T0348592

The Enzymes

VOLUME XXI

PROTEIN LIPIDATION

Third Edition

THE ENZYMES

Edited by

Fuyuhiko Tamanoi

Department of Microbiology
and Molecular Genetics, and
Molecular Biology Institute
University of California
Los Angeles, California

David S. Sigman

Department of Biological Chemistry
and Molecular Biology Institute
University of California
School of Medicine
Los Angeles, California

Volume XXI

PROTEIN LIPIDATION

THIRD EDITION

ACADEMIC PRESS

San Diego   New York   Boston
London   Sydney   Tokyo   Toronto

Academic Press
A Harcourt Science and Technology Company
525 B Street, Suite 1900, San Diego, California 92101-4495, USA
http://www.academicpress.com

Academic Press
Harcourt Place, 32 Jamestown Road, London NW1 7BY, UK
http://www.academicpress.com

Library of Congress Catalog Card Number: 00-107123

International Standard Book Number: 0-12-122722-7

Printed and bound in the United Kingdom
Transferred to Digital Printing, 2011

Contents

4. Farnesyltransferase Inhibitors

JACKSON B. GIBBS

5. Protein Geranylgeranyltransferase Type I

KOHEI YOKOYAMA AND MICHAEL H. GELB

6. Biochemistry of Rab Geranylgeranyltransferase

MIGUEL C. SEABRA

7. Postisoprenylation Protein Processing: CXXX(CaaX) Endoproteases and Isoprenylcysteine Carboxyl Methyltransferase

STEPHEN G. YOUNG, PATRICIA AMBROZIAK, EDWARD KIM, AND STEVEN CLARKE

8. Reversible Modification of Proteins with Thioester-Linked Fatty Acids

MAURINE E. LINDER

9. Biology and Enzymology of Protein N-Myristoylation

RAJIV S. BHATNAGAR, KAVEH ASHRAFI, KLAUS FUTTERER, GABRIEL WAKSMAN, AND JEFFREY I. GORDON

Preface

Posttranslational modification by the addition of lipids such as isoprenoids and fatty acids is critical for the function of a number of proteins involved in cell growth, differentiation, and morphology. Many proteins involved in signal transduction such as Ras superfamily G proteins as well as heterotrimeric G proteins undergo these lipidation events. Furthermore, recent results suggest that protein lipidation plays an important role in maintaining transformed phenotypes of cancer cells.

This past decade has seen a major development in the identification and characterization of enzymes that catalyze protein lipidation. A family of protein prenyltransferases as well as enzymes involved in fatty acid acylation have been characterized. In addition to enzymological and mutational studies of these enzymes, there has been a remarkable advance in elucidating three-dimensional structures of these enzymes as well as developing inhibitors of these enzymes. Inhibitors of protein prenyltransferases have shown promise as anticancer drugs. This volume reviews these developments by bringing together different areas of research that have been instrumental in elucidating the structure and function of these enzymes.

The first six chapters deal with enzymes catalyzing protein prenylation. Protein farnesyltransferase (abbreviated as FTase or PFT in this volume) is responsible for catalyzing the addition of a farnesyl group to substrate proteins. This enzyme recognizes a special motif (CaaX motif; cysteine followed by two aliphatic amino acids, and X is the C-terminal amino acid) present at the C termini of proteins such as Ras, transducin γ subunit, and nuclear lamins. Protein geranylgeranyltransferase type I (abbreviated as GGTase-I or PGGT-I) recognizes a similar but distinct motif, the CaaL motif (similar to the CaaX motif except that the C-terminal residue is leucine or phenylalanine), found at the C termini of proteins such as Rho, Rac, Cdc42, and the γ subunit of heterotrimeric G proteins. GGTase-I transfers a geranylgeranyl group to these proteins. Finally, protein geranylgeranyltransferase type II (abbreviated as GGTase-II or RGGT) recognizes a CC or CXC motif present at the C termini of Rab family G proteins involved in protein secretion. Proteins modified by FTase or GGTase-I undergo further C-terminal processing events that involve proteolytic cleavage of the aaX residues and carboxymethylation of the C-terminal cysteine. CXXX endoproteases and isoprenylcysteine carboxyl methyltransferase, which catalyze these associated modifications, are the topics of Chapter 7.

Chapters 8 and 9 deal with enzymes involved in the modification of proteins with fatty acids. Chapter 8 is concerned with thioacylation, a unique modification that can be reversed by deacylation; the thioacylated proteins undergo a dynamic cycle of acylation and deacylation. Both acylation and deacylation enzymes are discussed in this chapter. Chapter 9 deals with N-myristoyltransferase, which catalyzes the irreversible addition of myristic acid at the N termini of proteins such as Src family proteins as well as the α subunit of heterotrimeric G proteins.

There are other protein lipidation events that are not covered in this volume. An important example is the addition of a glycosylphosphatidyl-inositol (GPI) group. Although basic molecular mechanisms for this modification, which involve transamidation reactions, have been elucidated, we did not include this topic in this volume because it deserves a volume of its own. Other lipidation events such as protein retinoylation and the modification of hedgehog protein by cholesterol are not included in this volume. These areas represent emerging fields and could become topics of future volumes.

This volume is the first publication to provide a comprehensive review of enzymes involved in protein lipidation. We are grateful to all the chapter authors, who contributed exciting, up-to-date reviews. We are also grateful to Ms. Shirley Light of Academic Press, who has worked with us since the conception of this project until its completion.

Fuyuhiko Tamanoi
David S. Sigman

1

Mechanism of Catalysis by Protein Farnesyltransferase

REBECCA A. SPENCE • PATRICK J. CASEY

Departments of Pharmacology and Cancer Biology and of Biochemistry
Duke University Medical Center
Durham, North Carolina 27710

I. Introduction to Protein Prenyltransferases: Protein Farnesyltransferase and Protein Geranylgeranyltransferase Types I and II

Prenylation is a class of protein lipidation involving covalent addition of either farnesyl (15-carbon) or geranylgeranyl (20-carbon) isoprenoids to

1

THE ENZYMES, Vol. XXI

cysteine residues at or near the C terminus of a select group of intracellular proteins. The isoprenoid lipid is attached via a thioether linkage, and is generally required for proper function of the modified protein, participating either as a mediator of membrane association or as a determinant of specific protein–protein interactions. Protein prenylation is ubiquitous in the eukaryotic world, and the majority of prenylated proteins are involved in cellular signaling and/or regulatory events that occur at or near the cytoplasmic surfaces of cellular membranes (1, 2). This chapter focuses almost exclusively on the enzymology of addition of the farnesyl isoprenoid to cellular proteins; the reader is directed to other chapters in this volume, and several other reviews (2–5), for more detailed description of aspects of the biochemistry and biology of this process not covered here.

The enzymes responsible for isoprenoid addition to proteins, termed protein prenyltransferases, were identified in the early 1990s and have been extensively characterized at a molecular level both in mammalian systems and in *Saccharomyces cerevisiae* (3, 6). The three distinct protein prenyltransferases can be classified in two functional classes: the CaaX prenyltransferases, classified by their lipid substrate and termed protein farnesyltransferase (FTase) and protein geranylgeranyltransferase type I (GGTase-I); and protein geranylgeranyltransferase type II (GGTase-II, also called Rab geranylgeranyltransferase). FTase and GGTase-I have been designated CaaX prenyltransferases because their protein substrates invariably contain a cysteine residue precisely fourth from the C terminus of the protein; this motif is commonly referred to as a "CaaX box" (3, 7). Substrates of FTase include Ras proteins and several Ras-related GTP-binding proteins, nuclear lamins A and B, several proteins involved in visual signal transduction, and fungal mating factors (7–9). Known targets of GGTase-I include most γ subunits of heterotrimeric G proteins and a multitude of Ras-related GTPases such as most members of the Rac and Rho subfamilies (2, 7). GGTase-II attaches geranylgeranyl groups to two carboxyl-terminal cysteines of members of a single subfamily of Ras-related proteins termed Rab proteins (also termed Ypt/Sec4 proteins in fungi); these proteins generally terminate in Cys-Cys or Cys-X-Cys motifs (2, 10).

II. General Features of Protein Farnesyltransferase, a CaaX Prenyltransferase

FTase was the first protein prenyltransferase to be identified. The enzyme was originally isolated from rat brain cytosol, using an assay that followed the incorporation of radiolabel from ^{3}H-labeled farnesyl diphosphate ([^{3}H]FPP) into a recombinant Ras protein (11). Mammalian FTase was

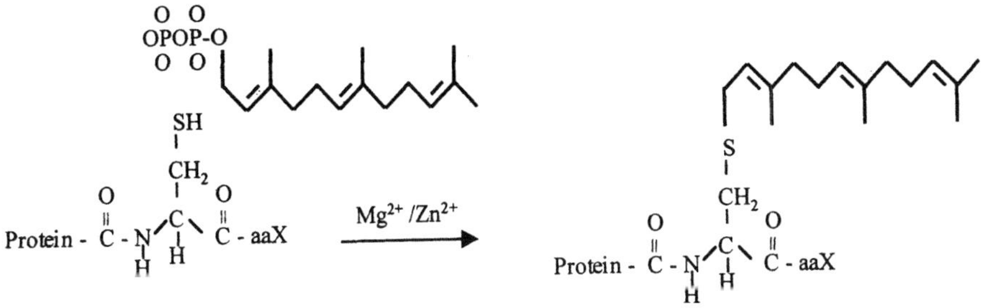

FIG. 1. The reaction catalyzed by FTase. Shown are the FPP and protein substrates; in the latter only the C-terminal CaaX region is detailed.

found to be a heterodimer that contained a 48-kDa α subunit (later determined also to be present in GGTase-I) and a 46-kDa β subunit (*11–13*). The cDNA cloning of FTase and GGTase-I confirmed that the α subunits were the products of the same gene and revealed that the β subunits shared ~35% sequence identity at the amino acid level (*14–16*). Although the significance of the two enzymes sharing a common subunit is not yet clear, the finding that both contained an identical α subunit provided the initial evidence that discrete segment(s) of the β subunit would be responsible for the substrate specificities of the enzymes. A role for the α subunit in catalysis was first suggested by the finding that catalytic activity was severely compromised by altering residue Lys-164 in the α subunit (*17*).

Two distinct divalent metals are required for catalysis by FTase (Fig. 1). First, FTase is a zinc metalloenzyme (*18, 19*). In addition, FTase activity is dependent on millimolar levels of Mg^{2+} (*13, 18*). As described in further detail in the next section, the zinc atom is not required for isoprenoid substrate binding but is required for protein substrate binding by FTase; on the other hand, Mg^{2+} does not seem to be required for binding of either substrate but instead may exert its influence directly on the step involving product formation.

Structural information has just begun to emerge on the CaaX prenyltransferases. The first structural data have come from analysis of mammalian FTase, whose X-ray crystal structure was determined at 2.2-Å resolution (*20*). In this structure, which was of the free (i.e., unliganded) enzyme, the α subunit was found to be folded into a crescent-shaped domain composed of seven successive pairs of coiled coils, termed helical hairpins, that contact a significant portion of the β subunit. The existence of repeat motifs in this subunit was first predicted from sequence alignments of mammalian and fungal α subunits (*21*). The β subunit was also found to consist largely of helical domains, with the majority of the helices arranged into an α–α

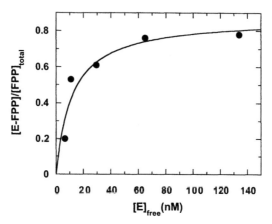

Fig. 2. High-affinity binding of FPP to FTase. Data are from an equilibrium binding experiment.

barrel structure. One end of the barrel was open to the solvent, while the other end was blocked by a short stretch of residues near the C terminus of the β subunit. This arrangement results in a structure containing a deep hydrophobic cleft in the center of the barrel that possesses all the features expected for the binding site of farnesyl diphosphate (FPP) with the preposed catalytic zinc ion at the opening of this cleft. The metal ion was coordinated by Cys-299, Asp-298, and His-362 in the β subunit; these residue have also been implicated as zinc ligands on the basis of biochemical studies (22–24). All three protein ligands to the metal are conserved in the known sequences of all three protein prenyltransferases. Additional crystal structures of the FTase complexes, including those of the binary complexes with bound FPP and also ternary complexes containing an FPP analog and CaaX-type peptides, have been reported that provide a detailed snapshot of the binding sites for substrates on the enzyme (25–28).

III.   Substrate Recognition by Protein Farnesyltransferase

A.   Recognition of Isoprenoid Substrates

FTase binds FPP with high affinity; K_D values in the low nanomolar range have been measured (Fig. 2) (29–31). Such high-affinity binding was first realized when it was discovered that the FTase–FPP complex could be isolated by gel filtration (32). Studies using a photoactivatable analog of FPP revealed specific cross-linking to the β subunit of FTase on photoactivation (33, 34), providing the initial evidence that the FPP-binding site for the enzyme was, as expected, predominantly associated with the β subunit. The aforementioned crystal structures determined for complexes of FTase with its substrates have provided the formal proof of this hypothesis, as well as delineating the structural features of the enzyme that comprise the binding site for FPP (see also [2] in this volume (34a)).

Early studies with FTase indicated that the enzyme could bind both FPP and geranylgeranyl diphosphate (GGPP) with relatively high affinity, although only FPP could be utilized as a substrate in the reaction (32). This property has been analyzed in greater detail, and GGPP binding to the enzyme was determined to be ~15-fold weaker than that of FPP (31), although this still translates to an apparent affinity of ~100 nM for GGPP binding to FTase. Elucidation of the structure of the FTase–FPP complex led to the development of a hypothesis concerning why the FTase–GGPP complex is essentially catalytically inactive (26); this hypothesis is discussed in detail in [2] in this volume (34a). In addition, analogs of FPP have been synthesized that retain high-affinity binding to FTase but that cannot participate in catalysis (35, 36). These analogs have proved extraordinarily useful in mechanistic studies of the enzyme, as they allow formation of an inactive FTase–isoprenoid binary complex (see below). It has been suggested that this extremely tight isoprenoid binding is to ensure that sufficient FPP is available for protein prenylation *in vivo*, even during fluctuations in isoprenoid biosynthesis (37).

B. RECOGNITION OF PROTEIN SUBSTRATES

Eukaryotic cells express a wide variety of proteins that are processed by FTase. Substrates of the enzyme in mammalian cells include H-, K-, and N-Ras, several Ras-related proteins such as specific subtypes of Rho and Rap, nuclear lamins A and B, at least two γ-subunit trimeric G proteins, and a number of kinases and phosphatases (5, 38–42). All these protein substrates contain the invariant cysteine residue in the CaaX motif at the C terminus. One important finding from early studies of FTase and GGTase-I was that the C-terminal residue of the CaaX motif determined which of the two enzymes will process a particular protein. FTase prefers proteins containing serine, methionine, alanine, or glutamine at the "X" position, whereas leucine at this position produces a preferred substrate of GGTase-I (3, 7). Because of this property of the two enzymes, it can be relatively accurately predicted whether the farnesyl or geranylgeranyl isoprenoid will be found on a CaaX protein in the cell.

An important property of FTase is that the enzyme can recognize, and use as a substrate, short peptides containing appropriate CaaX sequences

(*11, 43, 44*). Early studies using a competition-type assay resulted in a rather detailed analysis of the ability of FTase to recognize specific Ca_1a_2X sequences (*44, 45*). In these studies, a wide variety of residues were found to be allowed at the a_1 position, while variability at a_2 was more restricted. Basic and aromatic side chains were tolerated better at a_1 than at a_2, while acidic residues did not fare well at either position. In addition, placement of phenylalanine or tyrosine at the a_2 position in the context of a CaaX tetrapeptide created a molecule that was reported to be inactive as a substrate for FTase but functioned as a quite potent competitive inhibitor (*46*). One such peptide, CVFM, has served as the basis for design of peptidomimetic inhibitors of FTase (*47–49*). Interestingly, and for reasons not yet clear, N-acetylation of the cysteine residue in the tetrapeptide CVFM restored the ability of the peptide to be utilized as a substrate by FTase (*50*).

Although FTase (and GGTase-I) in general exhibits high selectivity toward specific protein substrates, some degree of cross-utilization of substrates has been reported (*43, 45*). Whether this so-called alternate prenylation has any biological relevance, however, remains an open question. In this regard, the most convincing evidence comes from studies of *S. cerevisiae*, in which the growth defects of yeast lacking FTase could be suppressed by overexpression of GGTase-I, suggesting that GGTase-I could prenylate substrates of FTase, at least partially restoring their function (*51, 52*). In terms of mammalian proteins, two specific Ras isoforms, these being K-Ras4B and N-Ras, have been found to serve as substrates for both FTase and GGTase I *in vitro*, although k_{cat}/K_m values are much greater for their modification by FTase (*53, 54*). In the context of a normal cell, these two Ras isoforms seem to be modified solely by the farnesyl group; however, geranylgeranylation of the proteins can be detected in cells treated with FTase inhibitors (*55, 56*). Both K-Ras4B and N-Ras terminate in methionine, and this C-terminal residue seems to "predispose" a protein to possible cross-prenylation, although the existence of multiple basic residues just upstream of the CaaX motif (e.g., the polybasic region found in K-Ras4B) also seems to confer some capacity of a protein to be processed by GGTase-I (*53*). In addition, phenylalanine as the C-terminal residue of a protein (as found in the Ras-related protein TC-21) seems to allow modification by either FTase or GGTase-I (*57*). An unusual situation occurs with another Ras-related protein, RhoB. This protein is modified by both farnesyl and geranylgeranyl groups, even though its C-terminal residue is a leucine (*58, 59*), and its farnesylation is apparently due to an ability to be processed by FTase (*60*).

As noted above, the zinc ion in FTase is required for high-affinity binding of the protein substrate and for catalysis (*18*). When the structure of the enzyme was solved, the zinc ion was found in the β subunit near the interface

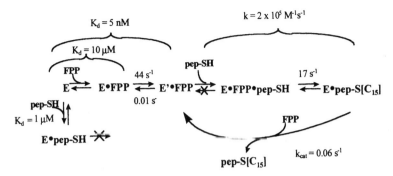

Fᴵɢ. 3. Kinetic mechanism of FTase. Shown is the overall kinetic scheme for the FTase reaction. The protein substrate is designated pep-SH; the farnesyl moiety attached to the protein is designated [C_{15}]. See text for details.

with the α subunit (20). This location was consistent with data indicating that both protein and peptide substrates could be cross-linked to the β subunit of FTase (32, 61), and that short peptide substrates containing divalent affinity groups label both the α and β subunits on photoactivation (61). As detailed below, studies have revealed a direct coordination of the thiolate of the cysteine residue of the protein substrate with the metal ion in the ternary complex of FTase–FPP–CaaX peptide that appears to be an integral component of the catalytic mechanism of the enzyme.

IV. Kinetic Mechanism of Protein Farnesyltransferase

Mammalian FTase is a slow enzyme, with k_{cat} values measured in the range of 0.05 sec^{-1} (Fig. 3) (11, 62). Steady state kinetics of mammalian FTase were initially interpreted as indicating a random-order binding mechanism, in which either substrate could bind first (62). However, the failure to trap enzyme-bound protein or peptide substrate in transient kinetic experiments suggested one of two things: either substrate binding is actually ordered, with FPP binding first; or the dissociation rate constant of the protein/peptide substrate from the free enzyme is so fast, and the affinity so weak, that FPP binding first is the kinetically preferred pathway (29, 62). Consistent with this functionally ordered mechanism, the affinity of FPP for FTase is in the low nanomolar range (29, 31), whereas the affinity of the peptide substrate in the absence of bound FPP is relatively weak (63) (see Fig. 3). However, affinity for CaaX peptide substrate is increased several hundredfold by the binding of FPP analogs (63). The transient kinetic studies also revealed that the association of the peptide substrate

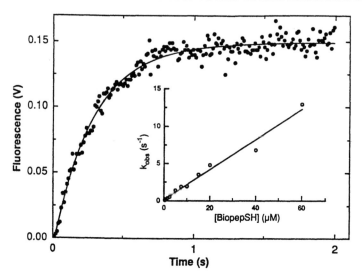

FIG. 4. Rapid association of a CaaX peptide substrate with the E·FPP complex measured by following intrinsic tryptophan fluorescence of the enzyme. The peptide substrate used in the experiment is designated BiopepSH. [Reprinted with permission from Furfine, E. S., Leban, J. J., Landavaso, A., Moomaw, J. F., and Casey, P. J. (1995). *Biochemistry* **34**, 6857; see text and cited reference for details.]

with the FTase·FPP binary complex was effectively irreversible, with a k_{assoc} of $2 \times 10^5\ M^{-1}\ sec^{-1}$ (Figs. 3 and 4) (*29*). While the rate constant for product formation could not be accurately determined in these studies, a lower limit of 12 sec^{-1} was established by a method exploiting protein fluorescence (*29*). The rate constant for product formation has been accurately determined for cobalt-substituted FTase (Co-FTase) from measurements of changes in the optical absorption spectrum during the catalytic process (*64*), in which a value of 17 sec^{-1} was obtained (Fig. 5). In any case, the rate constant for the product formation by mammalian FTase is clearly much greater than the steady state k_{cat}, indicating that the rate-limiting step is manifest after formation of the thioether product, i.e., product dissociation.

The kinetics of yeast FTase differ from mammalian FTase in several aspects. For the yeast enzyme, steady state kinetics clearly indicated that the mechanism was ordered with FPP binding first, because peptide bound to free FTase in a manner that precluded the productive binding of FPP (*37*). Substrate inhibition has also been observed to a lesser extent with the mammalian enzyme (*54*). Yeast FTase also has a markedly lower affinity for FPP (a decrease of some 30-fold) as well as a moderately reduced

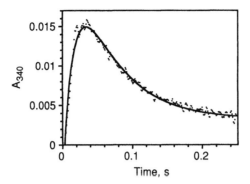

FIG. 5. Transient absorbance of the Co-FTase–FPP–CaaX peptide ternary complex. The absorbance increase is due to association of the CaaX peptide substrate with the Co-FTase–FPP complex, and the subsequent decrease is due to formation of the thioether product. [Reprinted with permission from Huang, C.-C., Casey, P. J., and Fierke, C. A. (1997). *J. Biol. Chem.* **272**, 20; see text and cited reference for details.]

affinity for protein/peptide substrates (decrease of ~4-fold) (*37*). Additional mechanistic observations of yeast FTase include the finding that the dissociation rate constant of the peptide substrate from the ternary complex ($k = 33$ sec^{-1}) is 70-fold faster than k_{off} for FPP from the enzyme (*65*). Thus, E·FPP can be efficiently trapped by added peptide and converted to product in spite of an unfavorable partitioning of the ternary complex between dissociation of peptide and product formation. Finally and most importantly, the chemical step for the yeast enzyme (10.5 sec^{-1}) is only threefold faster than its product release step; therefore, product dissociation is not the sole rate-limiting step for steady state turnover at saturating substrate concentrations (*65*). Nonetheless, the much greater steady state rate constant ($k_{cat} = 3.5$ sec^{-1}) for the yeast enzyme relative to the mammalian enzyme ($k_{cat} = 0.05$ sec^{-1}) is primarily a reflection of the faster product dissociation of the yeast enzyme.

Mammalian FTase exhibits a rather unusual property in terms of the product dissociation step. In the absence of excess substrates, the product dissociation rate constant is so slow that the adduct of FTase with bound product can be isolated (*66*). However, product dissociation can be enhanced by the addition of either substrate, with FPP being the most efficient in this regard (Fig. 6). However, the affinity of mammalian FTase for the thioether product is weak ($>10 \mu M$), indicating the presence of two distinct binding conformations for the product on the enzyme. The preceding observations suggest that product dissociation is kinetically controlled by an associated step (i.e., a conformational change of the enzyme) that is triggered by substrate binding. It has not been determined whether this phe-

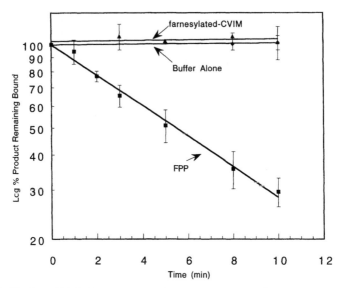

FIG. 6. Binding of FPP triggers release of product from FTase. [Adapted with permission from Tschantz, W. R., Furfine, E. S., and Casey, P. J. (1997). *J. Biol. Chem.* **272,** 9989; see text and cited reference for details.]

nomenon is also manifest in the yeast enzyme; however, farnesylated peptides are poor inhibitors of this enzyme, indicating weak binding of performed product similar to that seen with the mammalian FTase (*37*). Although the physiological significance of this substrate-triggered product release is unclear, the most intriguing possibility is that the farnesylated protein product remains bound to FTase until the enzyme encounters a specific site (e.g., a membrane compartment where FPP is located) to which it delivers its product (Fig. 7).

V. Chemical Mechanism of Protein Farnesyltransferase

A. A NOVEL TYPE OF ZINC METALLOENZYME

In proteins containing zinc, the metal ion can play either a catalytic or a structural role; specifically, a catalytic zinc participates directly in the chemical reaction whereas the role of a structural zinc is to stabilize protein structure (*67, 68*). As noted above, the finding that the zinc ion in FTase was required for catalytic activity, and for high-affinity binding of the protein substrate, led to the initial hypothesis that the metal ion would be involved in the catalytic mechanism of the enzyme (*18*). The first direct evidence in

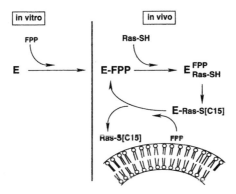

FIG. 7. Hypothetical *in vivo* kinetic scheme for FTase. In the scenario depicted, FTase delivers the product of its reaction, in this case farnesylated Ras (designated Ras-S[C_{15}]) to the intracellular membrane where further processing of the CaaX sequence (i.e., proteolysis and methylation) occur. [Adapted with permission from Tschantz, W. R., Furfine, E. S., and Casey, P. J. (1997). *J. Biol. Chem.* **272**, 9989; see text and cited reference for details.]

support of this hypothesis came from the demonstration of metal–thiol bond formation in the ternary complex of FTase–isoprenoid–CaaX peptide (64). Additional support for a catalytic role for the zinc ion came from the initial crystal structure of FTase, which revealed that the zinc-binding polyhedron contained a water molecule in addition to protein ligands forming an open coordination sphere. This is a unique feature of a catalytic zinc site, in contrast to structural zinc atoms, which are coordinated by protein ligands alone (20, 69). Furthermore, the three protein-derived zinc ligands in FTase (Asp-297, Cys-299, and His-362) were found in a spacing characteristic of many catalytic zinc sites (20, 22, 68). More recent crystal structures of FTase containing bound FPP analog and peptide substrates revealed that the reactive thiol of the peptide substrate cysteine was within 2.5 Å of the zinc ion (27, 28), which is highly suggestive of direct coordination.

As detailed in the following section, FTase is one of the first members of a new class of zinc metalloproteins that exhibit a novel catalytic function of the zinc ion, that being to enhance the nucleophilicity of a thiol group at neutral pH (68, 70). The primary function proposed for the zinc ion in this class of enzymes is to coordinate the cysteine thiol of the protein substrate, lowering the pK_a of the –SH group and thereby enhancing its reactivity by promoting thiolate anion formation at neutral pH (71). The list of these zinc-containing enzymes that catalyze S-alkylation reactions includes, in addition to FTase (and GGTase-I), the *Escherichia coli* Ada repair enzyme (O^6-methylguanine-DNA methyltransferase), cobalamin-dependent and -independent methionine synthases, methanol:coenzyme M methyltransferase, and betaine-homocysteine methyltransferase (70–77). Evidence from work on both Ada and FTase suggests that the thioether product remains coordinated to zinc (64, 78); these interactions may be an important factor for the slow product release step discussed above. Although the overall catalytic mechanism of the sulfur-alkylating metalloproteins is probably similar, there is no evidence of an electrophilic contribution to the mechanism of any of the other metalloproteins except for the prenyltransferases (see below).

In addition to the zinc ion, FTase also requires millimolar levels of magnesium to achieve efficient catalysis (18). Although it is unclear as to the role of this metal in the mechanism of FTase, no effect is observed on the affinity of either the FPP or peptide substrate in the absence of magnesium (18, 30, 63), indicating that this metal affects a step after formation of the ternary complex. Possibilities for a role for magnesium are that the metal could bind and thereby stabilize the pyrophosphate leaving group or stabilize carbocation formation of the C-1 of FPP in the transition state.

B. NUCLEOPHILIC VERSUS ELECTROPHILIC MECHANISMS FOR PROTEIN FARNESYLTRANSFERASE

Both electrophilic and nucleophilic mechanisms have been proposed for FTase (79, 80). The initial proposals for a nucleophilic mechanism were primarily prompted by the findings that the bound zinc ion in the enzyme was required both for catalysis as well as for the high affinity of peptide and protein substrates (13, 18). The first direct evidence in support of a mechanism with significant nucleophilic character came from a study in which optical absorbance spectroscopy of cobalt-substituted FTase revealed that the metal ion coordinated the thiol(ate) of the peptide substrate in the FTase·isoprenoid·CaaX peptide ternary complex (64). More compelling evidence in this regard has come from an examination of the pH dependence of peptide substrate binding to FTase, which revealed that the peptide thiol coordinates the metal as the thiolate anion, and that the pK_a of the SH group is shifted from 8.3 in the free peptide to ≤ 6.5 in the enzyme-bound peptide (63).

Metal substitution studies have also been used to investigate the contribution of a metal-coordinated substrate thiolate role in catalysis by mammalian FTase (80a). Cadmium is a softer metal than zinc, hence it has higher affinity for sulfur atom. In model studies using small thiol compounds, Cd^{2+} was found to bind the thiolate 1–2 orders of magnitude more tightly than Zn^{2+} (81). In studies of FTase containing Cd-for-Zn substitution, the affinity of the FTase·isoprenoid binary complex for peptide substrates was increased >5-fold for Cd-FTase compared with Zn-FTase, a finding consistent with the notion that Cd^{2+} enhances the binding of peptide substrate through stronger coordination of the substrate thiolate. Furthermore, in single-turnover experiments the rate of product formation catalyzed by Cd-FTase was decreased approximately fivefold compared with that of Zn-FTase, suggesting that the metal–thiolate nucleophile is important in the transition state.

Stereochemical studies using FPP containing chiral deuterium-for-hydrogen substitutions at its C-1 carbon have demonstrated that both the mammalian and yeast forms of FTase carry out the reaction with inversion of configuration at the C-1 farnesyl center (82, 83). Although this finding is consistent with an S_N2-type nucleophilic displacement, it could also result from an S_N1-type mechanism in which the configuration of carbocation formed is subject to steric hindrance. Secondary kinetic isotope effects have been used to further probe the transition state of the yeast enzyme (84). In this study, deuterium substitution at C-1 of FPP produced a kinetic isotope effect on the rate of product formation, consistent with an associative (S_N2) mechanism. However, because the value of the isotope effect was greater than those reported for reactions known to proceed via S_N2 mechanisms, the authors concluded that the yeast FTase may catalyze farnesylation with partial, but not complete, associative character (84).

Evidence has also been obtained of an electrophilic component in the mechanism of FTase. This initial evidence came from an examination of the ability of the yeast enzyme to use FPP substrate analogs that contained fluorine substitutions at the C-3 methyl position under steady state conditions. Because fluorine is a strongly electron-withdrawing atom, its presence at this C-3 position would be expected to destabilize a developing carbocation in the transition state (79). It was the use of such fluoromethyl isoprenoid analogs that provided some of the most compelling data for electrophilic mechanisms for the prenyl diphosphate synthases; reactions catalyzed by these enzymes proceed predominantly through formation of a carbocation at the site of bond formation (85–87). Decreased steady state activity of yeast FTase with the C-3 fluoromethyl-FPP analogs was in fact observed (79). Similar results have been obtained for transient kinetic studies of mammalian FTase, using the same fluoromethyl FPP analogs (80a). These findings strongly suggested that the transition state for FTase has carbocation character. Importantly, the decreases in reactivity resulting from fluorine substitution in the FPP analogs were significantly smaller than the effects of fluorine substitution on either the solvolysis of dimethylallyl-p-methoxybenzene sulfonates or the reactivity of farnesyl pyrophosphate synthase, both of which proceed via an S_N1 reaction (79, 88). In fact, the effects of the fluorine substitution on FTase reactivity more closely parallel

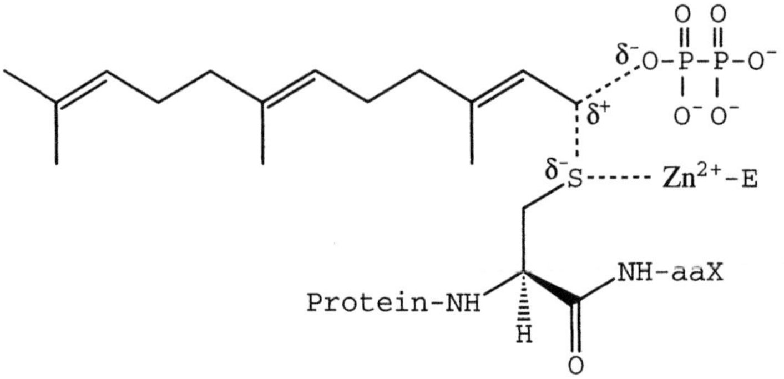

FIG. 8. Proposed structure of the transition state for FTase. [Adapted with permission from Hightower, K. E., and Fierke, C. A. (1998). *Curr. Opin. Chem. Biol.* **3,** 176; see text and cited reference for details.]

the effects on solvolysis reactions in the presence of a potent nucleophile such as azide, which proceed with significant S_N2 character through what has been termed an open, "exploded" transition state (*89*).

Taken together, the data summarized above suggest that the mechanism of FTase is a carbocation–nucleophile combination reaction (Fig. 8). It seems that neither the nucleophilic nature of the metal-thiolate nor the carbocation character of the C-1 atom of FPP can be ignored. For such a carbocation–nucleophile combination reaction, whether the reaction proceeds through an S_N1 or S_N2 mechanism depends on both the stability of the carbocation in the transition state and the strength of the nucleophile (*90, 91*). When the carbocation intermediate has a long enough lifetime, the reaction can proceed via a stepwise S_N1 mechanism. On the other hand, if the lifetime of the intermediate is short, the reaction may occur through an enforced preassociation mechanism, where the reactants are assembled before the first bond-making or -breaking step occurs. This preassociation mechanism could either be concerted with no intermediate or stepwise with an intermediate (*90*).

VI. Concluding Remarks

There has been major progress in understanding catalysis by FTase since it was first isolated in 1990. As detailed in other chapters in this volume, efforts to design selective cell-active inhibitors of the enzyme for evaluation as anticancer agents have been successful; the acquisition of mechanistic and structural information on FTase has greatly aided these drug discovery programs. Remaining questions on this enzyme are primarily centered on definition of the precise chemical mechanism of the reaction and the structural basis for the ability of the enzyme to discriminate between similar protein substrates. While the close similarities in both structure and function of FTase and GGTase-I make it is likely that both enzymes will use a similar catalytic mechanism, it will still be important to conduct detailed studies with GGTase-I to determine whether there are important differences between these enzymes. When studying these enzymes, it is best to keep in mind that the actual step of product formation has little influence on their steady state rate, and thus extreme caution must be exercised in the use of steady state data to draw conclusions on such parameters as specificity in recognition of CaaX sequences. Complete understanding of the mechanism of substrate specificity and other parameters of these enzymes requires examination of the individual steps in the catalytic process by presteady state kinetics and direct physical methods.

REFERENCES

1. Casey, P. J. (1995). *Science* **268**, 221.
2. Glomset, J. A., and Farnsworth, C. C. (1994). *Annu. Rev. Cell Biol.* **10**, 181.
3. Zhang, F. L., and Casey, P. J. (1996). *Annu. Rev. Biochem.* **65**, 241.
4. Gibbs, J. B., and Oliff, A. (1997). *Annu. Rev. Pharmacol. Toxicol.* **37**, 143.
5. Cox, A. D., and Der, C. J. (1997). *Biochim. Biophys. Acta* **1333**, F51.
6. Schafer, W. R., and Rine, J. (1992). *Annu. Rev. Genet.* **25**, 209.
7. Clarke, S. (1992). *Annu. Rev. Biochem.* **61**, 355.
8. Caldwell, G. A., Naider, F., and Becker, J. M. (1995). *Microbiol. Rev.* **59**, 406.
9. Omer, C. A., and Gibbs, J. B. (1994). *Mol. Microbiol.* **11**, 219.
10. Seabra, M. C., Goldstein, J. L., Sudhof, T. C., and Brown, M. S. (1992). *J. Biol. Chem.* **267**, 14497.
11. Reiss, Y., Goldstein, J. L., Seabra, M. C., Casey, P. J., and Brown, M. S. (1990). *Cell* **62**, 81.
12. Seabra, M. C., Reiss, Y., Casey, P. J., Brown, M. S., and Goldstein, J. L. (1991). *Cell* **65**, 429.
13. Moomaw, J. F., and Casey, P. J. (1992). *J. Biol. Chem.* **267**, 17438.
14. Chen, W.-J., Andres, D. A., Goldstein, J. L., Russell, D. W., and Brown, M. S. (1991). *Cell* **66**, 327.
15. Kohl, N. E., Diehl, R. E., Schaber, M. D., Rands, E., Soderman, D. D., He, B., Moores, S. L., Pompliano, D. L., Ferro-Novick, S., Powers, S., Thomas, K. A., and Gibbs, J. B. (1991). *J. Biol. Chem.* **266**, 18884.
16. Zhang, F. L., Diehl, R. E., Kohl, N. E., Gibbs, J. B., Giros, B., Casey, P. J., and Omer, C. A. (1994). *J. Biol. Chem.* **269**, 3175.
17. Andres, D. A., Goldstein, J. L., Ho, Y. K., and Brown, M. S. (1993). *J. Biol. Chem.* **268**, 1383.
18. Reiss, Y., Brown, M. S., and Goldstein, J. L. (1992). *J. Biol. Chem.* **267**, 6403.
19. Chen, W.-J., Moomaw, J. F., Overton, L., Kost, T. A., and Casey, P. J. (1993). *J. Biol. Chem.* **268**, 9675.

20. Park, H.-W., Boduluri, S. R., Moomaw, J. F., Casey, P. J., and Beese, L. S. (1997). *Science* **275,** 1800.
21. Boguski, M. S., Murray, A. W., and Powers, S. (1992). *New Biol.* **4,** 408.
22. Fu, H.-W., Moomaw, J. F., Moomaw, C. R., and Casey, P. J. (1996). *J. Biol. Chem.* **271,** 28541.
23. Fu, H.-W., Beese, L. S., and Casey, P. J. (1998). *Biochemistry* **37,** 4465.
24. Kral, A. M., Diehl, R. E., deSolms, S. J., Williams, T. M., Kohl, N. E., and Omer, C. A. (1997). *J. Biol. Chem.* **272,** 27319.
25. Dunten, P., Kammlott, U., Crowther, R., Weber, D., Palermo, R., and Birktoft, J. (1998). *Biochemistry* **37,** 7907.
26. Long, S. B., Casey, P. J., and Beese, L. S. (1998). *Biochemistry* **37,** 9612.
27. Strickland, C. L., Windsor, W. T., Syto, R., Wang, L., Bond, R., Wu, Z., Schwartz, J., Le, H. V., Beese, L. S., and Weber, P. C. (1998). *Biochemistry* **37,** 16601.
28. Long, S. B., Casey, P. J., and Beese, L. S. (2000). *Structure* **8,** 209.
29. Furfine, E. S., Leban, J. J., Landavazo, A., Moomaw, J. F., and Casey, P. J. (1995). *Biochemistry* **34,** 6857.
30. Zhang, F. L., and Casey, P. J. (1996). *Biochem. J.* **320,** 925.
31. Yokoyama, K., Zimmerman, K., Scholten, J., and Gelb, M. H. (1997). *J. Biol. Chem.* **272,** 3944.
32. Reiss, Y., Seabra, M. C., Armstrong, S. A., Slaughter, C. A., Goldstein, J. L., and Brown, M. S. (1991). *J. Biol. Chem.* **266,** 10672.
33. Omer, C. A., Kral, A. M., Diehl, R. E., Prendergast, G. C., Powers, S., Allen, C. M., Gibbs, J. B., and Kohl, N. E. (1993). *Biochemistry* **32,** 5167.
34. Yokoyama, K., McGeady, P., and Gelb, M. H. (1995). *Biochemistry* **34,** 1344.
34a. Terry, K., Long, S. B., and Beese, L. S. (2000). *In* "The Enzymes" (F. Tamanoi and D. S. Sigman, eds.), 3rd Ed., Vol. XXI: "Protein Lipidation," Chap. 2. Academic Press, San Diego, California (this volume).
35. Gibbs, J. B., Pompliano, D. L., Mosser, S. D., Rands, E., Lingham, R. B., Singh, S. B., Scolnick, E. M., Kohl, N. E., and Oliff, A. (1993). *J. Biol. Chem.* **268,** 7617.
36. Patel, D. V., Schmidt, R. J., Biller, S. A., Gordon, E. M., Robinson, S. S., and Manne, V. (1995). *J. Med. Chem.* **38,** 2906.
37. Dolence, J. M., Cassidy, P. B., Mathis, J. R., and Poulter, C. D. (1995). *Biochemistry* **34,** 16687.
38. Inglese, J., Glickman, J. F., Lorenz, W., Caron, M., and Lefkowitz, R. J. (1992). *J. Biol. Chem.* **267,** 1422.
39. Cox, A. D., and Der, C. J. (1992). *Crit. Rev. Oncogenesis* **3,** 365.
40. James, G. L., Goldstein, J. L., Pathak, R. L., Anderson, R. G. W., and Brown, M. S. (1994). *J. Biol. Chem.* **269,** 14182.
41. Heilmeyer, L. M. G., Serwe, M., Weber, C., Metzger, J., Hoffmann-Posorske, E., and Meyer, H. E. (1992). *Proc. Natl. Acad. Sci. U.S.A.* **89,** 9554.
42. Smed, F. D., Boom, A., Pesesse, X., Schiffmann, S. N., and Erneux, C. (1996). *J. Biol. Chem.* **271,** 10419.
43. Yokoyama, K., Goodwin, G. W., Ghomashchi, F., Glomset, J. A., and Gelb, M. H. (1991). *Proc. Natl. Acad. Sci. U.S.A.* **88,** 5302.
44. Reiss, Y., Stradley, S. J., Gierasch, L. M., Brown, M. S., and Goldstein, J. L. (1991). *Proc. Natl. Acad. Sci. U.S.A.* **88,** 732.
45. Moores, S. L., Schaber, M. D., Mosser, S. D., Rands, E., O'Hara, M. B., Garsky, V. M., Marshall, M. S., Pompliano, D. L., and Gibbs, J. B. (1991). *J. Biol. Chem.* **266,** 14603.

46. Goldstein, J. L., Brown, M. S., Stradley, S. J., Reiss, Y., and Gierasch, L. M. (1991). *J. Biol. Chem.* **266**, 15575.
47. Garcia, A. M., Rowell, C., Ackermann, K., Kowalczyk, J. J., and Lewis, M. D. (1993). *J. Biol. Chem.* **268**, 18415.
48. Kohl, N. E., Mosser, S. D., deSolms, S. J., Giuliani, E. A., Pompliano, D. L., Graham, S. L., Smith, R. L., Scolnick, E. M., Oliff, A., and Gibbs, J. B. (1993). *Science* **260**, 1934.
49. Nigam, M., Seong, C.-M., Qian, Y., Hamilton, A. D., and Sebti, S. M. (1993). *J. Biol. Chem.* **268**, 20695.
50. Brown, M. S., Goldstein, J. L., Paris, K. J., Burnier, J. P., and Marsters, J. C. (1992). *Proc. Natl. Acad. Sci. U.S.A.* **89**, 8313.
51. Powers, S., Michaelis, S., Broek, D., Santa-Ana, A. S., Field, J., Herskowitz, I., and Wigler, M. (1986). *Cell* **47**, 413.
52. Trueblood, C. E., Ohya, Y., and Rine, J. (1993). *Mol. Cell. Biol.* **13**, 4260.
53. James, G. L., Goldstein, J. L., and Brown, M. S. (1995). *J. Biol. Chem.* **270**, 6221.
54. Zhang, F. L., Kirschmeier, P., Carr, D., James, L., Bond, R. W., Wang, L., Patton, R., Windsor, W. T., Syto, R., Zhang, R., and Bishop, W. R. (1997). *J. Biol. Chem.* **272**, 10232.
55. Rowell, C. A., Kowalczyk, J. J., Lewis, M. D., and Garcia, A. M. (1997). *J. Biol. Chem.* **272**, 14093.
56. Whyte, D. B., Kirschmeier, P., Hockenberry, T. N., Nunez-Oliva, I., James, L., Catino, J. J., Bishop, W. R., and Pai, J.-K. (1997). *J. Biol. Chem.* **272**, 14459.
57. Carboni, J. M., Yan, N., Cox, A. D., Bustelo, X., Graham, S. M., Lynch, M. J., Weinmann, R., Seizinger, B. R., Der, C. J., Barbacid, M., and Manne, V. (1995). *Oncogene* **10**, 1905.
58. Adamson, P., Marshall, C. J., Hall, A., and Tilbrook, P. A. (1992). *J. Biol. Chem.* **267**, 20033.
59. Armstrong, S. A., Hannah, V. C., Goldstein, J. L., and Brown, M. S. (1995). *J. Biol. Chem.* **270**, 7864.
60. Lebowitz, P. F., Casey, P. J., Prendergast, G. C., and Thissen, J. A. (1997). *J. Biol. Chem.* **272**, 15591.
61. Ying, W., Sepp-Lorenzino, L., Cai, K., Aloise, P., and Coleman, P. S. (1994). *J. Biol. Chem.* **269**, 470.
62. Pompliano, D. L., Rands, E., Schaber, M. D., Mosser, S. D., Anthony, N. J., and Gibbs, J. B. (1992). *Biochemistry* **31**, 3800.
63. Hightower, K. E., Huang, C.-C., Casey, P. J., and Fierke, C. A. (1998). *Biochemistry* **37**, 15555.
64. Huang, C.-C., Casey, P. J., and Fierke, C. A. (1997). *J. Biol. Chem.* **272**, 20.
65. Mathis, J. R., and Poulter, C. D. (1997). *Biochemistry* **36**, 6367.
66. Tschantz, W. R., Furfine, E. S., and Casey, P. J. (1997). *J. Biol. Chem.* **272**, 9989.
67. Vallee, B. L., and Auld, D. S. (1990). *Biochemistry* **29**, 5647.
68. Hightower, K. E., and Fierke, C. A. (1998). *Curr. Opin. Chem. Biol.* **3**, 176.
69. Vallee, B. L., and Auld, D. S. (1992). *Faraday Discuss.* **93**, 47.
70. Matthews, R. G., and Goulding, C. W. (1998). *Curr. Opin. Chem. Biol.* **1**, 332.
71. Myers, L. C., Terranova, M. P., Ferentz, A. E., Wagner, G., and Verdine, G. L. (1993). *Science* **261**, 1164.
72. Gonzalez, J. C., Peariso, K., Penner-Hahn, J. E., and Matthews, R. G. (1996). *Biochemistry* **35**, 12228.
73. LeClerc, G. M., and Grahame, D. A. (1996). *J. Biol. Chem.* **271**, 18725.
74. Goulding, C. W., and Matthews, R. G. (1997). *Biochemistry* **36**, 15749.
75. Millian, N. S., and Garrow, T. A. (1998). *Arch. Biochem. Biophys.* **356**, 93.
76. Jarrett, J. T., Choi, C. Y., and Matthews, R. G. (1997). *Biochemistry* **36**, 15739.
77. Sauer, K., and Thauer, R. K. (1997). *Eur. J. Biochem.* **249**, 280.
78. Myers, L. C., Cushing, T. D., Wagner, G., and Verdine, G. L. (1994). *Chem. Biol.* **1**, 91.

79. Dolence, J. M., and Poulter, C. D. (1995). *Proc. Natl. Acad. Sci. U.S.A.* **92**, 5008.
80. Casey, P. J., and Seabra, M. C. (1996). *J. Biol. Chem.* **271**, 5289.
80a. Huang, C.-C., Hightower, K. E., and Fierke, C. A. (2000). *Biochemistry* **39**, 2593.
81. Li, N. C., and Manning, R. A. (1955). *J. Am. Chem. Soc.* **77**, 5225.
82. Mu, Y., Omer, C. A., and Gibbs, R. A. (1996). *J. Am. Chem. Soc.* **118**, 1817.
83. Edelstein, R. L., Weller, V. A., and Distefano, M. D. (1998). *J. Org. Chem.* **63**, 5298.
84. Weller, V. A., and Distefano, M. D. (1998). *J. Am. Chem. Soc.* **120**, 7975.
85. Gebler, J. C., Woodside, A. B., and Poulter, C. D. (1992). *J. Am. Chem. Soc.* **114**, 7354.
86. Chen, A., Kroon, P. A., and Poulter, C. D. (1994). *Protein Sci.* **3**, 600.
87. Lesburg, C. A., Zhai, G., Cane, D. E., and Christianson, D. W. (1997). *Science* **277**, 1820.
88. Poulter, C. D., Wiggins, P. L., and Le, A. T. (1981). *J. Am. Chem. Soc.* **103**, 3926.
89. Richard, J. P., and Jencks, W. P. (1984). *J. Am. Chem. Soc.* **106**, 1383.
90. Jencks, W. P. (1980). *Acc. Chem. Res.* **13**, 161.
91. Richard, J. P. (1995). *Tetrahedron* **51**, 1535.

2

Structure of Protein Farnesyltransferase

KIMBERLY L. TERRY · STEPHEN B. LONG ·
LORENA S. BEESE
Department of Biochemistry
Duke University Medical Center
Durham, North Carolina 27710

THE ENZYMES, Vol. XXI

I. Introduction

Posttranslational modifications are essential to the proper functioning of many proteins. One important modification involves the addition of a 15-carbon farnesyl isoprenoid group, a reaction catalyzed by protein farnesyltransferase (FTase). Protein substrates for this enzyme include Ras guanosine triphosphatases (GTPases), nuclear lamins, and several proteins involved in visual signal transduction (*1, 2*). These proteins require prenylation for localization to cellular membranes and subsequent interaction with other proteins. This localization is necessary for the mutated activity of Ras proteins found to be associated with a large percentage of human cancers, making FTase of great interest in anticancer treatments.

Prenylation by FTase and its related enzyme, protein geranylgeranyltransferase type I (GGTase-I), covalently attaches an isoprenoid group to a conserved cysteine residue of the substrate protein. This cysteine occurs in the fourth position from the C terminus in what is commonly known as the CaaX motif, where "C" indicates cysteine, "a," a small, aliphatic residue, and "X," the uncharged, specificity-determining residue. When X is serine, methionine, alanine, or glutamine, the protein is farnesylated by FTase, but if this position is occupied by a leucine, the protein is geranylgeranylated by GGTase-I. Farnesyl diphosphate (FPP) is the prenyl donor in the FTase reaction, which requires both zinc (*3–6*) and magnesium for full activity (*3, 4*). Prenylation is followed by cleavage of the aaX residues by an endoprotease and methylation of the carboxyl group of the modified cysteine by a methyltransferase (*1*).

High-resolution crystal structures of protein FTase complexed with substrates and inhibitors provide a framework for understanding the molecular basis of substrate specificity and mechanism, facilitating the development of improved chemotherapeutics. The 2.25-Å resolution crystal structure of rat FTase provided the first structural information on any protein prenyltransferase enzyme (*7*). Rat FTase shares 93% sequence identity with the human enzyme and is predicted to be nearly indistinguishable from human FTase in the active site region, a prediction recently confirmed by the determination of the human FTase structure (*8*). A cocrystal structure of rat FTase with bound FPP revealed the location of the isoprenoid binding and gave insight into the molecular basis of isoprenoid substrate specificity (*9*). Multiple cocrystal structures of rat FTase with bound substrates or nonreactive analogs identified the location of both the peptide and isoprenoid binding sites in a ternary enzyme complex (*10, 11*). Recently, a complex of farnesylated K-Ras4B peptide product bound to rat FTase was determined (*12*), that increases our understanding of the FTase mechanism and specificity. In this chapter, we describe the crystal structures of FTase and

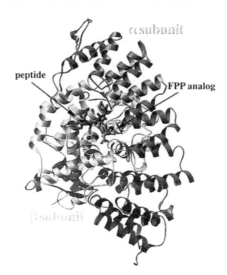

Fig. 1. Overview of the FTase structure with bound FPP analog and K-Ras4B-derived peptide substrate. FTase is a heterodimer of α (shown in red) and β (shown in yellow) subunits. As demonstrated here, the secondary structure of both subunits is composed primarily of α helices, with 15 in the α subunit and 14 in the β subunit. Both substrates (FPP in green and peptide in yellow) bind in the hydrophobic cavity of the β subunit. (See color plate.)

discuss their implications on our understanding of substrate specificity, mechanism, and inhibitor design.

II. Structure

A. PROTEIN ARCHITECTURE

FTase exists as a heterodimer composed of α and β subunits (48 and 46 kDa in molecular mass, respectively) (3, 13, 14), the former also being a component of the heterodimeric protein GGTase-I. The secondary structure of both subunits is composed primarily of α helices (Fig. 1), with 15 in the α subunit and 14 in the β subunit. Helices 2 to 15 of the α subunit are folded into seven successive helical pairs, forming a series of right-handed antiparallel coiled coils. These "helical hairpins" are arranged in a right-handed superhelix resulting in a crescent-shaped subunit that envelops a portion of the β subunit, an unusual structure also observed in the crystal structures of lipovitellin-phosvitin (15) and bacterial muramidase (16). Twelve α helices of the β subunit are folded into an α–α barrel (Fig.

FIG. 2. The α–α barrel of the β subunit. This domain, composed of helices 2β to 13β, forms the hydrophobic cavity in which the substrates bind. The aromatic residues lining this cavity are shown in yellow, with the bound zinc in magenta. This view is a 90° clockwise rotation from the orientation shown in Fig. 1. (See color plate.)

2), similar to those found in bacterial cellulase (17), endoglucanase CelA (18), and glycoamylase (19, 20). Six of these helices (3β, 5β, 7β, 9β, 11β, and 13β) are arranged in parallel, constituting the core of the barrel. The remaining six helices (2β, 4β, 6β, 8β, 10β, and 12β) form the outside of the barrel. These peripheral helices are parallel to each other, but antiparallel with respect to the core helices. One end of the α–α barrel is obstructed by a loop containing residues 399β to 402β. The opposite end is open to the solvent, creating a deep funnel-shaped cavity in the center of the barrel, with an inner diameter of 15 to 16 Å and a depth of 14 Å. This cavity is hydrophobic in nature and is lined with conserved aromatic residues (Fig. 2). The active site of FTase is located in this cavity, as are those of other α–α barrel proteins.

The N-terminal proline-rich domain of the α subunit (residues 1 to 54) is disordered in the crystal structure (7). Deletion of this domain does not affect the catalytic activity of FTase (21). Additionally, a crystal structure of a truncated form of rat FTase lacking the proline-rich domain has been determined to 2.75-Å resolution (22). Not surprisingly, the deletion of this domain had no significant effect on the structure of the rest of the protein. These observations suggest that the proline-rich domain may interact with other cellular factors, perhaps functioning in enzyme localization.

Multiple sequence alignments of mammalian and yeast α subunits revealed five tandem sequence repeats (23). Each repeat consists of two highly conserved regions separated by a divergent region of fixed length. These motifs appear in the first five "helical hairpins" of the FTase structure.

The second α helix of each helical pair contains an invariant tryptophan residue, which, together with other hydrophobic residues, forms the hydrophobic core of the "hairpin." The conserved sequence Pro-X-Asn-Tyr (where X is any amino acid) (23) is found in the turns connecting two helices of the coiled coil. These turns form part of the interface with the β subunit. Internal repeats of glycine-rich sequences have also been identified in the β subunits of other protein prenyltransferases (23). These repeats correspond to the loop regions that connect the C termini of the peripheral helices with the N termini of the core helices in the barrel.

The interface between the two subunits of FTase is quite extensive, burying 3322 Å^2 or 19.5% of accessible surface area of the α subunit and 3220 Å^2 or 17.2% of accessible surface area of the β subunit (7). Such surfaces are commonly composed of 65% nonpolar residues, but the α/β interface of FTase is 54% nonpolar, 29% polar, and 17% charged residues. As a result of the unusually high polar and charged residue content, there are nearly double the normal number of hydrogen bonds. Each of these hydrogen bonds contributes 0.5–1.8 kcal/mol to the free energy of stabilization when the donor and acceptor are neutral and 3–6 kcal/mol when the donor and acceptor are charged (23). The number of hydrogen bonds found in the FTase subunit interface may contribute to the unusual stability of the FTase heterodimer, which cannot be dissociated unless denatured (13).

B. ZINC-BINDING SITE

FTase is a zinc metalloenzyme, binding one zinc ion per protein dimer (4, 5, 7). A wealth of experimental evidence indicates that the zinc ion is required for catalytic activity (4, 25–27). Additionally, the zinc ion is important for the binding of peptide, but not for isoprenoid substrates (4). A direct involvement of zinc in catalysis is supported by several studies (4, 6, 25–27); the most compelling evidence is that the metal ion is coordinated by the thiol of the CaaX cysteine residue in the ternary substrate complexes (6, 10, 11). The crystal structures of FTase show a single zinc ion bound to the β subunit, near the subunit interface, marking the location of the active site. In the FTase apoenzyme crystal structures (7, 22) the zinc ion is coordinated by residues D297β (a bidentate ligand), C299β, H362β, and a well-ordered water molecule (Fig. 3). Crystal structures of FTase with bound substrates (10, 11) show that the cysteine thiol of the CaaX protein substrate also coordinates the zinc, displacing the well-ordered water observed in the apoenzyme structures.

Each of the protein ligands is strictly conserved in the β subunits of protein prenyltransferase enzymes. Mutational analysis and biochemical studies had previously identified the significance of C299β, demonstrating

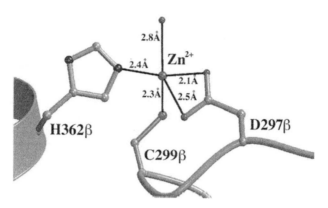

FIG. 3. The zinc-binding site. The zinc ion is coordinated by three enzyme side chains (D297β, C299β, and H362β) and a well-ordered water molecule in the apoenzyme structure. Carbon is shown in coral, oxygen in red, nitrogen in blue, sulfur in green, zinc in magenta, and polypeptide in cyan. The coordinating hydrogen bonds are depicted in black. (See color plate.)

its ability to affect zinc binding and abolish catalytic activity when mutated (*28*). Additional mutagenesis studies confirm a zinc-binding role for residues D297β, C299β, and H362β (*29–31*), and suggest that D359β also has a role in the binding of zinc (*30*). In the crystal structure, D359β forms a hydrogen bond to H362β, possibly stabilizing a required conformation for binding zinc.

C. FARNESYL DIPHOSPHATE-BINDING SITE

In the cocrystal structure of rat FTase complexed with FPP, the isoprenoid moiety binds in an extended conformation along one side of the hydrophobic cavity of the α–α barrel (*9*) (Fig. 4). Lining this cavity are conserved aromatic residues including W102β, W303β, Y205β, Y251β, and Y200α, each making hydrophobic interactions with the FPP molecule. The strictly conserved residue R202β, which has an alternate conformation with respect to the apostructure, participates in a hydrophobic interaction with the isoprenoid. The conformation of R202β is stabilized by interactions with D200β and M193β. Conserved residues C254β and G250β also contribute to the isoprenoid-binding site.

In the binary complex, the diphosphate moiety of the FPP molecule binds in a positively charged cleft near the subunit interface at the rim of the α–α barrel, adjacent to the catalytic zinc ion. This moiety forms hydrogen bonds with the strictly conserved residues H248β, R291β, Y300β, K294β, and K164α and lies directly above helix 9β, whose N-terminal

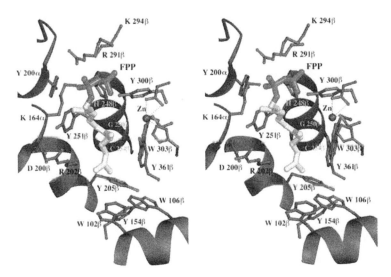

Fig. 4. Stereoview of the active site with bound FPP. Backbone ribbon diagrams indicate portions of the α and β subunits (in red and blue, respectively) forming the binding surface of the FPP molecule. The diphosphate moiety of FPP binds in a positively charged cleft near the subunit interface, with lining enzyme residues shown in pink. The nearby zinc ion is liganded by FTase residues illustrated in gray. The isoprenoid moiety descends into the hydrophobic cavity inside the α–α barrel of the β subunit, whose lining residues are colored in green. (See color plate.)

positive dipole contributes to the strong positive charge of this pocket (Fig. 4).

Site-directed mutagenesis of conserved residues that lie in the FTase active site support the observed location of FPP binding (30). Of the 11 residues investigated, 5 increased $K_{D(FPP)}$ when mutated. Each of these residues (H248β, R291β, K294β, Y300β, and W303β) interacts with the FPP molecule in the cocrystal structure. None of the mutants investigated demonstrated an increased affinity for FPP.

A similar isoprenoid-binding cavity is observed in the crystal structure of farnesyl diphosphate synthase (FPP synthase) (32). This enzyme, a homodimer of 44-kDa subunits, catalyzes the synthesis of FPP using isopentenyl diphosphate and dimethylallyl diphosphate as initial substrates. This enzyme binds the diphosphate moiety of its substrates through two Mg^{2+} to conserved aspartate residues (33), in contrast to FTase, which binds the diphosphate moiety of FPP by interaction with positively charged residues. It is possible that at some stage in the reaction Mg^{2+} may interact with the

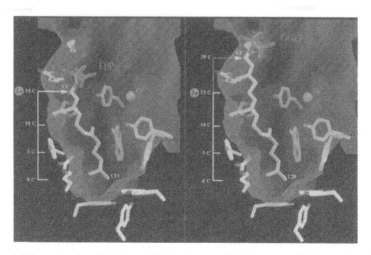

Fig. 5. The molecular ruler hypothesis for isoprenoid substrate specificity. *Left:* Solvent-accessible surface demonstrating the observed FPP-binding location in the binary complex. The catalytic zinc is in register with the C-1 atom of the FPP molecule, the atom that forms a bond with the CaaX cysteine during prenyl transfer. *Right:* Solvent-accessible surface with the modeled GGPP-binding location (based on the FPP position). The extra five carbon atoms of the GGPP isoprenoid chain positions the C-1 atom out of register with the catalytic zinc ion, demonstrating the structural basis for the specificity of FTase for FPP over GGPP. (See color plate.)

diphosphate moiety, as this metal ion is required for the full catalytic activity of FTase (*3, 4*).

D. STRUCTURAL BASIS FOR ISOPRENOID SPECIFICITY

From the cocrystal structure of FTase with bound FPP (*7*), it was hypothesized that the depth of the hydrophobic cavity may act as the primary determinant for the specificity of the 15-carbon FPP molecule over the 20-carbon geranylgeranyl pyrophosphate (GGPP) molecule, the prenyl substrate of GGTase-I. From this structure, it appears that the cavity functions as a molecular ruler, a hypothesis supported by the observation that FPP and GGPP bind to FTase in a competitive manner, but only FPP serves as an effective substrate. One end of this "ruler" is marked by the bottom of the hydrophobic cavity, and the other by the positive cleft at the top of the α–α barrel, where the diphosphate moiety binds. The distance between the catalytic zinc ion and the bottom of the hydrophobic cavity is equal to that between the C-1 atom of the FPP molecule and the bottom of the cavity in the binary complex. This "molecular ruler" hypothesis is illustrated in Fig. 5, a comparison of the observed FPP-binding conformation with a modeled GGPP (created by superimposing the isoprenoid portions of the two molecules). The C-1 atom of the FPP molecule is in register with the catalytic zinc, while the longer geranylgeranyl isoprenoid of GGPP would put the C-1 atom out of register. The isoprenoid-binding cavity of GGTase-I may be deeper, allowing the longer GGPP molecule to bind in a catalytically competent conformation. A deeper binding of GGPP would allow the conserved diphosphate-binding residues to interact with the longer molecule.

FPP synthase may employ a similar mechanism for controlling isoprenoid product chain length. A double mutation (F112A/F113S) was introduced into the hydrophobic cavity of FPP synthase, thereby increasing the depth of the cavity by 5.8 Å, a distance that roughly corresponds to the difference in length between FPP and the longer GGPP. The mutant FPP synthase was capable of producing isoprenoid products longer than FPP (33).

E. PEPTIDE- AND ISOPRENOID-BINDING SITES IN TERNARY COMPLEX

Three crystal structures of rat FTase with bound peptide substrates and nonreactive isoprenoid analogs have been determined (10, 11). One complex (11) shows the peptide substrate acetyl-CVI-selenoM-COOH bound in conjunction with the nonreactive isoprenoid analog α-hydroxyfarnesylphosphonic acid (α-HFP) (34). The two other complexes (10), determined to 2.1- and 2.0-Å resolution, respectively, contain the peptide substrates TKCVIM and KKKSKTKCVIM bound with the nonreactive FPP analog (E,E)-2-[2-oxo-2-[[(3,7,11-trimethyl-2,6,10-dodecatrienyl)oxy]amino]ethyl]phosphonic acid (FPT-II) (35). The chemical structures of both isoprenoid analogs are illustrated in Fig. 6. The peptides used in these studies were derived from the K-Ras4B protein sequence, representing the carboxyl-terminal 4, 6, or 11 residues, respectively. The structures of these three complexes are essentially identical, both to each other and to the previously determined structures. There are, however, a number of active-site side-chain rearrangements when compared with the observed unliganded FTase structure.

In the ternary complexes, the peptide Ca_1a_2X motif binds in an extended conformation in the hydrophobic cavity of the β subunit with the C-terminal carboxylate near the bottom of this cleft (Fig. 7). The C-terminal carboxylate of the methionine residue (X position) hydrogen bonds with Q167α and a buried water molecule coordinated by R202β, H149β, and E198β. The methionine residue is also in van der Waals contact with residues A98β, S99β, W102β, H149β, A151β, P152β, and Y131α. These interactions bury more than 90% of the methionine residue accessible surface area. The isoleucine side chain of the peptide substrate (a_2) is positioned in close proximity to the isoprenoid (Fig. 8), making a hydrophobic interaction,

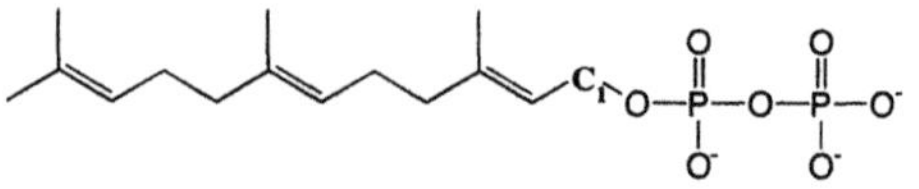

Farnesyl diphosphate (FPP)

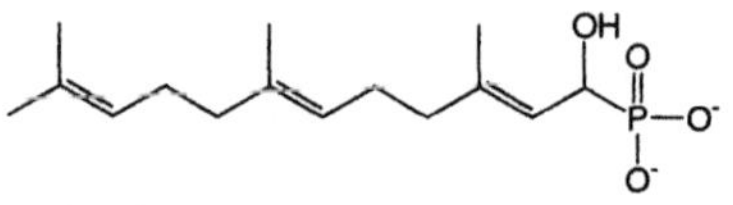

α-Hydroxyfarnesylphosphonic acid (α-HFP)

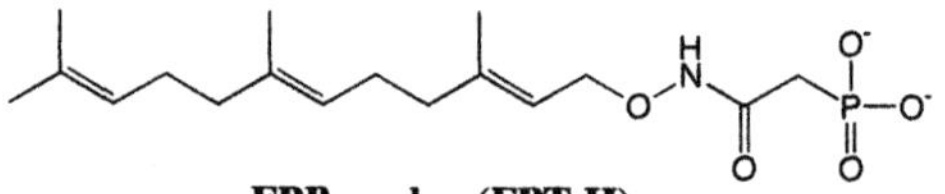

FPP analog (FPT II)

FIG. 6. Isoprenoid molecules used in the ternary complex cocrystal structures.

while the backbone carbonyl oxygen of this residue participates in a hydrogen bond with R202β. The side chain of the peptide valine residue (a_1) points into solvent without making any hydrophobic interactions with the isoprenoid. As predicted from spectroscopic studies (6), the cysteine residue of the peptide Ca_1a_2X sequence directly coordinates the zinc ion. As noted, the only directional, bonded interactions between the peptide CaaX sequence and the protein are the zinc-to-cysteine thiolate interaction and the hydrogen bonds involving the carbonyl oxygen of the a_2 isoleucine residue and the C-terminal carboxylate. Together, these interactions anchor the CaaX peptide into its observed substrate-binding conformation.

In the longer peptide, the polybasic region immediately adjacent to the CaaX motif forms a type-I β turn and binds at the rim of the α–α barrel of the β subunit. This turn is formed by residues $S_4K_5T_6K_7$ of the $K_1K_2K_3S_4K_5T_6K_7C_8V_9I_{10}M_{11}$ sequence (Fig. 9), with peptide backbone angles of K_5, $\phi = -65°$, $\psi = -22°$; T_6, $\phi = -81°$, $\psi = -3°$ (an ideal type I β turn has angles of $\phi = -60°$, $\psi = -30°$; $\phi = -90°$, $\psi = 0°$). The turn is stabilized by a characteristic intrapeptide hydrogen bond between the carbonyl oxygen of S_4 and the backbone nitrogen of T_6. K_3 forms a hydrogen bond with the carbonyl oxygen of C_8, further stabilizing the conformation. The polybasic region participates in several hydrogen bonds with the enzyme as it reaches out of the active pocket, although most of these interactions are water mediated. K_7 forms an ion pair with the conserved residue

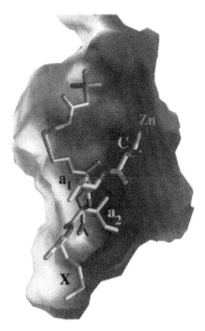

Fig. 7. A portion of the solvent-accessible surface of FTase with bound isoprenoid and peptide substrate in the ternary complex. The isoprenoid binds along one wall of the hydrophobic cavity, in the same position as in the binary complex. The peptide binds in an extended conformation, adjacent to the isoprenoid in the cavity. The COOH terminus binds near the bottom of the cavity, with the cysteine of the CaaX motif coordinating the catalytic zinc ion. The isoprenoid comprises a large portion of the peptide substrate-binding surface and makes a hydrophobic interaction with the isoleucine residue of the peptide (a_2 of Ca_1a_2X). (See color plate.)

$D359\beta$. K_5 shares a hydrogen bond with the carbonyl oxygen of $E161\alpha$. K_1 forms an ion pair with $D91\beta$, and the amino terminus of the peptide sequence forms a hydrogen bond with $E94\beta$.

The isoprenoid forms a large part of the CaaX sequence binding surface, its second and third isoprene units in direct van der Waals contact with the CaaX motif (Fig. 8). The peptide substrate sandwiches the isoprenoid against the wall of the hydrophobic cavity, burying 115 $Å^2$ of accessible FPT-II surface area and completely sequestering the third isoprene unit from the bulk solvent. This structure is consistent with the observation that the reaction is functionally ordered, with FPP binding preceding peptide binding (36, 37). The location and conformation of the isoprenoid are similar to those observed in the binary complex with FPP, with the position of the terminal phosphate of the FPP analog (FPT-II) overlapping that of

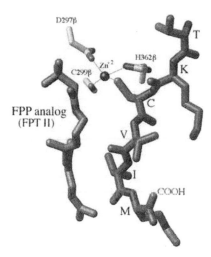

FIG. 8. The binding conformation of the peptide and nonreactive FPP analog FPT-II in the ternary complex. The catalytically active zinc ion coordinates the cysteine residue of the peptide CaaX motif, along with residues D297β, C299β, and H362β. (See color plate.)

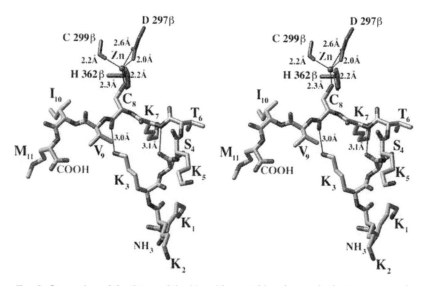

FIG. 9. Stereoview of the β turn of the 11-residue peptide substrate in the ternary complex. This illustration shows the binding conformation of the entire 11-residue peptide substrate (in yellow) in the active site. The N-terminal residues of the long peptide form a type I β turn, with intrapeptide hydrogen bonds shown in black. (See color plate.)

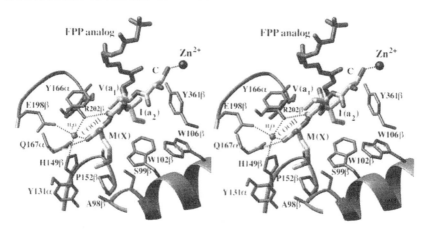

FIG. 10. Stereoview of the active site binding of the peptide CaaX motif (yellow) and FPP analog (purple). Enzyme residues within 4 Å of the CaaX box and the buried water molecule (red), which coordinates the C-terminal carboxylate, are shown. Also illustrated is the catalytically active zinc ion, shown hydrogen bonded with the CaaX cysteine. (See color plate.)

the β-phosphate of FPP. Furthermore, the FTase residues that interact with the FPP molecule (Fig. 4) were also observed to interact with the FPP analog.

Figure 10 illustrates the FTase residues within 4 Å of the CaaX motif of the peptide in the ternary complex. Three α-subunit and eight β-subunit residues are found within this distance: Y131α, Y166α, Q167α, A98β, S99β, W102β, H149β, P152β, E198β, R202β, and Y361β. Of these interactions, there are only two direct FTase–peptide hydrogen bonds: Q167α side chain to the C-terminal carboxylate and R202β to the isoleucine a_2 backbone carbonyl oxygen. All other hydrogen bonds between the peptide and the enzyme are water mediated. Residues H149β, E198β, and R202β coordinate a well-ordered water molecule, which in turn participates in a hydrogen bond with the C-terminal carboxylate.

Recent insight into the role of zinc in peptide substrate binding is obtained from a crystal structure (10) of zinc-depleted rat FTase with bound FPP analog and the K-Ras4B-derived peptide KKKSKTKCVIM. The overall structure of the enzyme is essentially identical to the other FTase structures (rms deviation of 0.23 Å for Cα atoms), indicating that zinc is not required for maintaining the proper three-dimensional fold of FTase. Furthermore, the conformations of the zinc ligands (D297β, C299β, and H362β) are largely undisturbed (rms deviation of 0.37 Å for side-chain atoms). Both the FPP analog and peptide are seen bound in the active site without zinc. The FPP analog conformation is unchanged from the other ternary

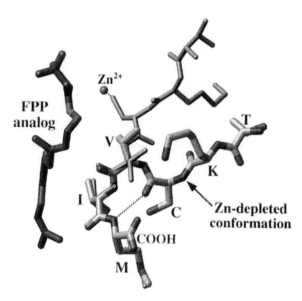

FIG. 11. The β-turn conformation of the CaaX motif of the peptide (yellow) bound to the zinc-depleted FTase. The zinc-coordinated conformation (light blue) is superimposed. Also shown is the FPP analog (FPT-II) (dark blue). (See color plate.)

complexes, supporting the observation that zinc is not required for FPP binding (*4*), while the peptide binds quite differently.

A comparison of the zinc-bound and zinc-depleted peptide conformations is shown in Fig. 11. In the absence of zinc, the conformation of the CaaX motif of the bound peptide closely approximates a type I β turn, with backbone angles of $\phi = -64°$, $\psi = 4°$ for the valine (a_1) residue and $\phi = -112°$, $\psi = 7°$ for the isoleucine (a_2) residue. The residues in the a_2 and X positions bind in the same manner as in the presence of zinc, while the cysteine and valine residues bind much differently. The rms deviation for the a_2X portion of the peptide is 0.36 Å between the structures with and without zinc, while the rms deviation for the Ca_1 portion is 4.8 Å. The cysteine sulfer has shifted 9.2 Å from its position in coordination of the zinc ion. Interactions between the peptide and the FPP substrate are unchanged, as are the interactions between the a_2X residues of the peptide and the active site, indicating that these two C-terminal residues bind independently of zinc and are held in position by hydrogen bonds and van der Waals contacts with the isoprenoid and enzyme.

The binding conformation of the peptide in the absence of zinc indicates that the zinc-thiol(ate) interaction anchors the cysteine residue, holding the peptide in an extended conformation. This observation is consistent with the finding that peptide binding is greatly enhanced when zinc is bound to the enzyme (4). This structure shows that while zinc is not strictly required for peptide binding, it is necessary for the stabilization of a productive peptide substrate conformation.

F. STRUCTURAL BASIS FOR PEPTIDE SUBSTRATE SPECIFICITY

FTase is selective for protein substrates that have a cysteine residue at the fourth position from the C terminus. Peptide competition studies have shown that the cysteine residue must be at this position and not the third or fifth position from the C terminus (3, 38). The structural basis for this length specificity has been proposed from the binding of Ca_1a_2X peptides (10). Peptide binding to zinc-depleted FTase indicates that the conformation of the a_2X portion of the peptide is independent of thiol(ate)-zinc coordination. Hydrogen bonds made with the C-terminal carboxylate and the carbonyl oxygen of the a_2 residue help orient this part of the peptide. These hydrogen bonds may function as a peptide C-terminal anchor. The dramatic difference in how the cysteine binds in the absence of zinc indicates that interaction with zinc anchors this residue. FTase may distinguish its CaaX protein substrates from proteins that have a cysteine residue at other positions according to the distance between the cysteine residue and the C-terminus using a molecular ruler. The ends of the molecular ruler on the protein correspond to the zinc ion and the molecular interactions that anchor the a_2X portion of the peptide.

The selectivity of FTase for Ca_1a_2X peptide substrates has largely been assigned to the X residue. If X is methionine, serine, glutamine, or alanine, the protein substrate is modified by FTase under most circumstances, whereas if the X residue is leucine, GGTase-I attaches a geranylgeranyl isoprenoid to the protein acceptor (1, 39, 40). There are exceptions to this general basis of selectivity, such as the observation that K-Ras4B can act as a moderately efficient substrate for GGTase-I, despite having a methionine as the C-terminal residue (2, 41–43).

The protein substrate selectivity of FTase is consistent with modeled interactions with other Ca_1a_2X peptides based on the observed binding of K-Ras4B-derived peptides (10). In this modeling study, the observed conformation of the peptide backbone was held fixed and the ability of FTase to accommodate different CaaX sequences assessed. The results of this modeling study were consistent with the experimentally observed peptide selectivity of FTase (1, 38). The a_2X residues are in extensive van der Waals contact with the enxyme–FPP analog complex (Fig. 7). The GGTase–I peptide substrate CVIL was modeled in the FTase active site and found to have steric clashes with the C-terminal leucine. On the other hand, the FTase peptide substrates CVIS, CVIA, and CVIQ fit without steric clashes. Although glutamate and lysine can be modeled without steric hindrance at the C-terminus, their charged nature would not be accommodated.

The a_2 position of the Ca_1a_2X peptide was analyzed in the same manner. Leucine and isoleucine in position a_2 of the peptide fit without causing steric clashes. Phenylalanine, tyrosine, and tryptophan at position a_2 have only minor steric overlap that could be accommodated by a shift of the peptide backbone away from Trp102β and Trp106β on the order of 0.5 Å. Charged residues such as aspartic acid would not be accommodated because of the highly hydrophobic nature of the a_2 binding surface. These results are consistent with FTase peptide substrate specificity (38). The side-chain of the a_1 residue of the Ca_1a_2X peptide is exposed to solvent (Figs. 7 and 10) and can easily accommodate a variety of amino acids. FTase shows little selectivity for this residue, but Ca_1a_2X peptides with amino acids that tend to disrupt backbone structure, such as proline and glycine, at the a_1 position have lower affinity (38). From the crystallographic studies of FTase with bound Ca_1a_2X peptides, it appears that interactions with both the a_2 and X residues are important for peptide substrate specificity.

G.   MAGNESIUM-BINDING SITE

A structure of rat FTase complexed with FPT-II and TKCVIM in the presence of both zinc and manganese has been determined to 2.1-Å resolution (10). As manganese is capable of substituting for Mg^{2+} in catalysis (4), yet is more easily detectable by X-ray crystallography because of its anomalous diffraction, the purpose of the experiment was to determine the location of the magnesium-binding site.

The peptide and FPP analog bind similarly as compared with the previously described ternary complex. The portion of the analog corresponding to the α-phosphate of FPP is slightly less ordered in this structure, however. The manganese ion was observed bound above the diphosphate-like moiety of FPT-II (Fig. 12), coordinating with the α- and β-phosphate mimics of the analog molecule. The ion does not form any ligands with the enzyme or peptide substrate. Of interest, however, is the movement of Lys164α. This residue was observed liganded with the α-phosphate of FPP in the binary complex and the corresponding carbonyl of FPT-II in the ternary complexes without manganese or magnesium. In the presence of the manganese ion, this residue has shifted away from the FPP analog, and is no longer within hydrogen-bonding distance.

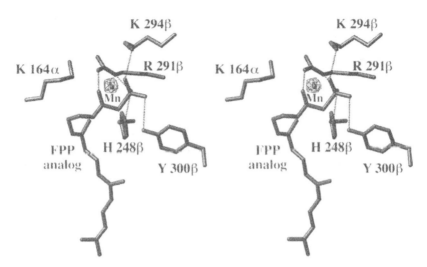

FIG. 12. Stereoview of Mn^{2+} bound to the enzyme. An anomalous electron density map was calculated from 10 to 2.5 Å, using phases from the final model and contoured at 3.5σ, indicating the location of the Mn^{2+} ion. The enzyme ligands (gray) that interact with the diphosphate of FPP are shown forming hydrogen bonds with the FPP analog (brown). (See color plate.)

H. FARNESYLATED PRODUCT COMPLEX

Kinetic studies indicate that binding of the FPP molecule to FTase is the first step in the reaction pathway followed by peptide/protein substrate binding and catalysis (37, 44, 45). Interestingly, the rate-limiting step in the reaction is the release of the farnesylated product (37, 45) and is facilitated by the binding of additional substrate (46). The crystal structures of the ternary complexes already described show FTase substrates bound in the active site of the enzyme. In these structures, C-1 atom of the isoprenoid and the cysteine sulfur of the CaaX peptide are separated by approximately 8 Å, much too far for catalysis. The substrates must move closer to one another for the reaction to occur. A recent crystal structure of rat FTase with bound farnesylated peptide product strongly suggests that it is the isoprenoid that moves to bridge this gap (12).

The overall structure of the product-bound enzyme is virtually identical to that of the ternary complex. The only active site residues that have shifted significantly are those that interact with the diphosphate moiety of FPP, which is reasonable as the pyrophosphate by-product is not observed in this complex. The largest motion is seen in the diphosphate ligands H248β and Y300β, which have moved approximately 1.6 and 1.1 Å, respectively.

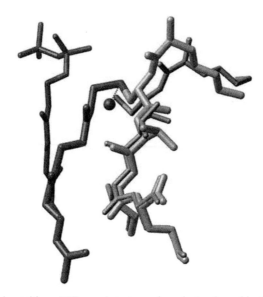

FIG. 13. Superimposition of FTase substrates and product as bound in the active site. The peptide portion of the product (yellow) binds in the same position as the corresponding peptide substrate (blue). The isoprenoid portion of the product (brown) binds as the FPP molecule (pink) does in the ternary complex through the third isoprene unit, while the first and second units shift and rotate, closing the distance between the C-1 atom of the FPP molecule and the cysteine sulfur of the CaaX peptide. (See color plate.)

A comparison of the bound product with the substrates (Fig. 13) demonstrates little change in the positioning of the CaaX motif. This observation validates the peptide-binding conformation observed in previous ternary complexes and suggests that these complexes represent a peptide substrate discrimination step in the reaction. In this product complex, the C-1 atom of the isoprenoid is approximately 6 Å closer to the catalytic zinc ion than the corresponding atom of the FPP substrate. The third isoprene unit of the farnesylated product retains its substrate-binding position, while the first and second isoprene units have shifted and rotated, suggesting that catalysis involves the movement of the isoprenoid toward the peptide.

III. Mutagenesis Studies

Site-directed mutagenesis studies of rat, human, and yeast FTase have identified many residues linked to substrate binding (*29, 30, 47–51*), substrate specificity (*52–54*), and catalytic activity (*21, 29, 30, 47, 51*). Comparison of this list of residues with the enzyme structure reveals that most of these residues are positioned near the site of FPP or peptide binding or stabilize residues in these regions.

A. MUTATIONS AFFECTING SUBSTRATE BINDING

Kinetic and biochemical analyses of site-directed mutants in human FTase β-subunit mutants suggest H248β, R291β, K294β, and W303β involvement in the binding and utilization of the FPP substrate (30). In another study (29), the E256A mutation in the β subunit of yeast FTase (corresponding to residue E246 in rat FTase) resulted in a 130-fold higher $K_{m(FPP)}$. In the binary complex with FPP, H248β, R291β, K294β, and W303β are observed to interact with the FPP molecule (Fig. 5), while E246β stabilizes the side-chain conformation of R291β. Further study (51) indicates the significance of residues Y200α, H201α, and H248β in FPP binding. The crystal structures of binary and ternary complexes of FTase show that Y200α and Y201α contribute to the hydrophobic binding pocket for the first isoprene unit of FPP (Fig. 4).

Analyses of the β subunit of yeast FTase (39, 40) and GGTase-I (41) have identified three mutations (D209N, G259V, and G328S) that increase the K_m of protein substrates. The corresponding substitutions (D200N, G249V, and G349S) were introduced into the human enzyme and the functional consequences examined (47). The D200N and G349S mutations increased the $K_{m(peptide)}$ only (29, 47) while the G249V muatation resulted in an increase in both $K_{m(peptide)}$ and $K_{m(FPP)}$. In a separate study, mutation of R202β in human FTase to alanine resulted in a >400-fold elevation in K_m for protein substrate (30). Further analysis of this mutant, using peptide-derived inhibitors, suggested that R202β interacts with the COOH terminus of CaaX peptide substrates (30). G249β is adjacent to H248β, which interacts with the diphosphate moiety of the FPP. R202β makes a hydrophobic interaction with the isoprenoid molecule and coordinates the C-terminal carboxylate of the peptide substrate through a well-ordered water molecule. D200β stabilizes the conformation of R202β.

B. MUTATIONS AFFECTING SUBSTRATE SPECIFICITY

Random screening identified three mutants (S159N, Y362H, and Y366N) in the β subunit of yeast FTase with relaxed protein substrate specificity, having gained the ability to farnesylate GGTase-I protein substrates (52). A similar effect can be observed by substitution of Y362β with a smaller side chain (52). These observations suggest that these residues are in close proximity to the protein substrate-binding site. In the ternary complex Y362β (Y361β in rat) is within 4 Å of the peptide substrate, and Y366β (Y365β in rat) is adjacent. Although not conserved, rat P152β (S159β in yeast) contributes to the surface that interacts with the C-terminal methionine side chain of the peptide substrate in the ternary complex with isoprenoid (Fig. 9). In another study on yeast FTase, residues conserved among FTase β subunits were mutated to their conserved counterparts in GGTase-I and the substrate specificities of the mutant yeast FTase enzymes were examined (53). Three regions in the β subunit of FTase were identified as participating in CaaX peptide substrate selectivity: residues 74, 206–212, and 351–354 (corresponding to β-subunit residues 67, 197–203, and 350–353 in rat FTase, respectively). In the crystal structure, L67β is located on a loop connecting helices 1β and 2β and is distant from the active site. Residues G197β–S203β make up part of the loop connecting helices 6β and 7β and are located near the bottom of the hydrophobic cavity in the β subunit. R202β, on this loop, interacts with both the isoprenoid and peptide substrates in the ternary complex, and its conformation is stabilized by hydrogen bonds made with D200β (Figs. 4 and 10). Residues L350β–K353β are on a loop connecting helices 12β and 13β and lie above the location of the catalytic zinc ion.

Interestingly, these same mutations affected the isoprenoid selectivity of the mutant enzymes. Moreover, the isoprenoid selectivity was observed to be affected by the CaaX peptide substrate used. These observations of interdependent selectivity in yeast FTase are consistent with the observation that the substrates directly interact in a ternary complex (10, 11).

C. Mutations Affecting Catalytic Turnover

Eleven conserved residues in the β subunit of human FTase, which lie in the active site region, were mutated and their steady state kinetic properties analyzed (30). Six mutants, R202A, D297A, C299A, Y300F, D359A, and H362A (corresponding to the same residues in rat FTase), had lower k_{cat} values than the wild-type enzyme. Mutation of the corresponding residues in yeast FTase, with the exception of D359β, which was not studied, also reduced k_{cat} (29). A more recent study (42) confirms the importance of residues K164α and Y300β through steady state, presteady state, and pH studies. In the crystal structures of FTase, D297β, C299β, and H362β coordinate the catalytic zinc ion, and D359β stabilizes the conformation of H362β. R202β forms a hydrophobic interaction with the isoprenoid and hydrogen bonds with the peptide (Figs. 4 and 10). Y300β makes a hydrogen bond with the terminal phosphate of the FPP molecule (Fig. 4).

Five conserved residues in the α subunit of rat FTase (K164N, Y166F,

R172E, N199D, and W203H) have also been analyzed (*21*). The K164N mutation significantly decreases FTase activity (*21, 51*). In the cocrystal structure of FTase with FPP, K164α lies directly above the C-1 atom of the FPP molecule, the site of prenyl transfer, consistent with a direct role in catalysis (*9*). The four other mutations reduced FTase activity from 30 to 75% of wild-type values. Y166α and R172α are on helix 5α. Y166α contributes to the surface that binds both substrates in the ternary complex (*10*). Y200α also forms part of the isoprenoid-binding surface, which may explain why mutation of the nearby residues N199α or W203α affected catalytic turnover.

IV. Inhibition of Catalytic Activity

Three types of genetic mutations have been determined to give rise to the pathogenesis of nearly all cancers. These mutations occur on oncogenes, tumor suppressor genes, and genes controlling DNA replication fidelity. Among the oncogenes currently under study, *ras* oncogenes are prime targets for cancer treatments. Mutations in *ras* genes are associated with approximately 30% of all human cancers, most commonly in pancreatic cancer, colorectal cancer, and adenocarcinoma of the lung.

In normal cell growth, Ras cycles between the GTP-bound and GDP-bound states, with the GDP-bound form considered the inactive state. Interaction with guanine nucleotide exchange factors converts Ras to its GTP-bound state, activating its signaling function, and "turning on" cell division. Normally, the Ras protein itself cleaves the GTP, returning to an inactive state, but mutations in the *ras* gene often inactivate the GTPase activity, locking Ras in an active state. With constant Ras signaling, the cell undergoes continuous cell division, resulting in cancerous proliferation (*55–57*).

Ras proteins require posttranslational prenylation for cellular activity. In particular, K-Ras, the most commonly mutated form of Ras found in human cancerous cell lines, is a high-affinity substrate for FTase. Consequently, specific inhibition of FTase may prove to be an effective anticancer strategy. This possibility has prompted extensive research aimed at the discovery of potent FTase inhibitors. Both peptide and FPP-derived inhibitors, in addition to other classes of FTase inhibitors, have been studied kinetically, structurally, and clinically and continue to show great promise as chemotherapeutic agents (reviewed in *56, 58*),

A. PEPTIDE-DERIVED INHIBITORS

Early studies of substrate specificity (*2, 38, 59*) indicate that FTase activity is inhibited by tetrapeptides with cysteine in the fourth position from the

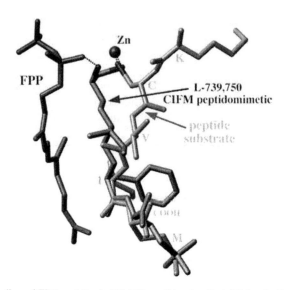

FIG. 14. Binding of FPP and the L-739,750 peptidomimetic inhibitor in the active site of human FTase. L-739,750 (light blue), as shown with the CVIM peptide substrate (yellow) superimposed, binds much more closely to the FPP molecule (blue). This conformation may hinder the movement of FPP toward the reactive cysteine, preventing catalysis. (See color plate.)

C terminus. It was noted that the greatest inhibition occurred when the a_1 and a_2 positions were occupied by nonpolar aliphatic or aromatic amino acids, with the a_2 position being more stringent. Substitution of an aromatic residue such as phenylalanine in the a_2 position results in a purely inhibitory complex that is not farnesylated. Interestingly, the removal or substitution of the amino group on the cysteine of a tetrapeptide inhibitor restores farnesylation of the peptide (*59, 60*). However, tetrapeptides as clinical molecules have three main disadvantages: poor cellular uptake, rapid degradation, and FTase farnesylation (in some cases), which results in inactivation of the inhibitor. A number of nonpeptide inhibitors that compete with peptide substrates have been developed and characterized (*8, 61–70*). The structures of two classes of inhibitors bound to FTase have been determined by X-ray crystallography (*8, 70*).

One of the first examples of a peptidomemetic inhibitor is the Merck compound L-739,750 (*69, 71*). This compound, modeled after the tetrapeptide CIFM, is the metabolic product of the prodrug L-744,832, which, when administered in rats, resulted in tumor regression without systemic toxicity (*69*). Shown complexed with human FTase and the FPP substrate in Fig. 14 (*8*), L-739,750 binds to the β-subunit active site in an extended conformation, overlapping the peptide-binding site observed in the rat FTase ternary complex with the FPP analog and peptide substrate. The cysteine moiety directly coordinates the catalytic zinc ion, with a sulfur-to-zinc distance of 2.3 Å, similar to that observed in the peptide ternary complex. The carboxyl terminus of the inhibitor binds in the same position as the peptide substrate, forming hydrogen bonds with Q167α and a buried water that is coordinated by R202β, E198β, and H149β. R202β also hydrogen bonds with the carbonyl oxygen of the phenylalanine-derived group, as it does in the equivalent position of the peptide complex. The N-terminal portion of the peptidomimetic inhibitor, corresponding to the cysteine and a_1 residues of the peptide substrate, binds in a different conformation than do the analogous residues in the peptide complex. This region of the inhibitor binds adjacent to the isoprenoid, with a 3.4-Å distance between the cysteine C_α atoms of the inhibitor and substrate, while still allowing the cysteine of the inhibitor to coordinate the zinc ion. This conformation is stabilized in part by a hydrogen bond between the amino terminus of the inhibitor and an oxygen on the α-phosphate of the FPP molecule, in addition to hydrophobic interactions between the phenylalanine-derived side chain of the inhibitor and residues W102β, W106β, and Y361β. A portion of the inhibitory activity of this compound may be due to the binding of its amino terminus. This region of the molecule binds (Fig. 14) between the isoprenoid and the substrate-binding location. In this conformation, L-739,750 may interfere with the movement of the isoprenoid toward the peptide during catalysis, thereby preventing FPP from reacting with the cysteine.

A class of nonpeptide inhibitors discovered at Schering-Plough includes a number of tricyclic compounds that also compete with the peptide substrate (72, 73). High-resolution co-crystal structures of ten related tricyclic compounds have been determined, complexed with FTase and FPP (70). These compounds are composed of three connected rings, with a tail containing two more rings extending from the central seven-carbon ring. One of these compounds, SCH 66336 (4-[2-[4-(3,10-dibromo-8-chloro-6,11-dihydro-5H-benzo-[5,6]cyclohepta[1,2-b]pyridin-11-yl)-1-piperidinyl]-2-oxoethyl]-1-piperidine-carboxamide), is currently undergoing clinical trials as an anticancer agent. Crystal structures show that the binding site of these compounds partially overlaps with the peptide binding site. The compounds bind with the tricyclic moiety positioned near the zinc and the tail descending into the cavity of the α subunit. Each compound participates in extensive hydrophobic interactions with both the enzyme and the FPP molecule. These inhibitors also form hydrogen bonds with FTase, most being water-mediated (70).

B.   FARNESYL DIPHOSPHATE-DERIVED INHIBITORS

A number of FPP analogs have been developed that are competitive inhibitors of FTase with respect to FPP (*34, 35, 74–78*). The selectivity of these inhibitors for FTase over other cellular enzymes, including GGTase-I, has stimulated their investigation as possible antitumor agents. The crystal structure of FTase with FPP and FPP analogs provide a structural framework from which to interpret the inhibitory potency of these inhibitor molecules. Not surprisingly, among the FPP-based inhibitor molecules, the most effective retained a hydrophobic farnesyl group and a negatively charged moiety mimicking the diphosphate (*75*). For one of these potent and selective FTase inhibitors [designated compound 3, IC_{50} of 75 nM, (*75*)], a systematic structure–activity analysis was carried out that indicated that the most potent inhibitors contained a terminal phosphate group rather than other types of negatively charged groups. This is consistent with the interactions seen with the terminal phosphate in the FTase complexes, specifically with W300β, which appears to interact with this phosphate. The length of the hydrophobic chain also had a dramatic effect on the activity in this study. Homoelongation of the farnesyl group by one carbon resulted in a decrease in IC_{50} of more than 200-fold, consistent with the crystal structure presented here and the molecular ruler hypothesis for prenyl substrate specificity.

Two FPP analogs, FPT-II and α-HFP (Fig. 6), have been crystallized with FTase in a number of complexes with K-Ras peptides (*10, 11*). Several of these structures have already been discussed in this chapter. Both the α-HFP and the FPT-II mimic the binding of FPP. These analogs possess a farnesyl isoprenoid moiety that interacts extensively with the peptide substrate, as well as the enzyme. The diphosphate mimic binds in the diphosphate site and forms hydrogen bonds with the enzyme.

V.   Summary

The high-resolution crystal structures of FTase complexed with substrates and inhibitors described in this chapter provide a framework for understanding the molecular basis of substrate specificity and mechanism that may facilitate the development of improved chemotherapeutics. The co-crystal structure of FTase with a FPP substrate suggests a "molecular ruler" mechanism for isoprenoid substrate specificity whereby the length of the isoprenoid moiety is selected based on the depth of the hydrophobic pocket into which it binds. Multiple structures of FTase with nonreactive FPP analogs and K-Ras peptides identified the substrate binding sites and, together with a recently determined farnesylated product complex, suggest how critical features of the CaaX peptide motif are recognized. Additionally, complexes with and without zinc illustrate the critical role of this ion both in peptide binding and in the catalytic reaction. The structural studies are consistent with mutagenesis data on this enzyme. The structures of FTase inhibitor complexes are assisting in the design and optimization of cancer chemotherapeutics agents. Several FTase inhibitors are currently in clinical trials as cancer treatments, and many more are expected soon to join them. However, understanding the detailed enzymatic mechanism and testing the hypotheses for the molecular basis of substrate specificity will require further biochemical, genetic, and structural analyses.

REFERENCES

1. Zhang, F. L., and Casey, P. J. (1996). *Annu. Rev. Biochem.* **65,** 241.
2. Gibbs, J. B., and Oliff, A. (1997). *Annu. Rev. Pharmacol. Toxicol.* **37,** 143.
3. Reiss, Y., Goldstein, J. L., Seabra, M. C., Casey, P. J., and Brown, M. S. (1990). *Cell* **62,** 81.
4. Reiss, Y., Brown, M. S., and Goldstein, J. L. (1992). *J. Biol. Chem.* **267,** 6403.
5. Chen, W.-J., Moomaw, J. F., Overton, L., Kost, T. A., and Casey, P. J. (1993). *J. Biol. Chem.* **268,** 9675.
6. Huang, C.-C., Casey, P. J., and Fierke, C. A. (1997). *J. Biol. Chem.* **272,** 20.
7. Park, H.-W., Boduluri, S. R., Moomaw, J. F., Casey, P. J., and Beese, L. S. (1997). *Science* **275,** 1800.
8. Long, S. B., and Beese, L. S. (in preparation).
9. Long, S., Casey, P. J., and Beese, L. S. (1998). *Biochemistry* **37,** 9612.
10. Long, S. B., Casey, P. J., and Beese, L. S. (2000). *Structure Fold. Des.* **8**(2), 209.
11. Strickland, C. L., Windsor, W. T., Syto, R., Wang, L., Bond, R., Wu, Z., Schwartz, J., Le, H. V., Beese, L. S., and Weber, P. C. (1998). *Biochemistry* **37,** 16601.
12. Long, S. B., Terry, K. L., and Beese, L. S. (in preparation).
13. Chen, W.-J., Andres, D. A., Goldstein, J. L., Russell, D. W., and Brown, M. S. (1991). *Cell* **66,** 327.
14. Chen, W.-J., Andres, D. A., Goldstein, J. L., and Brown, M. S. (1991). *Proc. Natl. Acad. Sci. U.S.A.* **88,** 11368.
15. Raag, R., Appelt, K., Xuong, N. H., and Banaszak, L. (1988). *J. Mol. Biol.* **200,** 553.
16. Thunnissen, A. M., Dijkstra, A. J., Kalk, K. H., Rozeboom, H. J., Engel, H., Keck, W., and Dijkstra, B. W. (1994). *Nature (London)* **367,** 750.
17. Juy, M., Amit, A. G., Alzari, P. M., Poljak, R. J., Claeyssens, M., Beguin, P., and Aubert, J. P. (1992). *Nature (London)* **357,** 89.
18. Alzari, P. M., Souchon, H., and Dominguez, R. (1996). *Structure* **4,** 265.
19. Aleshin, A., Golubev, A., Firsov, L. M., and Honzatko, R. B. (1992). *J. Biol. Chem.* **267,** 19291.
20. Aleshin, A. E., Hoffman, C., Firsov, L. M., and Honzatko, R. B. (1994). *J. Mol. Biol.* **238,** 575.
21. Andres, D. A., Goldstein, J. L., Ho, Y. K., and Brown, M. S. (1993). *J. Biol. Chem.* **268,** 1383.
22. Dunten, P., Kammlott, U., Crowther, R., Weber, D., Palermo, R., and Birktoft, J. (1998). *Biochemistry* **37,** 7907.

23. Boguski, M., Murray, A. W., and Powers, S. (1992). *New Biol.* **4,** 408.
24. Janin, J., Miller, S., and Chothia, C. (1988). *J. Mol. Biol.* **204,** 155.
25. Casey, P. J., and Seabra, M. C. (1996). *J. Biol. Chem.* **271,** 5289.
26. Moomaw, J. F., and Casey, P. J. (1992). *J. Biol. Chem.* **267,** 17438.
27. Zhang, F. L., Moomaw, J. F., and Casey, P. J. (1994). *J. Biol. Chem.* **269,** 23465.
28. Fu, H.-W., Moomaw, J. F., Moomaw, C. R., and Casey, P. J. (1996). *J. Biol. Chem.* **271,** 28541.
29. Dolence, J. M., Rozema, D. B., and Poulter, C. D. (1997). *Biochemistry* **36,** 9246.
30. Kral, A. M., Diehl, R. E., deSolms, S. J., Williams, T. M., Kohl, N. E., and Omer, C. A. (1997). *J. Biol. Chem.* **272,** 27319.
31. Fu, H.-W., Beese, L. S., and Casey, P. J. (1998). *Biochemistry* **37,** 4465.
32. Tarshis, L. C., Yan, M., Poulter, C. D., and Sacchettini, J. C. (1994). *Biochemistry* **33,** 10871.
33. Tarshis, L. C., Proteau, P. J., Kellogg, B. A., Sacchettini, J. C., and Poulter, C. D. (1996). *Proc. Natl. Acad. Sci. U.S.A.* **93,** 15018.
34. Manne, V., Ricca, C. S., Brown, J. G., Tuomari, A. V., Yan, N., Patel, D., Schmidt, R., Lynch, M. J. Ciosek, C. P., Jr., Carboni, J. M., Robinson, S., Gordon, E. M., Barbacid, M., Seizinger, B. R., and Biller, S. A. (1995). *Drug Dev. Res.* **34,** 121.
35. Gibbs, J. B., Pompliano, D. L., Mosser, S. D., Rands, E., Lingham, R. B., Singh, S. B., Scolnick, E. M., Kohl, N. E., and Oliff, A. (1993). *J. Biol. Chem.* **268,** 7617.
36. Pompliano, D. L., Schaber, M. D., Mosser, S. D., Omer, C. A., Shafer, J. A., and Gibbs, J. B. (1993). *Biochemistry* **32,** 8341.
37. Furfine, E. S., Leban, J. J., Landavazo, A., Moomaw, J. F., and Casey, P. J. (1995). *Biochemistry* **34,** 6857.
38. Reiss, Y., Stradley, S. J., Gierasch, L. M., Brown, M. S., and Goldstein, J. L. (1991). *Proc. Natl. Acad. Sci. U.S.A.* **88,** 732.
39. Yokoyama, K., Goodwin, G. W., Ghomashchi, F., Glomset, J. A., and Gelb, M. H. (1991). *Proc. Natl. Acad. Sci. USA* **88,** 5302.
40. Roskoski, R., Jr., and Ritchie, P. (1998). *Arch. Biochem. Biophys.* **356**(2), 167.
41. Trueblood, C. E., Ohya, Y., and Rine, J. (1993). *Mol. Cell. Bio.* **13,** 4260.
42. Whyte, D. B., Kirschmeier, P., Hockenberry, T. N., Nunez-Oliva, I., James, L., Catino, J. J., Bishop, W. R., and Pai, J.-K. (1997). *J. Biol. Chem.* **272,** 14459.
43. Rowell, C. A., Kowalczyk, J. J., Lewis, M. D., and Garcia, A. M. (1997). *J. Biol. Chem.* **272,** 14093.
44. Dolence, J. M., Cassidy, P. B., Mathis, J. R., and Poulter, C. D. (1995). *Biochemistry* **34,** 16687.
45. Mathis, J. R., and Poulter, C. D. (1997). *Biochemistry* **36,** 6367.
46. Tschantz, W. R., Furfine, E. S., and Casey, P. J. (1997). *J. Biol. Chem.* **272,** 9989.
47. Omer, C. A., Kral, A. M., Diehl, R. E., Prendergast, G. C., Powers, S., Allen, C. M., Gibbs, J. B., and Kohl, N. E. (1993). *Biochemistry* **32,** 5167.
48. Goodman, L. E., Judd, S. R., Farnsworth, C. C., Powers, S., Gelb, M. H., Glomset, J. A., and Tamanoi, F. (1990). *Proc. Natl. Acad. Sci. U.S.A.* **87,** 9665.
49. Powers, S., Michaelis, S., Broek, D., Santa Anna, S., Field, J., Herskowitz, I., and Wigler, M. (1986). *Cell* **47,** 413.
50. Ohya, Y., Goebl, M., Goodman, L. E., Petersen-Bjorn, S., Friesen, J. D., Tamanoi, F., and Anraku, Y. (1991). *J. Biol. Chem.* **266,** 12356.
51. Wu, Z., Demma, M., Strickland, C. L., Radisky, E. S., Poulter, C. D., Le, H. V., and Windsor, W. T. (1999). *Biochemistry* **38,** 11239.
52. Del Villar, K., Mitsuzawa, H., Yang, W., Sattler, I., and Tamanoi, F. (1997). *J. Biol. Chem.* **272,** 680.
53. Caplin, B. E., Ohya, Y., and Marshall, M. S. (1998). *J. Biol. Chem.* **273,** 9472.

54. Del Villar, K., Urano, J., Guo, L., and Tamanoi, F. (1999). *J. Biol. Chem.* **274,** 27010.
55. Bourne, H. R., Sanders, D. A., and McCormick, F. (1991). *Nature (London)* **349,** 117.
56. Boguski, M. S., and McCormick, F. (1993). *Nature (London)* **366,** 643.
57. Quilliam, L. A., Khosravi-Far, R., Huff, S. Y., and Der, C. J. (1995). *Bioessays* **17,** 395.
58. Oliff, A. (1999). *Biochim. Biophys. Acta* **1423,** C19.
59. Goldstein, J. L., Brown, M. S., Stradley, S. J., Reiss, Y., and Geirasch, L. M. (1991). *J. Biol. Chem.* **266,** 15575.
60. Brown, M. S., Goldstein, J. L., Paris, K. J., Burnier, J. P., and Marsters, J. C. (1992). *Proc. Natl. Acad. Sci. U.S.A.* **89,** 8313.
61. James, G. L., Goldstein, J. L., Brown, M. S., Rawson, T. E., Somers, T. C., McDowell, R. S., Crowley, C. W., Lucas, B. K., Levinson, A. D., and Marsters, J. C., Jr. (1993). *Science* **260,** 1937.
62. Kohl, N. E., Mosser, S. D., deSolms, S. J., Giuliani, E. A., Pompliano, D. L., Graham, S. L., Smith, R. L., Scolnick, E. M., Oliff, A., and Gibbs, J. B. (1993). *Science* **260,** 1934.
63. Garcia, A. M., Rowell, C., Ackermann, K., Kowalczyk, J. J., and Lewis, M. D. (1993). *J. Biol. Chem.* **268,** 18415.
64. Qian, Y., Blaskovich, M. A., Saleem, M., Seong, C. M., Wathen, S. P., Hamilton, A. D., and Sebti, S. (1994). *J. Biol. Chem.* **269,** 12410.
65. Vogt, A., Qian, Y., Blaskovich, M. A., Fossum, R. D., Hamilton, A. D., and Sebti, S. M. (1995). *J. Biol. Chem.* **270,** 660.
66. deSolms, S. J., Giuliani, E. A., Graham, S. L., Koblan, K. S., Kohl, N. E., Mosser, S. D., Oliff, A. I., Pompliano, D. L., Rands, E., Scholz, T. H., Wiscount, C. M., Gibbs, J. B., and Smith, R. L. (1998). *J. Med. Chem.* **41,** 2651.
67. Shen, W., Fakhoury, S., Donner, G., Henry, K., Lee, J., Zhang, H., Cohen, J., Warner, R., Saeed, B., Cherian, S., Tahir, S., Kovar, P., Bauch, J., Ng, S.-C., Marsh, K., Sham, H., and Rosenberg, S. (1999). *Bioorg. Med. Chem. Lett.* **9,** 703.
68. Liu, R., Dong, D. L.-Y., Sherlock, R., and Nestler, H. P. (1999). *Bioorg. Med. Chem. Lett.* **9,** 847.
69. Kohl, N. E., Omer, C. A., Conner, M. W., Anthony, N. J., Davide, J. P., deSolms, S. J., Giuliani, E. A., Gomez, R. P., Graham, S. L., Hamilton, K., Handt, L. K., Hartman, G. D., Koblan, K. S., Kral, A. M., Miller, P. J., Mosser, S. D., O'Neill, T. J., Rands, E., Schaber, M. D., Gibbs, J. B., and Oliff, A. (1995). *Nature Med.* **1,** 792.
70. Strickland, C. L., Weber, P. C., Windsor, W. T., Wu, Z., Le, H. V., Albanese, M. M., Alvarez, C. S., Cesarz, D., del Rosario, J., Deskus, J., Mallams, A. K., Njoroge, F. G., Piwinski, J. J., Remiszewski, S., Rossman, R. R., Taveras, A. G., Vibulbhan, B., Doll, R. J., Girijavallabhan, V. M., and Ganguly, A. K. (1999). *J. Med. Chem.* **42,** 2125.
71. Kohl, N. E., Wilson, F. R., Mosser, S. D., Giuliani, E., deSolms, S. J., Conner, M. W., Anthony, N. J., Holtz, W. J., Gomez, R. P., Lee, T. J., and et al. (1994). *Proc. Natl. Acad. Sci. U.S.A.* **91**(19), 9141.
72. Njoroge, F. G., Taveras, A. G., Kelly, J., Remiszewski, S., Mallams, A. K., Wolin, R., Afonso, A., Cooper, A. B., Rane, D. F., Liu, Y., Wong, J., Vibulbhan, B., Pinto, P., Deskus, J., Alvarez, C. S., del Rosario, J., Connolly, M., Wang, J., Desani, J., Rossman, R. R., Bishop, W. R., Patton, R., Wang, L., Kirschmeier, P., Bryant, M. S., Nomeir, A. A., Lin, C., Liu, M., McPhail, A. T., Doll, R. J., Girijavallabhan, V. M., and Ganguly, A. K. (1998). *J. Med. Chem.* **41,** 4890.
73. Bishop, W. R., Bond, R., Petrin, J., Wang, L., Patton, R., Doll, R., Njoroge, G., Catino, J., Schwartz, J., Windsor, W., Syto, R., Carr, D., James, L., and Kirschmeier, P. (1995). *J. Biol. Chem.* **270,** 30611.
74. Cohen, L. H., Valentijn, A. R. P. M., Roodenburg, L., Van Leeuwen, R. E. W., Huisman,

R. H., Lutz, R. J., Van Der Marel, G. A., and Van Boom, J. H. (1995). *Biochem. Pharmacol.* **49,** 839.

75. Patel, D. V., Schmidt, R. J., Biller, S. A., Gordon, E. M., Robinson, S. S., and Manne, V. (1995). *J. Med. Chem.* **38,** 2906.
76. Yonemoto, M., Satoh, T., Arakawa, H., Suzuki-Takahashi, I., Monden, Y., Kodera, T., Tanaka, K., Aoyama, T., Iwasawa, Y., Kamei, T., Nishimura, S., and Tomimoto, K. (1998). *Mol. Pharm.* **54,** 1.
77. Aoyama, T., Satoh, T., Yonemoto, M., Shibata, J., Nonoshita, K., Arai, S., Kawakami, K., Iwasawa, Y., Sano, H., Tanaka, K., Monden, Y., Kodera, T., Arakawa, H., Suzuki-Takahashi, I., Kamei, T., and Tomimoto, K. (1998). *J. Med. Chem.* **41,** 143.
78. Eummer, J. T., Gibbs, B. S., Zahn, T. J., Sebolt-Leopold, J. S., and Gibbs, R. A. (1999). *Bioorg. Med. Chem.* **7,** 241.

3

Mutational Analyses of Protein Farnesyltransferase

JUN URANO · WENLI YANG · FUYUHIKO TAMANOI
Department of Microbiology and Molecular Genetics
Jonsson Comprehensive Cancer Center
University of California, Los Angeles
Los Angeles, California 90095

THE ENZYMES, Vol. XXI

I. Introduction

Mutational analyses of protein farnesyltransferase (FTase) (*1, 2*) have provided valuable information concerning how this enzyme functions. Insights into roles amino acid residues at the active site play in the recognition of substrate molecules have been obtained by mutational analyses. This understanding was facilitated by the availability of substantial amounts of knowledge concerning the three-dimensional structure of this enzyme. Crystal structures of the enzyme with and without substrates as well as with enzyme inhibitors have been characterized (*3–7*) (see [2] in this volume[7a]). In addition, mutational analyses led to the identification of FTase mutants with new properties such as altered substrate specificity and resistance to farnesyltransferase inhibitors.

Two different approaches have been taken to identify FTase mutants. In one approach, site-directed mutagenesis was applied to conserved residues in FTase. Such conserved residues were identified through the comparison of amino acid sequences of FTase and a related prenyltransferase, geranylgeranyltransferase type I (GGTase-I), from a number of different organisms. Characterization of these mutants uncovered the roles these conserved residues play in FTase function. The second approach utilized random mutagenesis followed by screening for mutants of interest. Use of the yeast system for this random mutagenesis approach enabled quick and extensive screens (*8, 9*), which led to the identification of mutants that exhibit altered specificity for the substrate protein.

In this chapter, we first discuss deletion studies that defined the minimal size of FTase subunits critical for the structural integrity and function of the enzyme. We then describe the identification and characterization of a variety of FTase mutants that exhibit altered zinc binding, substrate affinity, and catalysis. In addition, we discuss different approaches that led to the identification of FTase mutants possessing altered substrate specificity. Some FTase mutants obtained by mutational analyses may provide valuable tools with which to evaluate the physiological function of FTase. For example, mutants resistant to farnesyltransferase inhibitor (FTI) could be used to determine whether any biological effects of FTI are indeed due to the inhibition of FTase. In this review, we discuss both mammalian and yeast FTase mutants, as there are striking similarities in the behavior of yeast and mammalian enzymes carrying analogous mutations (see below). For ease of comparison, in our discussions of the yeast enzyme, we indicate the corresponding residues in mammalian enzymes.

II. Protein Farnesyltransferase

FTase is a heterodimeric enzyme consisting of α and β subunits (*1, 2*). The enzyme contains a tightly bound zinc atom, which plays critical roles in catalysis (*10*). FTase recognizes a tetrapeptide sequence called the CaaX motif (cysteine followed by two aliphatic amino acids; X is the C-terminal amino acid, which is usually serine, cysteine, glutamine, alanine, or methionine), present at the C termini of substrate proteins. This enzyme transfers a farnesyl group from farnesyl diphosphate (FPP) to the cysteine within the CaaX motif. Some examples of farnesylated proteins include Ras, RhoE, nuclear lamins, and transducin. A closely related enzyme that is relevant for the discussion in this chapter is GGTase-I, which recognizes a related C-terminal CaaL motif (a motif similar to the CaaX motif, but containing leucine or phenylalanine as the C-terminal residue). This CaaL motif is found in proteins such as Cdc42, RhoA, and Rac proteins. GGTase-I transfers a geranylgeranyl group from geranylgeranyl diphosphate (GGPP) to these substrate proteins.

Figures 1 and 2 show sequences of FTase α and β subunits isolated from *Saccharomyces cerevisiae* (*11, 12*), *Schizosaccharomyces pombe* (*13, 14*),

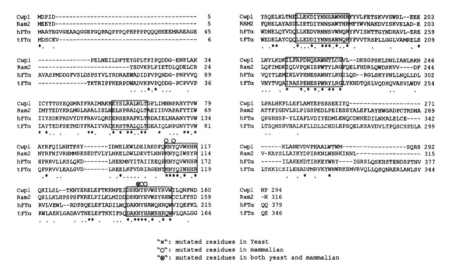

"x": mutated residues in Yeast
"O": mutated residues in mammalian
"⊗": mutated residues in both yeast and mammalian

FIG. 1. Alignment of FTase α-subunit sequences from *S. pombe* (Cwp1), *S. cerevisiae* (Ram2), human (hFTα), and tomato (tFTα). Identical residues among these subunits are shown by asterisks (*) and similar residues are shown by dots (.). Boxes indicate highly conserved stretches. Residues that have been mutated in mutational analyses described in this chapter are indicated by symbols defined at the bottom.

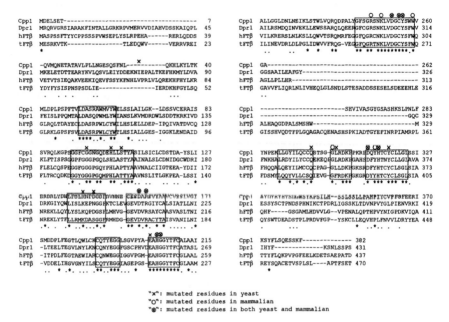

"x": mutated residues in yeast
"O": mutated residues in mammalian
"⊗": mutated residues in both yeast and mammalian

FIG. 2. Alignment of FTase β-subunit sequences from *S. pombe* (Cpp1), *S. cerevisiae* (Dpr1), human (hFTβ), and tomato (tFTβ). Identical residues and similar residues are indicated as in Fig. 1. Boxes indicate highly conserved stretches. Residues that have been mutated in mutational analyses are indicated by symbols defined at the bottom.

tomato (*15*), and human (*16*). In addition to these, FTase subunit genes were identified in other organisms such as rat (*17, 18*), cow (*19*), *Arabidopsis* (*20*), *Xenopus laevis* (*21*), and *Trypanosoma brucei* (*22*). The α subunit contains about 300 amino acid residues, whereas the average size of the β subunit is about 400 amino acids. The β subunit of tomato FTase contains an extra stretch inserted in the middle of the molecule. There is significant sequence similarity among these β subunits. For example, 37% identity is detected even between *S. cerevisiae* and rat β subunits (*17*). Significant similarity is also detected among α subunits from different organisms. Stretches with high sequence similarity are indicated in Figs. 1 and 2.

III. Deletion Studies

Deletion analysis was employed to examine whether any regions at the N terminus and C terminus of FTase are dispensable. The three-dimensional structure of FTase demonstrates contacts between the two subunits oc-

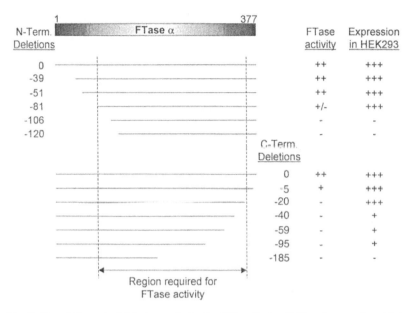

Fig. 3. N- and C-terminal deletion analysis of rat FTase β subunit. Based on results published by Andres *et al.* (*23*).

curring along a wide stretch of each subunit, suggesting that a significant portion of each subunit is required to form an active enzyme (*3*). However, deletion studies suggest that a portion of the α and β subunits at their N termini can be deleted.

A. DELETION OF α SUBUNIT OF RAT PROTEIN FARNESYLTRANSFERASE

Andres *et al.* (*23*) performed deletion analyses of the α subunit of rat FTase (Fig. 3). A series of N-terminal deletions was constructed by placing a start codon in the cDNA at positions corresponding to residues 39, 51, 81, 106, and 120. These cDNA constructs were introduced into HEK293 cells together with the β subunit, and the expression and FTase activity of the deletions were examined. Because the endogenous level of FTase is low in HEK293 cells, it is possible to examine the activity of the expressed enzyme. The truncated α subunits that initiate at positions 39, 51, and 81 all produced proteins at levels comparable to that of the intact subunit. In contrast, the truncated α subunits that initiate at positions 106 and 120 were not detectable. When FTase activity of extracts was examined, it was found that truncations 39 and 51 produced appreciable activity, whereas mutant 81 produced less activity. Thus, it appears that the deletion of up to 51 amino acids at the N terminus of the α subunit does not affect enzyme function.

A series of C-terminal deletions was constructed by placing a stop codon at positions corresponding to 192 (-185), 282 (-95), 318 (-59), 337 (-40), 357 (-20), and 372 (-5). These truncations were expressed in HEK293 cells together with the intact β subunit and their expression and FTase activity were examined. All truncation mutants were expressed except for the truncation terminating at residue 192 (-185); however, the level of expression of the -95, -59 and -40 deletions was significantly reduced compared with the intact protein. In terms of FTase activity, cells transfected with the -20 mutant had no FTase activity. This -20 mutant was also incapable of forming a complex with farnesyl diphosphate (FPP). The only mutant that produced activity was the -5 mutant. This activity, however, was significantly reduced compared with that of the intact protein. Thus, deletion of even five amino acids at the C terminus results in a significant reduction in FTase activity.

The structure of the α subunit of rat FTase showed that this polypeptide is highly helical, containing 15 α helices, four 3_{10} helices, and a β strand (*3*). However, these structures are not present within the N-terminal 55 residues. This lack of structural feature may be consistent with the deletion analyses, which showed that the N-terminal 51 amino acids are dispensable. In contrast, α-helix 15α is located close to the C terminus, which may explain why even a short deletion at the C terminus is not tolerated.

B.   DELETION STUDY OF α AND β SUBUNITS OF YEAST PROTEIN FARNESYLTRANSFERASE

Urano and Tamanoi (*24*) carried out similar deletion analyses with *S. cerevisiae* FTase. They focused on the interaction between the α and β subunits by using the yeast two-hybrid assay. In this experiment, the α subunit (Ram2) was fused with the activation domain of Gal4p, while the β subunit (Dpr1/Ram1) was fused with the DNA-binding domain of Gal4p. Both constructs were transformed into yeast strain Y190 and the expression of β-galactosidase was examined by the method described by Poullet and Tamanoi (*25*). This assay was used to assess the effects on the Dpr1–Ram2 interaction of a series of N-terminal and C-terminal deletions of Dpr1 as well as Ram2. Figure 4 summarizes the results of these experiments.

Deletions of 41 or 62 residues at the N terminus of the α subunit (Ram2) did not destory interaction with the β subunit (Dpr1/Ram1), whereas an α subunit deleted for 109 residues no longer interacted with the β subunit (Dpr1/Ram1). This dispensability of a significant portion of N-terminal residues is similar to that found with the mammalian FTase discussed above

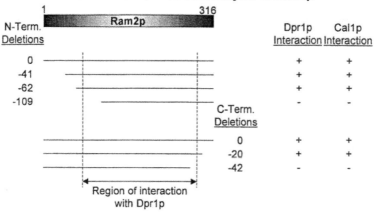

N- & C- Terminal Deletion Analysis of Ram2p

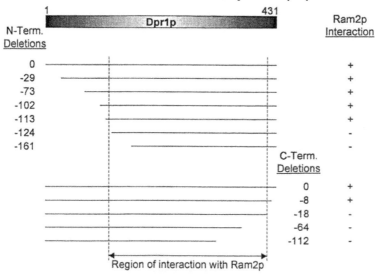

N- & C- Terminal Deletion Analysis of Dpr1p

FIG. 4. N- and C-terminal deletion analyses of the α and β subunits of *S. cerevisiae* FTase. Interactions detected by the use of the yeast two-hybrid assay are indicated as +. Ram2 protein was fused with the activation domain of Gal4 protein, while Dpr1 and Cal1 proteins were fused with the DNA-binding domain of Gal4p. These constructs were transformed into the yeast Y190 strain and β-galactosidase activity of the transformants was assayed as described by Poullet and Tamanoi (*25*).

(23). On the other hand, whereas deletion of 20 amino acids at the C terminus did not destory the interaction with the β subunit, the deletion of 42 amino acids at the C terminus destroyed the interaction with the β subunit (Dpr1/Ram1). Thus, the C-terminal 20 amino acids can be deleted. The α subunit of FTase is shared with GGTase-I. This raises an interesting question concerning whether deletions affecting interactions with the β subunit of FTase also affect interaction with Cdc43/Cal1, the β subunit of GGTase-I. This appears to be the case at least with the N-terminal 109 amino acid deletions and the C-terminal 42 amino acid deletions, which did not interact with the β subunit of GGTase-I (Cdc43/Cal1).

Figure 4 also shows deletion analyses of the β subunit of *S. cerevisiae* FTase, Dpr1/Ram1. Analyses of a series of deletions from the N terminus showed that deletions of 29, 73, 102, or 113 residues did not destroy the interaction with the α subunit (Ram2). However, a deletion of 124 amino acids was not tolerated. On the other hand, deletion of 18 amino acids from the C terminus destroyed the interaction with the α subunit (Ram2), whereas a deletion of 8 residues was tolerated. Thus, it appears that a significant region of the β subunit (the first 100 residues) is dispensable for the interaction with the α subunit, but the C-terminal residues are critical for the interaction.

Additional experiments were carried out to examine whether the largest N-terminal and C-terminal deletions of the β subunit that were still able to interact with the α subunit were functional. This assessment was performed by cloning the deletion constructs into a centromeric plasmid under the control of the *DPR1* promoter. The functionality of these constructs was assessed by assaying for their ability to complement the temperature-sensitive growth of *dpr1* mutant strains. In this experiment, the C-terminal deletion of 8 residues was able to efficiently complement this phenotype; however, the N-terminal 102-residue deletion was completely inactive. Therefore, the C-terminal eight residues are unnecessary for function. Furthermore, it appears that there exists a region within the N-terminal 102 residues that, although unnecessary for complex formation, is critical for function.

C. DELETION STUDY OF PLANT PROTEIN FARNESYLTRANSFERASE

An interesting deletion study was carried out with tomato FTase (15). The β subunit of tomato FTase (LeFTB) contains an insertion of 66 amino acids located in the middle of the molecule. A similar insertion is found in pea FTase but not in other FTases. This feature raised an interesting question concerning the role of the insertion, which was addressed by constructing a mutant LeFTB in which the insertion was deleted. The activity of the deletion mutant was assessed by examining the ability of the mutant tomato FTase to complement temperature-sensitive growth of the yeast FTase mutant, *dpr1/ram1*. For this experiment, both α and β subunits of the tomato FTase needed to be expressed, because the plant β subunit does not form a complex with the yeast α subunit. Reduction of the ability to complement was detected with the deletion mutant compared with wild type, suggesting that this insertion domain plays a role in the activity of tomato FTase. Further characterization of this insertion domain may be informative.

IV. Mutants Affecting Zinc Binding, Substrate Affinity, and Catalysis

Mutants affecting zinc binding, substrate affinity, and catalysis were obtained by site-directed mutagenesis of residues conserved among different FTase enzymes isolated from a number of different organisms (see Figs. 1 and 2). Sequence comparison of the α and β subunits from different organisms showed that, although the homology is scattered over the entire length of the protein, there are residues that are perfectly conserved among these proteins. A number of stretches can be noted that are perfectly conserved among different FTase enzymes; sequences containing these perfectly conserved stretches are shown in boxes in Figs. 1 and 2. In these figures, we also indicated the residues that were mutated in mutational analyses described in this chapter.

Many of the mutants characterized by site-directed mutagenesis of conserved residues have decreased FTase activity, which could be due to effects on (*1*) zinc binding, (*2*) affinity for CaaX peptide and/or FPP, and (*3*) catalysis. Kinetic analyses of these mutants were carried out to gain insights into their enzymatic properties. Table I lists the mammalian FTase mutants identified. Tables II and III lists yeast FTase mutants identified.

A. Zinc Affinity Mutants

Kral *et al.* (*26*) carried out an extensive site-directed mutagenesis of the β subunit of human FTase. In this study, 11 conserved residues whose side chains contained aromatic rings or heteroatoms were mutated to alanine because of its small size and lack of effects on secondary structure. The mutations created were R202A (i.e., the arginine at residue 202 was changed to alanine), H248A, C254A, R291A, K294A, D297A, C299A, Y300F, W303A, D359A, and H362A. The kinetic parameters, K_m values for farnesyl diphosphate (FPP) and for CaaX substrate (Ras-CVLS), as well as k_{cat} and apparent dissociation constant for FPP (K_d), were determined for each

TABLE I

MAMMALIAN PROTEIN FARNESYLTRANSFERASE MUTANTS[a]

Experiment	Mutation	k_{cat}	K_m CaaX	K_m FPP	Property	Ref.
1	Wild type	1	1	1		26
	βD297A	0.016	7.0	0.53	Defective zinc binding	26
	βC299A	0.00093	ND	ND	Defective zinc binding	26
	βH362A	0.11	8.1	0.74	Defective zinc binding	26
	βD359A	0.11	6.1	0.52	Defective zinc binding	26
	βY300F	0.21	0.65	0.53	Decreased catalysis	26
	βR202A	0.46	472	0.37	Decreased CaaX affinity	26
	βH248A	1.1	1.6	8.5	Decreased FPP affinity	26
	βR291A	7.1	8.1	36	Decreased CaaX and FPP affinity	26
	βK294A	3.0	5.6	37	Decreased CaaX and FPP afinity	26
	βW303A	2.0	11	19	Decreased CaaX and FPP affinity	26
2	αK164N				Decreased FTase activity, binds FPP and CaaX	23
3	Wild type	1	1	1		30
	αK164N	0.17	0.71	1.6	Decreased catalysis	30
	αY166F	1.2	0.4	1.3		30
	αY166A	0.028	1.7	2.4	Decreased catalysis	30
	αY200F	1.4	1.8	71	Decreased FPP affinity	30
	αH201A	0.63	2.1	2.1	Decreased CaaX and FPP afinity	30
4	Wild type	1	1	1		31
	βD200N	0.085	5.6	0.52	Decreased CaaX affinity	31
	βG349S	0.11	7.2	1.1	Decreased CaaX affinity	31
	βG249V	0.037	3.9	9.5	Decreased CaaX and FPP affinity	31
	αN199K	0.048	2.3	0.38	Decreased catalysis	31
5	βY361L				Increased utilization of CaaL motif, resistant to tricyclic FTI	42
	βY361M				Increased utilization of CaaL motif	42

[a] Kinetic parameters are standardized to the wild-type value.

TABLE II

YEAST MUTANTS DEFECTIVE IN FTase, GGTase-I

Mutant	Mutation	Corresponding residue in human enzyme	Ref.
Saccharomyces cerevisiae FTase			
ram1-1	βD209N	βD200	31
ram1-2	βG259V	βG249	31
ram2-1	αN143K	αN199	31
Saccharomyces cerevisiae GGTase-I			
cal1-1	βG328S	βG349	32
cdc43-2	βS85F	βN91	32
cdc43-3	βC138Y	βL154	32
cdc43-7	βR280C	βR263	32
cdc43-4,5,6	βA171V/T	βA178	32
Schizosaccharomyces pombe GGTase-I			
cwg2-1	βA202T	βA259	13

mutant enzyme. Properties of mutants exhibiting altered kinetic parameters are described in Table I, experiment 1.

Significant reduction in the catalytic activity was observed with five mutants: D297A, C299A, Y300F, D359A, and H362A. In particular, C299A had less than 0.1% of the k_{cat} value compared with the wild-type enzyme. The catalytic activities of D297A, Y300F, D359A, and H362A were 5- to 1000-fold lower than that of the wild-type enzyme. When all the mutants were assayed for zinc binding by the use of radioactive zinc (^{65}Zn), it was found that D297A, C299A, D359A, and H362A mutants were defective in binding zinc. Results obtained with three of these mutants, D297A, C299A, and H362A, were expected, because these residues are involved in coordinating the zinc ion in the crystal structure of FTase (3–6). The result with the D359A mutant is interesting, because the crystal structure does not identify this residue to be directly involved in the zinc coordination. In addition, the Y300F and H248A mutants showed a slight reduction in their ability to bind zinc. The mechanism by which these residues affect zinc binding is unclear. The H248A mutant also exhibits a significant decrease in binding FPP and is discussed below. Zinc dependence for FTase activity was examined in these mutants by comparing activity in the presence and absence of zinc. The mutants D297A, D359A, and H362A were found to exhibit a greater than 10-fold increase in their dependence on exogenously added ZnCl$_2$. These mutants exhibited K_m values for CaaX substrate that were elevated severalfold compared with the wild-type enzyme. Their K_m values for FPP and K_d (app) for FPP, on the other hand, were comparable to those of the wild-type enzyme. This lack of effect on FPP binding in

TABLE III

Yeast Protein Farnesyltransferase Mutants[a]

Experiment	Mutant	k_{cat}	K_m CaaX	K_m FPP	Property	Residue in human enzyme	Ref.
1	Wild type	1	1	1			
	βD307A	0.0027	1.2	0.15	Decreased zinc binding	D297	34
	βC309A	0.0022	0.5	0.43	Decreased zinc binding	C299	34
	βH363A	0.0015	1.4	0.39	Decreased zinc binding	H362	34
	βH156A	0.49	4.3	0.52	Fast dissociation of zinc	H149	34
	βY310F	0.0071	0.10	0.035	Increased CaaX and FPP affinity	Y300	34
	βR211Q	0.0018	1.8	0.56		R202	34
	βE256A	0.013	2.0	129	Decreased FPP affinity	E246	34
	βQ344A	0.14	1.9	4.6	Decreased FPP affinity	Q344	34
	βK193A	1.9	4.1	1.5	Decreased CaaX affinity	K185	34
	βN196A	1.0	3.6	1.1	Decreased CaaX affinity	D188	34
	βD209A	0.16	15	0.15	Decreased CaaX affinity	D200	34
2	Wild type	1	1	1			
	βH258N	0.29	2.6	3.0	Decreased CaaX and FPP affinity	H248	35
	βD360N				Decreased activity, increased FPP binding	H359	35
3	βS159N				Increased utilization of CaaL motif	P152	37,40
	βY362L (M, I)				Increased utilization of CaaL motif	Y361	40
	βY366N				Increased utilization of CaaL motif	Y365	40
4	βI74D				Increased specificity for CaaX motif	L67	44
	β351FSKN				Decreased CaaL recognition		44
5	βG149E (D)				Charge–charge interaction with CaaX ending with R or K	G142	45

[a] Kinetic parameters are standardized to the wild-type values.

these mutants with decreased zinc binding is in agreement with the idea that zinc is not required for the binding of FPP (27).

Fu et al. (28) further characterized two residues of rat FTase β subunit, D297 and H362. These residues are directly involved in coordinating the zinc ion. Five mutants, D297A, D297N, H362A, H362Q, and H362E, were purified and their kinetic properties were examined with H-Ras protein and FPP. This analysis showed that all these mutants have significantly reduced k_{cat} values ranging from 15-fold (H362E) to 200-fold (D297A). All five mutants had significantly reduced affinity for zinc. These mutants, however, still retained the ability to bind zinc, because increasing zinc concentration had a pronounced effect on the activity of the mutant enzymes. All the mutants bound FPP with affinity comparable to that of the wild-type enzyme. Single-turnover kinetics of the FTase mutants suggested that the rate-limiting step for three of the mutants, D297N, H362A, and H362Q, was changed from a product release step to a product formation step. This shift supports the idea that zinc is involved in catalyzing the product formation. On the other hand, with the D297A mutant, the rate-limiting step in steady state turnover remained as a product release step. For the H362E mutant, both the product release and product formation steps are involved in limiting its overall rate of catalysis.

B. MUTANTS AFFECTING CaaX, FARNESYL DIPHOSPHATE AFFINITY AND CATALYSIS

1. Site-Directed Mutagenesis of Conserved Residues in β Subunit

Of the 11 mutants Kral et al. constructed (26), R202A exhibited more than a 400-fold increase in the K_m value for CaaX peptide substrate, while retaining a K_m for FPP that is comparable to that observed with the wild-type enzyme. The k_{cat} value, as determined by the use of Ras-CVLS substrate, was 46% that of the wild-type enzyme. These results suggest that the R202 residue is critical for the recognition of the CaaX substrate. Further support for the idea that this residue is involved in the recognition of the CaaX motif was obtained by the use of peptidomimetic FTase inhibitors. Kral et al. (26) speculated that the loss of peptide affinity in the mutant was due to the loss of charge–charge interaction between the arginine and the negatively charged carboxylate of the terminal X of the CaaX peptide. They further speculated that if this were the case, the R202A mutant should no longer be sensitive to CaaX peptidomimetics that contain carboxylate (CVFM and L-739, 750), but should still be sensitive to CaaX peptidomimetics that do not contain carboxylate (L-739, 787 and L-745, 631). As expected, CVFM and L-739, 750, which contain carboxylate, had 100- to 200-fold higher IC_{50} (50% inhibitory concentration) values for the R202A mutant compared with the wild-type enzyme. On the other hand, the noncarboxylate inhibitors L739, 787 and L-745, 631 had IC_{50} values that were less than fourfold higher than that of the wild-type enzyme. Crystallographic studies revealed that the R202 residue makes direct contact with the protein substrate (6). The crystal structure of the rat FTase complexed with FPP and CVIM peptide revealed that R202 interacts with the isoleucine in the CVIM peptide through hydrogen bonding. Therefore, R202 does appear to be involved in interacting with the carboxylate of the CaaX substrate.

In contrast to the R202 mutant, the H248A mutant exhibited a K_m value for CaaX substrate, as determined using Ras-CVLS, comparable to that of the wild-type enzyme. However, this mutant had an approximately ninefold higher K_m value for FPP compared with the wild-type enzyme. Furthermore, this mutant had a k_{cat} value similar to that of the wild-type enzyme. Thus, the H248A mutation appears to affect primarily FPP affinity.

Several mutants (R291A, K294A, and W303A) exhibited increased K_m values for both CaaX and FPP substrates. These mutants had an approximately 10-fold increase in the K_m for CaaX substrate, Ras-CVLS. In addition, they had a 20- to 40-fold increase in the K_m for FPP. Interestingly, although these mutants have decreased affinity for both substrates, their k_{cat} values are slightly higher than that of the wild-type enzyme. In particular, the R291A mutant had a k_{cat} value that was seven times higher than that of the wild-type enzyme.

In agreement with the preceding results, the crystallographic structure of rat FTase complexed with FPP and CaaX peptide showed that FPP interacts with residues H248, R291, K294, and W303 of the β subunit (6). Mutations of these residues result in an enzyme with an increased K_m for FPP. The R291A, K294A and W303A mutants also exhibited an increase in K_m for CaaX substrates. Because FPP binds before the CaaX substrate, the increased K_m for CaaX substrate may be a reflection of the effects on FPP binding. In fact, the two substrates, FPP and CaaX peptide, are within van der Waals contact in the FPP–CaaX–FTase complex (6).

2. Site-Directed Mutagenesis of Conserved Residues in α Subunit

The role of specific residues of the α subunit in FTase activity was investigated in rat FTase by mutagenesis of conserved residues of the α subunit (23). Five residues located within five internal repeats in the α subunit (29) were chosen from the comparison of the α subunits of rat, human, yeast, and bovine FTases. These residues were mutated and the resulting mutants, K164N, Y166F, R172E, N199D, and W203H, were purified and characterized. The K164N mutant exhibited only a trace amount of FTase activity, while the other mutants maintained 30–75% activity of the wild-type enzyme. The deficiency of the K164N mutant was due to its inability to transfer the farnesyl group to the protein substrate as demonstrated by examining the transfer of farnesyl from a preformed enzyme–FPP complex to H-Ras. This mutant FTase transferred less farnesyl to H-Ras in 5 min than the wild-type FTase transferred in 10 sec. These results point to the idea that the K164 mutant is involved in catalysis. An interesting feature of this mutant is that it still retained the ability to bind FPP as well as CaaX substrate. However, kinetic analysis with purified preparation is needed to gain more concrete information about this mutant.

Wu *et al.* (*30*) characterized five α-subunit mutants, αK164N, αY166F, αY166A, αY200F, and αH201A, of rat FTase. The mutant enzymes were purified after their expression in *Escherichia coli* and FTase activity was assayed by using biotinylated KKSKTKCVIM as a protein substrate (Table I, experiment 3). Of the five mutants, Y200F and H201A appear to be mainly involved in substrate binding. The Y200F mutant exhibited a 71-fold increase in K_m for FPP. On the other hand, the H201A mutant exhibited an approximately twofold increase in K_m for both FPP and CaaX substrate. These results are consistent with structural data that show that both residues, Y200 and H201, contribute to the formation of a hydrophobic pocket for the first isoprene unit of FPP (*6*). In addition, the H201 residue interacts with the CaaX peptide through a water molecule (*6*). Two Y166 mutants show significantly different characteristics. Whereas the Y166F mutant exhibited kinetic values similar to those of the wild-type enzyme, the Y166A mutant exhibited increased K_m values for FPP and CaaX substrate. In addition, the k_{cat} value of Y166A was significantly decreased. These results suggest that the aromatic ring of tyrosine is important for substrate binding. A significant decrease in the k_{cat} value was detected with the K164N mutant. On the other hand, this mutant had K_m values for the Caax and FPP substrates comparable to those of the wild-type enzyme. These results, together with pre-steady-state kinetic studies of this mutant, suggest that residue K164 of the α subunit plays an important role in both the catalytic step and product release.

3. Mutations Analogous to Mutations Found in Prenylation-Deficient Yeast

Additional FTase mutants were identified by Omer *et al.* (*31*), who took advantage of the availability of yeast mutant strains that show drastically reduced FTase and GGTase-I activities. FTase and GGTase-I mutants of *S. cerevisiae* and *S. pombe,* as well as the position and amino acid changes found in these mutants, are described in Table II. The mutations are generally found at positions that are well conserved among FTases obtained from different organisms. The roles for these residues at these positions were analyzed by introducing analogous mutations into human FTase and characterizing the mutant enzymes (31). The mutants examined were D200N, G249V, and G349S of the β subunit of FTase as well as N199K of the α subunit of FTase. The FTase β-subunit mutations D200N and G249V correspond to the D209N and G259V mutations found in the yeast FTase β subunit, Dpr1/Ram1p (31). The G349S mutation corresponds to the G328S mutation found in the GGTase-I β subunit, Cal1/Cdc43p (32). Because FTase and GGTase-I share significant similarity, the expectation was that this residue would also be critical for the function of FTase. The N199K mutant of the α subunit was based on the ram2-1 (12) mutation of the yeast FTase α subunit.

Kinetic analysis showed that D200N and G349S exhibited increased K_m values for the CaaX substrate, Ras-CVLS (Table I, experiment 4). Similar results were obtained by using Ras-CVIM or Ras-CVLQ. These increases ranged from 4- to 30-fold. The K_m values for FPP, however, were unchanged. The G249V mutant, on the other hand, exhibited a significant increase in K_m values for both the CAAX and FPP substrates. The αN199K mutant exhibited a K_m value for Ras-CVLS that was 2.3-fold higher than that of the wild-type enzyme and a K_m value for FPP that was 38% relative to that of the wild-type enzyme. The k_{cat} values for the preceding mutants were generally 5- to 50-fold lower than for the wild-type enzyme (Table I).

An interesting observation concerning the D200N mutant was made when different CaaX motif proteins were used to determine the kinetic properties of this mutant enzyme (31). As shown in Table I, when Ras-CVLS was used as a substrate, the mutant exhibited a k_{cat} value 10-fold lower than that for the wild-type enzyme. However, when another substrate, Ras-CVIM, was used as a substrate, the k_{cat} value was 3-fold higher than that observed with the wild-type enzyme. This was seen despite a 40-fold increase in the K_m for the Ras-CVIM substrate. Omer et al. (31) speculated that the D200N mutation alters the CaaX-binding site to a conformation that is favorable for catalysis but is less favorable for binding Ras-CVIM.

The αN199K mutant requires further characterization. This mutant has an amino acid change in the α subunit that is shared with GGTase-I. Thus, it will be interesting to know what effect this mutation has on the biochemical properties of GGTase-I. It is interesting to note that, in yeast, the corresponding mutant (ram2-1) still exhibits approximately 30% of GGTase-I activity whereas the FTase activity is almost completely defective (31). Therefore, the ram2-1 mutation is expected to exert more pronounced effects on FTase than on GGTase-I. However, detailed characterization of this point has not been carried out. Another potentially interesting observation concerning the αN199K mutant is that this mutant may exhibit a dominant inhibitory effect when expressed in mammalian cells. This suggestion is based on experiments in which Prendergast *et al.* (*33*) attempted to express this mutant FTase in *ras*- and *mos*-transformed Rat1 cells. Whereas *mos*-transformed cells expressing the mutant FTase were readily obtained, *ras*-transformed cells expressing the mutant FTase could not be obtained. Prendergast *et al.* (*33*) suggested that the expression of the αN199K mutant results in a selection against cell growth by inhibiting Ras function. Further investigation is needed to assess the possibility that this mutant acts as dominant negative in mammalian cells.

C. YEAST PROTEIN FARNESYLTRANSFERASE

Mutational analyses of *S. cerevisiae* FTase were carried out by approaches similar to those taken for the mammalian enzyme. Table III lists yeast FTase mutants that were obtained. Dolence *et al.* (*34*) aligned sequences of the β subunits of FTase (yeast, human, rat, bovine, pea), GGTase-I (human, rat, *S. cerevisiae*, *S. pombe*), and GGTase-II (*S.cerevisiae*, rat) and identified five regions of high similarity. Of the similar and conserved residues in these regions, 13 polar or charged residues were mutated in the yeast FTase β subunit (Dpr1/Ram1). The mutants characterized were H156A, K193A, N196A, D209A, R211Q, E256A, R301Q, K304A, D307A, C309A, Y310F, Q344A, and H363A. Steady state kinetic properties of these mutant enzymes were analyzed with a peptide substrate, GCVIA (Table III, experiment 1). Of these, five mutants, R211Q, D307A, C309A, Y310F, and H363A, were found to exhibit drastically reduced k_{cat} values (1.5×10^{-3} to 7.1×10^{-3} of the wild-type values). Zinc-binding studies for these five yeast mutants showed that the D307A, C309A, and H363A mutants retained less than 0.5 atom of zinc per heterodimer, whereas R211Q and Y310F mutants retained one atom of zinc. The decreased abilities of the D307A, C309A, and H363A mutants to bind zinc were expected as these residues correspond to residues D297, C299, and H362 of the rat FTase β subunit, which coordinate the zinc ion. The mechanistic reason for the loss of FTase activity due to the R211Q mutation is unclear. The Y310F mutation is discussed further below. In addition to these five mutants, the H156A mutant was found to dissociate its zinc faster than the wild-type enzyme during incubation with EDTA and 1,10-phenanthroline. Thus, the H156 residue may help maintain the conformational integrity of the enzyme and affect the release and uptake of the metal ion.

Of the β-subunit mutants constructed by Dolence *et al.* (*34*) described above, a number of mutants exhibited altered affinity for FPP and CaaX substrates. Two mutants, E256A and Q344A, exhibited an increased K_m for FPP. In particular, the E256A mutant exhibited a 130-fold increase in K_m for FPP. This mutant, on the other hand, exhibited a K_m value for

CaaX substrate that was only 2-fold higher than that observed with the wild-type enzyme, whereas the k_{cat} was decreased 20-fold. Therefore, the E256A mutation appears to have an influence preferentially on FPP binding, and residue E256 may be involved in the binding of FPP. The corresponding residue in rat FTase is E246, which is located close to residues that make contacts with FPP in the crystal structure (6).

Four mutants, H156A, K193A, N196A, and D209A, exhibited increased K_m values for CaaX substrate. These increases ranged from approximately 3- to 15-fold. With the H156A, K193A, and N196A mutants, the K_m values for FPP were similar to that of the wild-type enzyme. The D209A mutant, which showed a 15-fold increase in K_m for CaaX protein substrate, exhibited a K_m value for FPP that was reduced approximately 7-fold. The increased K_m values for CaaX substrate suggest that these residues are involved in binding the CaaX motif. In addition, the kinetic behaviors of the D209A mutant also raise the possibility that this residue is involved in binding the FPP substrate.

One mutant that is particularly interesting is Y310F, which exhibited an approximately 10-fold decrease in K_m value for CaaX protein substrate as well as a 30-fold decrease in K_m for FPP. Thus, this mutant has acquired an increased affinity for both CaaX substrate and FPP. The residue Y310 corresponds to the residue Y300 in the human enzyme. It is interesting to note that the alteration of Tyr-300 to phenylalanine in the β subunit of the human enzyme also results in a slight increase in the affinity for Ras-CVLS and FPP (26). However, the yeast mutant (Y310F) shows an approximately 140-fold decrease in k_{cat} compared with the human mutant (Y300F), which exhibits a 5-fold decrease in the k_{cat} value. Although these results suggest that this residue is involved in binding both the CaaX substrate and FPP, it may play a more significant role in the yeast enzyme than in the human enzyme. Further mutational analysis of this residue as well as the elucidation of the three-dimensional structure of the yeast enzyme may shed light on this point.

Kurth et al. (35) also carried out site-directed mutagenesis of conserved residues of S. cerevisiae FTase. They identified five conserved blocks of sequences by comparing FTase β subunits from human, rat, and yeast, as well as the GGTase-I β subunit from yeast. Six yeast FTase β-subunit mutants with single amino acid changes, H156Q, H253N, E256Q, H258N, D360N, and H363Q, as well as two kinds of double mutants, E256Q/H258N and H253N/H258N, were constructed and characterized (Table III, experiment 2). Four mutations, E256Q, D307N, C309A, and H363Q, caused nearly a complete loss of FTase activity. As discussed above, the three residues D307, C309, and H363 correspond to residues in rat FTase that coordinate the bound zinc ion, which would explain the loss of FTase activity in the D307N, C309A, and H363Q mutants. The decreased FTase activity of the E256Q mutant agrees with the decreased activity of the E256A mutant described above.

Three mutations, D360N, H156Q, and H258N, caused a decrease in FTase activity, whereas the H253A mutation did not affect FTase activity. The mechanism by which D360N reduces FTase activity is unclear. The reduced activity seen with H156Q may be due to effects on zinc binding, as was seen with the H156A mutant described by Dolence et al. (34). The H258N mutant exhibited altered properties in addition to decreased activity. First, this mutant was more sensitive than the wild type to inhibition by 1,10-phenanthroline. Second, the mutant required more zinc to restore full activity to the mutant apoenzyme compared with the wild-type apoenzyme. These results point to changes related to zinc binding. This mutant also had reduced binding of FPP. Because FPP binding to the yeast enzyme requires zinc, this decrease may reflect the reduced affinity for the zinc ion. Further examination of the kinetic properties of the H258N mutant showed that this mutant also had decreased affinity for the CaaX protein, with a two- to sixfold increase in the K_m value. In addition, the k_{cat} value was decreased 3.5-fold. Interestingly, combining this mutation with the H253A mutation resulted in a dramatic decrease in FTase activity, although the H253A mutation by itself did not have any significant effect on FTase activity. Residue 258 corresponds to H248 of the human enzyme.

V. Conversion of Protein Substrate Specificity

Further insights into residues involved in recognition of the CaaX substrate were obtained by random mutagenesis of the yeast FTase β subunit. Yeast provided an ideal system with which to carry out random mutagenesis, because it is genetically amenable and a number of screens could be used to identify mutatnts of interest (8, 9, 36).

A. CONVERSION OF PROTEIN SUBSTRATE SPECIFICITY OF PROTEIN FARNESYLTRANSFERASE TO THAT OF GERANYLGERANYLTRANSFERASE TYPE I

1. Alteration of Protein Substrate Specificity of Yeast Protein Farnesyltransferase

Mitsuzawa et al. (37) obtained yeast FTase mutants that could recognize protein substrates ending with the CaaL sequence, a motif recognized by GGTase-I (Table III, experiment 3). A diagram of this type of mutant is

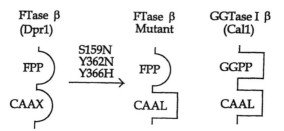

FIG. 5. Conversion of the CaaX recognition of FTase by single amino acid changes. [Reproduced, with permission, from Yang, W., Del Villar, K., Urano, J., Mitsuzawa, H., and Tamanoi, F. (1997). *J. Cell. Biochem.* **S27**, 12.] FTase binds FPP and CaaX, whereas GGTase-I binds GGPP and the CaaL motif. Mutants obtained by Del Villar *et al.* (*40*) exhibit increased recognition of the CaaL motif, but retain FPP binding.

shown in Fig. 5 (*37a*). The idea that it would be possible to convert the CaaX recognition specificity of FTase to the CaaL recognition of GGTase-I was based on two key observations. First, FTase and GGTase I are structurally similar; the α subunit is shared between the two enzymes and the β subunits of FTase and GGTase-I share significant similarity (*1, 2*). Second, these enzymes exhibit cross-specificities (GGTase-I can modify FTase substrates, albeit at low efficiency) (*38, 39*). The screen developed by Mitsuzawa *et al.* (*37*) took advantage of a *cal1* mutant strain, which is temperature sensitive for growth because of a defect in the GGTase-I β subunit, Cal1/Cdc43p. A library of randomly mutagenized FTase β-subunit gene, *DPR1/RAM1* [prepared by polymerase chain reaction (PCR) mutagenesis], was expressed in the *cal1* mutant, and transformants capable of growing at the nonpermissive temperature were isolated. If a mutant Dpr1/Ram1 protein was able to recognize the CaaL motif and functionally complement the defect in GGTase-I β subunit, the cells would be able to grow at the nonpermissive temperature. Three different mutants, S159N, Y362H, and Y366N, were obtained by this screen (*37, 40*). Expression of any one of these mutants was sufficient to suppress the temperature-sensitive growth of the *cal1* mutant. All three mutants also retained FTase activity, because they were able to complement the temperature-sensitive growth of the *dpr1/ram1* mutant.

To gain further insights into the kinds of amino acid changes at residues 159 and 362 that can cause increased utilization of the CaaL motif proteins, Del Villar et al. (*40*) changed each of these residues to all possible amino acids. These mutants were examined for their ability to suppress the temperature-sensitive growth of the *cal1* mutant. The results of the analysis at residue 159 showed that asparagine and aspartic acid were the only amino acids that could suppress the *call* phenotype. The suppression by aspartic acid, however, was much less efficient than that seen with asparagine. In the case of residue 362, three amino acids, leucine, methionine, and isoleucine, were found to confer increased utilization of the CaaL motif protein. One common characteristic among these amino acids that could confer increased CaaL utilization is that they are similar in size. Leucine, isoleucine, and methionine possess the identical van der Waals volume of 124 Å. Asparagine and aspartic acid, which could confer CaaL utilization when present at residue 159, also have similar van der Waals volumes. Thus, the mass of the amino acid introduced at these residues may be critical in increasing utilization of the CaaL motif proteins. It is also possible that the hydrophobic character of the amino acid residue at position 362 may contribute to the increased utilization of the CaaL motif, because methionine, leucine, and isoleucine are hydrophobic amino acids.

Del Villar *et al.* (*40*) also demonstrated that a mutant that can efficiently replace the GGTase-I β subunit can be generated by combining two mutations at residues 159 and 362, each of which alone shows only a minimal level of complementation. A variety of mutations at residue 159 (except asparagine) were combined with a variety of mutations at residue 362 (except methionine, leucine, or isoleucine). This led to the finding that a double mutant, S159D/Y362H, was effective in suppressing the temperature-sensitive growth of the *call* mutant. This suppression was as efficient as that obtained with strong suppressors such as Y362L. The additive effect of the two mutations may reflect a combined effect of conformational changes at two different regions within the CaaX-binding site of FTase.

K_m and k_{cat} values for the CaaX and CaaL substrates under steady state conditions were determined for the S159N and Y362L mutants and were compared with the wild-type enzyme. The S159N mutant had a threefold decrease in the K_m and an eightfold increase in the k_{cat} for the CaaL motif compared with the wild-type enzyme. Thus, the overall catalytic activity (k_{cat}/K_m) for the CaaL motif was increased 24-fold in the mutant. The mutant also exhibited a threefold increase in the K_m for the FTase substrate, CaaX motif. In the case of the Y362L mutant, there was an approximately 10-fold decrease in the K_m and a 2-fold increase in the K_{cat} for CaaL motif proteins, whereas for CaaX motif proteins the K_m is slightly increased and the k_{cat} is slightly decreased. It is important to note that these mutants transferred a farnesyl group rather than a geranylgeranyl group to the CaaL substrate. Little incorporation of radioactivity into the substrate protein was observed when radioactive GGPP was used. Thus, the *call/cdc43* phenotype can be suppressed when GGTase-I substrates are farnesylated. This finding agrees with a report by Ohya *et al.* (*41*), who showed that expression of mutant forms of both Rho 1 and Cdc42, which can be farnesylated instead of geranylgeranylated, enabled suppression of the temperature-sensitive growth of the *cal1* mutant.

2. Introduction of S159, Y362, Y366 Mutations into Human Protein Farnesyltransferase

Identification of yeast mutants with increased utilization of the CaaL motif raises a question concerning whether an analogous mutation introduced into the mammalian enzyme also causes altered substrate recognition. Residues Y362 and Y366 of the yeast enzyme correspond to residues Y361 and Y365 of the human enzyme, respectively. These residues are conserved among all FTase β subunits so far identified (see Fig. 2). In GGTase-I, either leucine or phenylalanine is found at these positions. On the basis of this information, Tyr-361 was changed to leucine. Similarly, Tyr-365 was changed to leucine. As for Ser-159, the corresponding residue in the human FTase is Pro-152. This proline was changed to methionine, which is the corresponding residue in human GGTase-I. The resulting human FTase mutants, P152M, Y361L, and Y365L, retained at least 50% FTase activity compared with the wild-type enzyme (Fig. 6A). The Y361L mutant exhibited significant ability to utilize CaaL substrates such as Cdc42 and GST-CIIL (Fig. 6B), whereas the P152M and Y365L mutants could not utilize the CaaL substrates. Kinetic studies were performed with the Y361L mutant. The k_{cat}/K_m value of the β361L mutant using GST-CIIL (CaaL motif) substrate and FPP was approximately 10-fold higher than that observed with the wild-type enzyme. On the other hand, the k_{cat}/K_m value of the mutant using GST-CIIS (CaaX motif) substrate was about 40% of that seen with the wild-type enzyme. Thus, the Y361L mutation appears to confer an increase in CaaL substrate utilization by human FTase. These results, however, represent steady state characteristics. Presteady state kinetics need to be examined to understand which step is affected by the mutation.

The CaaX substrate affinity of the Y361L mutant was further investigated by using a variety of substrates: GST-CIIS, GST-CIIL, GST-CIIM, GST-CIIC, and GST-CIIA. Utilization of CIIS, CIIM, CIIC, or CIIA remained similar to that of wild type, whereas there was a significant increase in the utilization of CIIL. The Tyr-361 residue of human FTase was also changed to other amino acids that possess the same van der Waals volume, i.e., methionine and isoleucine. Like the Y361L mutant, the Y361M mutant exhibited a significant increase in the utilization of the CIIL substrate. On the other hand, the Y361I mutant was not able to utilize the CIIL substrate.

Further support for the idea that the Y361L and Y361M mutants have increased CaaL recognition was obtained by the demonstration that these mutants are capable of functioning as GGTase-I in the yeast *cal1* mutant.

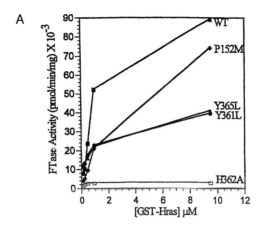

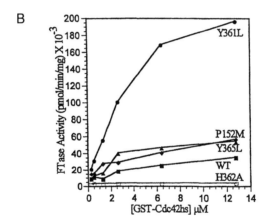

Fig. 6. Increased utilization of CaaL motif protein by the Y361L mutation. (A) Utilization of CaaX motif substrate, GST-Hras, by the wild-type and mutant forms of human FTase. P152M, Y361L, and Y365L mutants show more than 50% of the wild-type FTase activity. The H362A mutant exhibits dramatically reduced activity. (B) Utilization of CaaL motif protein, GST-Cdc42h, by wild-type and mutant FTases. [Reproduced, with permission, from Del Villar, K., Urano, J., Guo, L., and Tamanoi, F. (1999). *J. Biol. Chem.* **274,** 27010.]

For this experiment, it was necessary to express both the α and β subunits of human FTase. The Y361L and Y361M mutants were capable of suppressing temperature-sensitive growth of the *cal1* mutant, whereas the wild-type human enzyme was incapable of this suppression.

In addition to the increased affinity for CaaL substrates, the Y361L mutant exhibits other interesting properties. One striking property conferred by this mutation is increased resistance to tricyclic farnesyltransferase inhibitors (42). This is discussed below (see Section VI). Two additional properties conferred by the mutation have been identified. First, the Y361L mutant exhibits increased sensitivity to zinc chelators HPH-5 and HPH-6. HPH-5 is a tetradentate ligand that chelates divalent zinc ion through two nitrogen and two sulfur atoms (43); HPH-6 contains an extra carbon atom separating the nitrogen and sulfur atoms. Whereas the wild-type enzyme is inhibited by HPH-5 with an IC_{50} value of 100 μM, the IC_{50} value for the inhibition of the Y361L mutant is 2 μM. Thus, the Y361L mutant has approximately 50-fold increased sensitivity to this zinc chelator. This difference in zinc chelator sensitivity appears to reflect a subtle change in the coordination of zinc, because the sensitivity of the Y361L mutant to general metal chelators such as 1,10-phenanthroline, dipyridyl, or ethylenediamine was unchanged. Another interesting property of the Y361L mutant is its ability to utilize GGPP. Whereas the wild-type enzyme did not incorporate geranylgeranyl radioactivity into GST-CIIS substrate, significant incorporation of geranylgeranyl radioactivity was detected with the Y361L mutant. Incorporation of geranylgeranyl radioactivity by the Y361L mutant was also detected with other substrates, GST-CIIM and GST-CIIA.

In the crystal structure of the ternary complex of rat FTase–FPP–CaaX peptide (6), residue Y361 is located in a pocket that would be occupied by the second aliphatic amino acid in the FTase recognition motif (A2 position of Ca_1a_2X motif). This pocket is lined with the side chains of residues, W102, W106, and Y361. Alterations of multiple enzymatic properties caused by the Y361L mutation may reflect the critical location that this residue occupies within the CaaX recognition site of the enzyme. The C-terminal residue X of the CaaX motif, on the other hand, binds in a pocket defined by residues W102, H149, A151, and P152. It is interesting to note that one of these residues is Pro-152, which corresponds to Ser-159 of the yeast enzyme.

B. CREATING PROTEIN FARNESYLTRANSFERASE WITH STRICT SUBSTRATE SPECIFICITY

A different approach for creating FTase mutants with altered CaaX recognition was taken by Caplin et al. (44), who decided to change multiple amino acid residues in FTase to those in GGTase-I. Two types of conserved residues were identified from the comparison of a number of FTase and GGTase-I β-subunit sequences. One class of conserved residues is defined by residues that are conserved in both FTase and GGTase-I. Caplin et al. (44) speculated that these residues are important for maintaining the structure of these enzymes. Residues in the other class of conserved residues are specifically conserved in only one of the enzymes, and they may be involved in the distinct substrate specificity of each enzyme. By comparing Dpr1/Ram1p and Cdc43/Cal1p, seven regions that fit into this latter category were identified. The residues of Dpr1/Ram1p in these regions were changed to corresponding residues in Cal1/Cdc43p. Seven different mutants were constructed: 57HFFRH (R57H, V60F, L61F, S63R, and V64H), I74D, 206DDLF (G206D, V208D, R210L, and G212F), G308T, 351FSKN (L351F, R352S, D353K, and K354N), C367G, and S385K. The C367G and S385K mutants showed almost wild-type activity for each substrate using either FPP or GGPP. The 57HFFRH and G308T mutants showed general impairment of prenyltransferase activity. The other mutants exhibited interesting properties described below.

Three mutants, I74D, 206DDLF, and 351FSKN, are of particular interest, because they exhibit altered utilization of CaaX protein substrates. These mutants were examined for their ability to utilize three different CaaX substrates, Ras1-CIIS, Ras1-CIIM, and Ras1-CIIL, as well as prenyl substrates FPP and GGPP. The wild-type enzyme was able to geranylgeranylate Ras-CIIS and Ras-CIIM substrates. The I74D mutant farnesylated only Ras-CIIS substrate, but geranylgeranylated all three substrates as well as or better than the wild-type enzyme. The 206DDLF mutant farnesylated all three substrates at a wild-type level, but could not geranylgeranylate Ras1-CIIM or Ras1-CIIL substrate. The 351FSKN mutant farnesylated Ras1-CIIS and Ras1-CIIM but not Ras1-CIIL. Furthermore, this mutant lost the ability to geranylgeranylate any of the CaaX substrates. Thus, the 351FSKN mutant behaves as a "perfect" FTase, because it does not utilize the CIIL motif, which is a preferred motif for GGTase-I.

Prenyl specificity of the 351FSKN mutant was further examined. The K_m and K_d values for GGPP, using Ras-CIIM as a protein substrate, were 17.5 and 43.5 μM, respectively. With the wild-type enzyme, K_m and K_d values were 3.59 and 7.61 μM, respectively. Thus, the mutant exhibits a reduction in the affinity for GGPP of at least fourfold. The 351FSKN mutant has reduced affinity for Ras-CIIL when FPP is used as the prenyl donor, with a K_m value of 22.8 μM, whereas the K_m value for the wild-type enzyme is 10.2 μM. In addition, k_{cat} for Ras-CIIL, using FPP, is significantly decreased in the mutant compared with the wild-type enzyme.

The restricted ability of the mutant 351FSKN to use CaaX substrate and not CaaL substrate provided a useful tool with which to examine whether the ability of yeast FTase to utilize GGTase-I substrate is important for its function inside the cell. Examination of the *dpr1/ram1* mutant expressing the mutant FTase showed that the 351FSKN mutant is capable of complementing temperature-sensitive growth and mating defects of the FTase β subunit-deficient cells. Thus, the ability of FTase to utilize GGTase-I

substrates is not critical for its function to support growth at high temperatures as well as to mate.

The mutant I74D exhibited an interesting property *in vivo*. When the
above-described complementation experiment using the *dpr1/ram1* mutant
was carried out with the I74D mutant, it was found that the mutant was
incapable of complementing the temperature-sensitive growth but was capable of complementing the mating defect. This finding is interesting, because it raises the possibility that this mutant supports the function of only
a subset of farnesylated proteins. The altered substrate specificity of this
mutant may be causing farnesylation of only some substrates but not others.
In rat FTase, the analogous position is in a short loop between helix 1 and
helix 2 outside the CaaX- and FPP-binding pockets. This result also raises
an interesting structural question concerning whether any regions outside
the binding pocket can influence substrate specificity.

C. Mutants That Create Charge–Charge Interaction with CaaX Motif

Another type of interesting mutant was obtained from a screen based on
the function of the yeast mating factor peptide (*45*). These FTase mutants,
G149E and G149D, were capable of utilizing CaaX substrates ending with
the sequence CVIR or CVIK. Conversely, FTase mutants G149R and
G149K were capable of utilizing CaaX substrates ending with the sequence
CVID or CVIE. Thus, these FTase mutants acquired the ability to recognize
altered CaaX motifs. This ability to recognize new CaaX motifs appears
to be due to the establishment of charge–charge interactions between the
FTase enzyme and the C-terminal residue of the CaaX motif.

The mutants were identified from a screen based on a mating peptide,
a-factor, a peptide involved in the mating of *MATa* cells with *MATα* cells.
This mating peptide undergoes posttranslational modification events that
include farnesylation to produce a mature form (*46*). The level of mature
farnesylated **a**-factor production can be examined by the size of the zone
of growth inhibition ("halo") produced when *MATa* cells are spotted on
a lawn of *MATα* cells. This zone is due to **a**-factor secreted from *MATa*
cells and causing growth inhibition of *MATα* cells. Mutations in FTase that
cause decreased recognition of **a**-factor result in a decreased level of mature
a-factor, and hence a smaller halo. To increase the sensitivity of this assay,
a-factor ending with CAMQ was utilized. Because **a**-factor-CAMQ is less
active than the wild-type **a**-factor (CVIA), a slight change in the recognition
of **a**-factor-CAMQ can be detected. This screen led to the identification of

the G149E mutant of the β subunit of FTase (Dpr1/Ram1p). Cells expressing the G149E mutant produced little functional a-factor-CAMQ.

Utilization of different CaaX motifs by the G149E mutant was compared with that of the wild-type enzyme by constructing a variety of a-factor peptides ending with different amino acids (CVIC, CVIQ, CVIA, CVIG, CVIK, CVIW, CVIE, CVID, and CVIR). The results demonstrated that the CaaX recognition of the G149E mutant is different from that of the wild-type enzyme. The G149E mutant had better recognition of a-factor ending with CVIW than with a factor ending with CVIG. The wild-type enzyme, in contrast, favored the a-factor ending with CVIG. A surprising observation was that the a-factor ending with arginine could be farnesylated more effectively by the mutant enzyme than by the wild-type enzyme. Furthermore, the mutant enzyme as well as the wild-type enzyme was capable of farnesylating a-factor-CVIK. Thus, the mutant enzyme can recognize a-factor variants terminating with a positively charged amino acid at the C terminus (CVIR and CVIK). On the other hand, the mutant is incapable of farnesylating a-factor variants with negatively charged amino acids at the C terminus (CVIE and CVID). These results suggest that the residue at position 149 is involved in recognition of the C-terminal amino acid of the CaaX motif.

Potential interactions between residue 149 of FTase and the C-terminal residue of the CaaX motif were further examined with Ras2^{Val19} protein instead of the a-factor. Expression of this mutant Ras2 protein causes several dominant phenotypes including low viability upon nutrient starvation and heat shock sensitivity. Lack of Ras2^{Val19} farnesylation suppresses these phenotypes (45, 47, 48). Five different mutants of FTase (G149D, G149E, G149R, G149K, and G149A) were constructed and assessed for their ability to suppress the Ras2^{Val19} phenotypes. The results showed that FTase mutants having a negatively charged amino acid at residue 149 (G149D and G149E) preferentially farnesylated Ras2^{Val19} variants with positively charged C-terminal amino acids (CIIR and CIIK). On the other hand, FTase mutants with a positively charged amino acid at residue 149 (G149R and G149K) were most effective in utilizing Ras2^{Val19} variants with negatively charged C-terminal amino acids (CIID). One likely interpretation of these results is that charge–charge interaction is established on binding of the CaaX substrate to the mutant enzymes.

An interesting feature of the G149E mutant is that it exhibits altered substrate specificity *in vivo*. Cells expressing only this mutant FTase exhibited decreased sensitivity to growth arrest in response to the action of α-factor, the other mating factor that is not farnesylated, suggesting that the efficiency of the mating factor signal transduction is affected. In addition, the induction of a-factor synthesis by α-factor was decreased. These results suggest that the efficiency of the mutant FTase to farnesylate Ste18 (49), the γ subunit of heterotrimeric G protein involved in mating factor signal transduction, is lower than that of the wild-type cells. Cells expressing the G149E mutant were also temperature sensitive for growth. This appears to be due to a failure to farnesylate Ydj1 protein (50), a DnaJ homolog ending with the sequence CASQ, which is required for growth at 37°C but not at 30°C. Therefore, this mutant appears to be able to farnesylate only a subset of CaaX substrates.

VI. Mutants Resistant to Farnesyltransferase Inhibitors

A variety of compounds have been identified as inhibitors of farnesyltransferase (2, 51) (see [4] in this volume[51a]). These compounds have been shown to be effective in inhibiting growth of a variety of tumors in animal model systems and are being evaluated in clinical trials. Human FTase mutants resistant to farnesyltransferase inhibitor (FTI) should provide valuable materials to investigate how FTI interacts with FTase. In addition, such a mutant can be used to assess biological effects of FTI, because the expression of an FTI-resistant mutant is expected to render cells resistant to FTI. The first example of a mutant resistant to FTI has been identified (42).

Del Villar et al. (42) speculated that FTI-resistant mutants could be found among mutants that exhibited altered substrate recognition. Because many FTI compounds (peptidomimetics and tricyclic compounds) act as competitive inhibitors with respect to the CaaX substrate, mutations that affect CaaX affinity may have altered sensitivity to FTIs. Three human FTase mutants, P152M, Y361L, and Y365L, were examined for their sensitivity to FTIs (Fig. 7). This led to the finding that one of the mutants, Y361L, exhibits increased resistance to FTIs. The mutant enzyme was approximately 15- and 28-fold more resistant to the peptidomimetic-type inhibitors B1088 and BMS193269, respectively, compared with the wild-type enzyme. What was more remarkable was that the mutant exhibited a more than several-thousandfold increase in the resistance to a tricyclic type of FTI, SCH56582. SCH56582 inhibited the wild-type enzyme with an IC_{50} of 1.4 μM, whereas more than 10,000 μM SCH56582 was needed to obtain 50% inhibition of the mutant enzyme. Similarly, with another tricyclic FTI compound, SCH44342, a more than 200-fold increase in IC_{50} was observed with the mutant enzyme. Because Y361L mutant is able to recognize its normal CaaX substrate, this selective effect of the tricyclic inhibitor, which also binds at the CaaX-binding pocket, suggests that the Y361L mutant is capable of recognizing a subtle difference in the binding of the substrate and the inhibitor. This idea is further supported by the observation that the

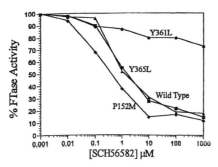

FIG. 7. Resistance of the human FTase mutant Y361L to inhibition by a tricyclic FTI compound, SCH56582. Increasing concentrations of SCH56582 were added to FTase reactions catalyzed by wild-type and mutant FTases. GST-CIIS was used as a protein substrate in this assay. Reproduced, with permission, from Del Villar, K., Urano, J., Guo, L., and Tamanoi, F. (1999). *J. Biol. Chem.* **274**, 27010.

mutant showed a more pronounced resistance to SCH56582 over another tricyclic compound, SCH44342. The only difference between the two tricyclic compounds is the presence of a bromine at position 3 of the tricyclic ring, and it appears that this slight difference can be detected by the mutant FTase.

The crystal structure of rat FTase complexed with FPP and the tricyclic FTI SCH44342 shows that the side chain of Y361 of the β subunit of FTase is within van der Waals contact of the piperidine ring of SCH44342 (7). Side chains of L96 and W106 of the FTase β subunit are also located close to the piperidine ring of SCH44342. The pyridinyl ring of SCH46432, on the other hand, binds in a pocket lined with several aromatic side chains. In addition, halogen atom in this FTI compound binds to a narrow groove in FTase. Overall, SCH44342, together with FPP, occupies about one-third of the active site cavity and there are extensive interactions between SCH44342 and FTase residues.

One of the important issues in the study of farnesyltransferase inhibitors is to critically evaluate that any biological effects seen with the inhibitors are due to the inhibition of FTase. As described in more detail in [4] in this volume (51a), FTI exhibits remarkable activity toward transformed cells. These biological effects include morphological changes, inhibition of anchorage-independent growth, enrichment of G_1-phase cells, and induction of apoptosis (51–55). Because a number of FTI compounds with different structures cause similar biological effects, it is believed that these effects are due to the inhibition of FTase. However, direct demonstration of this point has not been accomplished. FTI-resistant mutants such as Y361L may provide valuable materials to address this point. If the biological effects of

FTI are due to the inhibition of FTase, expression of FTI-resistant FTase should render cells resistant to FTI. Del Villar and Tamanoi (56) utilized this approach to show that morphological changes in vK-*ras*-transformed normal rat kidney (KNRK) cells induced by SCH56582 were due to the FTI. KNRK cells were transfected with FTase α subunit and either wild-type or Y361L FTase β subunit. FTase β subunit was fused with green fluorescent protein to selectively observe the transfected cells. The transfected cells were treated with SCH56582 and morphological changes were examined. KNRK cells expressing the wild-type FTase β subunit underwent morphological changes on treatment with SCH56582. In contrast, cells transfected with the Y361L mutant β subunit were resistant to morphological changes induced by the treatment with SCH56582. Thus, the expression of the Y361L mutant appears to correlate with the increased resistance to morphological changes induced by SCH56582. This approach could be applied to examine other biological effects of FTIs, such as induction of apoptosis.

VII. Future Prospects

A. FURTHER USE OF MUTANTS CREATED BY MUTAGENESIS

As described in this chapter, a variety of approaches have been used to identify a number of different classes of FTase mutants. Characterization of these mutants provided valuable information concerning the structure and function of the enzyme. Mutations of residues coordinating zinc ion caused a significant decrease in FTase activity and zinc binding. Interestingly, these mutants retained a normal level of FPP affinity, providing insights that FPP binding is independent of zinc binding. This finding agrees with the mechanistic studies that showed that zinc is involved in catalysis and is not required for FPP binding (27, 57). Insights into how substrates interact with the FTase residues were obtained by the identification of FTase mutants that affect the binding of substrates. Interestingly, some mutants affect the binding of one substrate without affecting the binding of another substrate. For example, the βR202A mutation in human FTase had a major effect on CaaX affinity but not on FPP affinity. On the other hand, the β248A mutation appeared to affect FPP affinity more than CaaX affinity. Thus, some residues are involved in the binding of only one of the substrates. In addition, some residues are involved in the binding of both substrates. Furthermore, mutants that affect catalysis but not substrate affinity should provide information concerning residues involved in the catalysis of FTase. So far, αK164N, αN199K, and βY300F mutants fit this characteristic. More mutants of this type should be valuable in gaining insights into the mechanism of catalysis of FTase enzyme.

Mutants with altered substrate specificity have been identified by random mutagenesis approaches. This approach was possible because of the powerful yeast genetics that could be applied to FTase mutagenesis. A number of clever genetic screens were devised and the screens led to the identification of two types of mutants: mutants with increased CaaL recognition and mutants that create charge–charge interactions with the CaaX motif.

Interestingly, the Y361L mutant of human FTase exhibits increased resistance to a tricyclic type of FTI. Farnesyltransferase inhibitors show promise as anticancer drugs, and they exhibit a number of biological effects that include inhibition of anchorage-independent growth, morphological changes, as well as effects on cell cycle and induction of apoptosis. FTase mutants that are resistant to FTIs may prove to be invaluable in evaluating whether any biological effects of FTIs are actually due to the inhibition of FTase. Although the Y361L mutant is the only mutant available at the moment, more mutants of this type are expected by further studies.

Because there are multiple substrates of FTase, it has been of importance to understand the function of farnesylated proteins. For example, a-factor is responsible for the mating of yeast cells, whereas Ydj1p is responsible for growth at high temperatures. Creating mutant FTase enzymes that prenylate a subset of substrate proteins is valuable in further investigations of this issue. In this regard, two yeast mutants, G149E and I74D, are noteworthy. The G149E mutant enzyme appears to modify a-factor and Ras2 protein; however, their utilization of Ste18p and Ydj1p proteins appears to be decreased. The I74D mutant appears to be capable of prenylating a-factor but incapable of prenylating Ydj1 protein. Thus, it is possible to construct FTase mutants that can farnesylate only a subset of substrate proteins. Characterization of yeast cells expressing these mutant forms of FTase may provide valuable information about the physiological roles of protein farnesylation.

B. MUTANTS THAT HAVE YET TO BE CREATED

There are a number of mutant types that have yet to be identified. One major class concerns mutants that can be called "substrate trap mutants." Such mutants have been identified with protein tyrosine phosphatase (58). Expression of the mutant followed by immunoprecipitation led to the identification of physiologically significant substrates of protein tyrosine phosphatase (58). More recently, a substrate trap mutant of caspase 3 has been identified. Use of this mutant in the yeast two-hybrid assay led to the demonstration that gelsolin is one of the substrates of caspase 3 (59). Similar mutants of FTase should be valuable in gaining insights into physiologically significant substrates of FTase. Such FTase mutants may have significantly increased substrate affinity (decrease in K_m). Alternatively, mutations that affect release of the product could show this phenotype, because product release has been shown to be the rate-limiting step during the catalysis by FTase (60).

Another set of mutants that have yet to be identified concerns subunit interaction mutants. Although regions of the α subunit that interact with the β subunit encompass a significant portion of the α-subunit molecule, it is possible that there are regions critical for the interaction. These mutants could be valuable in understanding the molecular basis of the interaction between the two subunits. An additional issue of interest concerns the interaction with the β subunit of GGTase-I. Because this subunit also interacts with the α subunit of FTase, it will be interesting to know the similarities and differences between the GGTase-I β-subunit/FTase α-subunit interaction and the FTase β-subunit/FTase α-subunit interaction.

Mutants that change from FPP recognition to GGPP recognition represent another class of mutants that is of interest. Such mutants have been obtained with farnesyl diphosphate (FPP) synthase of *Bacillus stearothermophilus* (61). The mutants were initially identified by random mutagenesis followed by a screen to find mutants that showed the activity of geranylgeranyl diphosphate (GGPP) synthase. Site-directed mutagenesis was further carried out to pinpoint the mutation, among multiple mutations found in initial mutants, that is responsible for the increased GGPP synthase activity. More recently, this group succeeded in converting archaeal geranylgeranyl diphosphate synthase to farnesyl diphosphate synthase (62). This type of analysis may be valuable with protein farnesyltransferase. Such mutants may be useful in addressing whether there are physiological differences between farnesylation and geranylgeranylation.

ACKNOWLEDGMENTS

This work was supported by NIH Grant CA41996. J.U. was also supported by U.S. Public Health Service National Research Service Award GM07185. W.Y. was also supported by an Edwin D. Pauley Foundation Fellowship.

REFERENCES

1. Zhang, F. L., and Casey, P. J. (1996). *Annu. Rev. Biochem.* **65**, 241.
2. Sattler, I., and Tamanoi, F. (1996). *In* "Regulation of the Ras Signaling Network" (H.

Maruta and A. W. Burgess, eds.), p. 95. Molecular Biology Intelligence Unit Series. R. G. Landes, Austin, Texas.

3. Park, H., Boduluri, S., Moomaw, J., Casey, P., and Beese, L. (1997). *Science* **275**, 1800.

4. Long, S. B., Casey, P. J., and Beese, L. S. (1998). *Biochemistry* **37**, 9612.

5. Dunten, P., Kammlott, U., Crowther, R., Weber, D., Palermo, R., and Birktoft, J. (1998). *Biochemistry* **37**, 7907.

6. Strickland, C. L., Windsor, W. T., Syto, R., Wnag, L., Bond, R., Wu, Z., Schwartz, J., Le, H. V., Beese, L., and Weber, P. C. (1998). *Biochemistry* **37**, 16601.

7. Strickland, C. L., Weber, P. C., Windsor, W. T., Wu, Z., Le, H. V., Albanese, M. M., Alvarez, C. S., Cesarz, D., del Rosario, J., Deskus, J., Mallams, A. K., Njoroge, G., Piwinski, J. J., Remiszewski, S., Rossman, R. R., Taveras, A. G., Vibulbhan, B., Doll, R. J., Girijavallabhan, V. M., and Ganguly, A. K. (1999). *J. Med. Chem.* **42**, 2125.

7a. Terry, K. L., Long, S., and Beese, L. S. (2000). *In* "The Enzymes," Vol. XXI, Chap. 2. Academic Press, San Diego, California. [This volume]

8. Tamanoi, F., Del Villar, K., Robinson, N., Urano, J., and Yang, W. (2000). *In* "Farnesyltransferase and Geranylgeranyltransferase I: Targets for Cancer and Cardiovascular Therapy" (S. M. Sebti and A. Hamilton, eds). Humana Press, Totoua, New Jersey.

9. Tamanoi, F., and Mitsuzawa, H. (1995). *Methods Enzymol.* **255**, 82.

10. Hightower, K. E., and Fierke, C. A. (1999). *Curr. Opin. Chem. Biol.* **3**, 176.

11. Goodman, L. E., Perou, C. M., Fujiyama, A., and Tamanoi, F. (1988). *Yeast* **4**, 271.

12. He, B., Chen, P., and Chen, S. Y. (1991). *Proc. Natl. Acad. Sci. U.S.A.* **88**, 11373.

13. Arellano, M., Coll, P. M., Yang, W., Duran, A., Tamanoi, F., and Perez, P. (1998). *Mol. Microbiol.* **29**, 1357.

14. Yang, W., Urano, J., and Tamanoi, F. (2000). *J. Biol. Chem.* **275**, 429.

15. Yalovsky, S., Trueblood, C. E., Callan, K. L., Narita, J. O., Jenkins, S. M., Rine, J., and Gruissem, W. (1997). *Mol. Cell. Biol.* **17**, 1986.

16. Andres, D. A., Milatovich, A., Ozcelik, T., Wenzlau, J. M., Brown, M. S., Goldstein, J. L., and Francke, U. (1993). *Genomics* **18**, 105.

17. Chen, W. J., Andres, D. A., Goldstein, J. L., Russell, D. W., and Brown, M. S. (1991). *Cell* **66**, 327.

18. Chen, W. J., Andres, D. A., Goldstein, J. L., and Brown, M. S. (1991). *Proc. Natl. Acad. Sci. U.S.A.* **88**, 11368.

19. Kohl, N. E., Diehl, R. E., Schaber, M. D., Rands, E., Soderman, D. D., He, B., Moores, S. L., Pompliano, D. L., Ferro-Novick, S., Powers, S., Thomas, K. A., and Gibbs, J. B. (1991). *J. Biol. Chem.* **266**, 18884.

20. Qian, D., Zhou, D., Ju, R., Cramer, C. L., and Yang, Z. (1996). *Plant Cell* **8**, 2381.

21. Goalstone, M. L., Diamond, C. L., and Sadler, S. E. (1996). *Biol. Reprod.* **54**, 675.

22. Yokoyama, K., Trobridge, P., Buckner, F. S., Van Voorhis, W. C., Stuart, K. D., and Gelb, M. H. (1998). *J. Biol. Chem.* **273**, 26497.

23. Andres, D. A., Goldstein, J. L., Ho, Y. K., and Brown, M. S. (1993). *J. Biol. Chem.* **268**, 1383.

24. Urano, J., and Tamanoi, F. (1999). Unpublished results.

25. Poullet, P., and Tamanoi, F. (1995). *Methods Enzymol.* **255**, 488.

26. Kral, A. M., Diehl, R. E., deSolmes, J., Williams, T. M., Kohl, N. E., and Omer, C. A. (1997). *J. Biol. Chem.* **272**, 27319.

27. Reiss, Y., Brown, M. S., and Goldstein, J. L. (1992). *J. Biol. Chem.* **267**, 6403.

28. Fu, H., Beese, L. S., and Casey, P. J. (1998). *Biochemistry* **37**, 4465.

29. Boguski, M. S., Murray, A. W., and Powers, S. (1992). *New Biol.* **4**, 408.

30. Wu, Z., Demma, M., Strickland, C. L., Radisky, E. S., Poulter, C. D., Le, H. V., and Windsor, W. T. (1999). *Biochemistry* **38**, 11239.

31. Omer, C. A., Kral, A. M., Diehl, R. E., Prendergast, G. C., Powers, S., Allen, C. M., Gibbs, J. B., and Kohl, N. E. (1993). *Biochemistry* **32,** 5167.
32. Ohya, Y., Caplin, B., Qadota, H., Tibbetts, M. F., Anraku, Y., Pringle, J. R., and Marshall, M. S. (1996). *Mol. Gen. Genet.* **252,** 1.
33. Prendergast, G. C., Davide, J. P., Kral, A., Diehl, R., Gibbs, J. B., Omer, C. A., and Kohl, N. E. (1993). *Cell Growth Differ.* **4,** 707.
34. Dolence, J. M., Rozema, D. B., and Poulter, C. D. (1997). *Biochemistry* **36,** 9246.
35. Kurth, D. D., Farh, L., and Deschenes, R. J. (1997). *Biochemistry* **36,** 15932.
36. Judd, S. R., and Tamanoi, F. (1991). "Methods: A Companion to *Methods in Enzymology*", Vol. 1, p. 246. Academic Press, San Diego, California.
37. Mitsuzawa, H., Esson, K., and Tamanoi, F. (1995). *Proc. Natl. Acad. Sci. U.S.A* **92,** 1704.
37a. Yang, W., Del Villar, K., Urano, J., Mitsuzawa, H., and Tamanoi, F. (1997). *J. Cell. Biochem.* **S27,** 12.
38. Trueblood, C. E., Ohya, Y., and Rine, J. (1993). *Mol. Cell. Biol.* **13,** 4260.
39. Yokoyama, K., Goodwin, G. W., Ghomashchi, F., Glomset, J. A., and Gelb, M. H. (1991). *Proc. Natl. Acad. Sci. U.S.A.* **88,** 5302.
40. Del Villar, K., Mitsuzawa, H., Yang, W., Sattler, I., and Tamanoi, F. (1997). *J. Biol. Chem.* **272,** 680.
41. Ohya, Y., Qadota, H., Anraku, Y., Pringle, J. R., and Botstein, D. (1993). *Mol. Biol. Cell* **4,** 1017.
42. Del Villar, K., Urano, J., Guo, L., and Tamanoi, F. (1999). *J. Biol. Chem.* **274,** 27010.
43. Fujita, M., Otsuka, M., and Sugiura, Y. (1996). *J. Med. Chem.* **39,** 503.
44. Caplin, B. E., Ohya, Y., and Marshall, M. S. (1998). *J. Biol. Chem.* **273,** 9472.
45. Trueblood, C. E., Boyartchuk, V. L., and Rine, J. (1997). *Proc. Natl. Acad. Sci. U.S.A.* **94,** 10774.
46. Caldwell, G. A., Naider, F., and Becker, J. M. (1995). *Microbiol. Rev.* **59,** 406.
47. Fujiyama, A., Matsumoto, K., and Tamanoi, F. (1987). *EMBO J.* **6,** 223.
48. Powers, S., Michaelis, S., Broek, D., Santa Anna, A., S., Field, J., Herskowitz, I., and Wigler, M. (1986). *Cell* **47,** 413.
49. Finegold, A. A., Schafer, W. R., Rine, J., Whiteway, M., and Tamanoi, F. (1990). *Science* **249,** 165.
50. Caplan, A. J., Tsai, J., Casey, P. J., and Douglas, M. G. (1992). *J. Biol. Chem.* **267,** 18890.
51. Gibbs, J. B., and Oliff, A. (1997). *Annu. Rev. Pharmacol. Toxicol.* **37,** 143.
51a. Gibbs, J. B. (2000). *In* "The Enzymes," Vol. XXI, Chap. 4. Academic Press, San Diego, California. [This volume]
52. Cox, A. D., and Der, C. J. (1997). *Biochim. Biophys. Acta.* **1333,** F51.
53. Lebowitz, P. F., Sakamuro, D., and Prendergast, G. C. (1997). *Cancer Res.* **57,** 708.
54. Suzuki, N., Urano, J., and Tamanoi, F. (1998). *Proc. Natl. Acad. Sci. U.S.A.* **95,** 15356.
55. Suzuki, N., Del Villar, K., and Tamanoi, F. (1998). *Proc. Natl. Acad. Sci. U.S.A.* **95,** 10499.
56. Del Villar, K., and Tamanoi, F. (1999). Unpublished results.
57. Huang, C. C., Casey, P. J., and Fierke, C. A. (1997). *J. Biol. Chem.* **272,** 20.
58. Flint, A. J., Tiganis, T., Barford, D., and Tonks, N. K. (1997). *Proc. Natl. Acad. Sci. U.S.A.* **94,** 1680.
59. Kamada, S., Kusano, H., Fujita, H., Ohtsu, M., Koya, R. C., Kuzumaki, N., and Tsujimoto, Y. (1998). *Proc. Natl. Acad. Sci. U.S.A.* **95,** 8532.
60. Furfine, E. S., Leban, J. J., Landavazo, A., Moomaw, J. F., and Casey, P. J. (1995). *Biochemistry* **34,** 6857.
61. Ohnuma, S., Nakazawa, T., Hommi, H., Halberg, A. M., Koyama, T., Ogura, K., and Nishino, T. (1996). *J. Biol. Chem.* **271,** 10087.
62. Ohnuma, S., Hirooka, K., Ohto, C., and Nishino, T. (1997). *J. Biol. Chem.* **272,** 5192.

4

Farnesyltransferase Inhibitors

JACKSON B. GIBBS

Department of Cancer Research
Merck Research Laboratories
West Point, Pennsylvania 19486

I. Introduction

The discovery of Ras farnesylation in 1989 (*1–3*) and farnesyltransferase (FTase) in 1990 (*4–7*) afforded the first pharmaceutically tractable approach toward inhibiting the cell-transforming properties of the *ras* oncogene. Interest in inhibiting the function of *ras* stems from the association of mutated forms of this oncogene with solid tumor types (e.g., colon, pancreatic, and lung), which are difficult to treat effectively with current therapeutic regimens (*8, 9*).

As GTP-binding proteins that cycle between active (GTP-bound) and

THE ENZYMES, Vol. XXI

inactive (GDP-bound) states, the Ras proteins (Harvey, Kirsten, and N-Ras) act as molecular switches to regulate the proliferative and survival signals of growth factors and their receptors (*10*). In normal cells, this switching mechanism is highly regulated, and the Ras protein is found bound predominantly in its inactive GDP state. Interaction with effector pathways is required to activate Ras in its GTP-bound form, followed by downregulation by proteins such as the GTPase-activating protein (GAP) or the neurofibromatosis gene product neurofibromin. However, in tumor cells, this regulation is lost and leads to an increase in the steady state intracellular levels of the Ras–GTP-bound state. Most often, activation of Ras occurs by mutations, in the Ras protein itself, that impair its intrinsic GTPase activity. However, changes in the activity of key regulatory proteins can also raise the intracellular levels of Ras–GTP. For example, a variety of growth factor receptor tyrosine kinases activate the function of nucleotide exchange factors that promote Ras into its GTP-bound state, and constitutive activation of tyrosine kinases subsequently leads to continuous exchange factor activity. Alternatively, loss of the GTPase-activating protein neurofibromin, which normally inactivates Ras, also raises Ras–GTP.

In the 1980s, efforts to develop inhibitors of Ras function were slowed by limited information about Ras functions. Thus, each new insight into Ras mechanisms was considered for the development of a small molecule inhibitor. However, these efforts failed because of practical limitations. For example, inhibitors of guanine nucleotide binding were unable to discriminate effectively between Ras and other GTP-binding proteins (*11*). This is because the active sites of GTP-binding proteins are highly conserved and the key residues that interact with guanine nucleotide are often identical. Perhaps the ideal Ras drug might derive from reversing the impaired GTPase phenotype, but knowledge of the Ras crystal structure and the mechanism of GAP-stimulated GTP hydrolysis is only slowly providing information relevant to this possibility, highlighting the difficulty of this approach. The first experimental evidence that reconstitution of mutant Ras GTPase activity is formally possible has been described by the Wittinghofer and Selinger laboratories (*12*). They have identified a modified GTP analog that can be cleaved effectively by Ras proteins having a mutation at either position 12 or 61. Further progress into this GTP hydrolytic activity will be an exciting area to monitor.

Prior to the identification of Ras effector proteins in mammalian systems, the only known targeting path was described in the yeast *Saccharomyces cerevisiae*. In yeast, Ras proteins activate adenylate cyclase (*13*). However, this effector interaction is not found in mammalian cells, and so inhibitors of adenylate cyclase catalytic activity were not considered. Instead, inhibitors of Ras protein interaction with the target molecule adenylyl cyclase were screened because it was hypothesized that Ras effector interactions would be conserved in yeast and mammalian systems based on mutagenesis studies of the residues in the effector-interacting domain (switch I region in the crystal structure of Ras) (*10*). While short peptides were identified that were capable of inhibiting both Ras–yeast adenylyl cyclase interactions and Ras–GAP interactions, these peptides were not active in cells and small molecule surrogates were not identified (*11*).

The lessons learned here highlighted the difficulty of finding small chemical molecules that could either restore a lost function (in this case GTP hydrolytic activity) or inhibit large protein–protein interaction domains. For investigators interested in discovering inhibitors of Ras functions, other approaches needed to be explored.

One of the earliest known properties of Ras was that it was localized in the plasma membrane of cells after posttranslational modification, and this localization was essential to the cell-transforming activity of Ras (*10*). However, the chemical nature of these posttranslational modifications was not identified until the carboxy-terminal structures of fungal mating factors and nuclear lamins (*14*), which share a small region of homology with Ras proteins, that is, the CaaX motif (C, cysteine; a, aliphatic amino acid; X, another amino acid), were identified by analytical techniques. As described in [1–3] in this volume (*14a*), FTase modifies the cysteine thiol of CaaX with a farnesyl group, using farnesyl diphosphate (FPP) as cosubstrate in the presence of Mg^{2+} and Zn^{2+} ions.

$$\text{Protein}-\text{CaaX} + \text{FPP} \xrightarrow[Mg^{2+}, Zn^{2+}]{\text{FTase}} \overset{\overset{\text{farnesyl}}{|}}{\text{protein}-\text{CaaX}} + \text{PP}_i$$

Interest in FTase as a pharmaceutical target was based on two broad reasons. First, there was genetic proof of concept that inhibition of farnesylation would lead to functionally inactive Ras. This was demonstrated both by mutating the CaaX sequence so that Ras could no longer accept a farnesyl group (*10*), and by experiments in yeast in which the FTase β subunit gene was mutated or disrupted (*15*). The yeast experiments were particularly interesting because the target enzyme for inhibitor development, FTase, was ablated and this disruption blocked Ras function; however, the yeast cells remained viable, suggesting that in yeast, at least, there was some level of therapeutic index between inhibition of the activated versus the normal phenotype of Ras. The second reason for pursuit of FTase as a target was the fact that it was an enzyme, and selective inhibitors of a variety of enzymes have been developed into highly effective drugs.

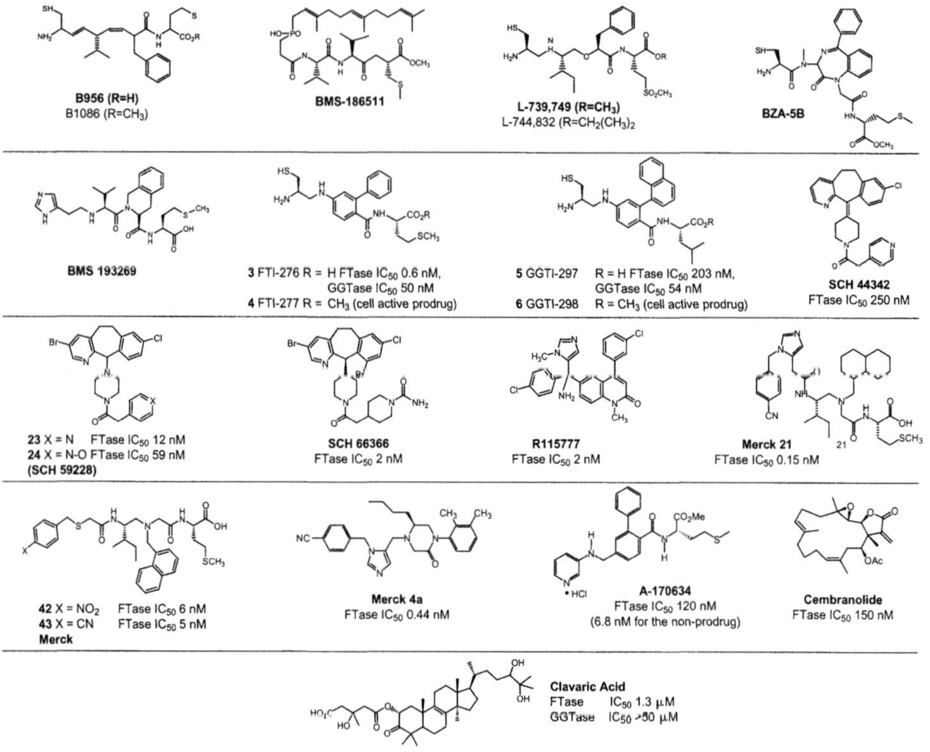

FIG. 1. Structures of FTIs competitive with protein substrate.

II. Early Inhibitors and Proof of Concept

The earliest described inhibitors of FTase were based on the substrates of the reaction, FPP and the CaaX tetrapeptide motif. These compounds were identified either by rational modification of substrate or by screening of chemical, natural product, and combinatorial libraries. Examples of these early structures are shown in Figs. 1 and 2 and include the thiol-based CaaX peptidomimetics L-744,832, BZA-5B, B-1086, and FTI-276; the non-thiol CaaX mimetic BMS-193269; the isoprenoid analogs FPT III, α-hydroxyfarnesyl phosphonic acid, Parke-Davis PD083176, RPR 130401, and H-D-Trp-D-Met-D-Fcl-Gla-NH; the bisubstrate compound BMS-186511; and natural products such as manumycin, chaetomellic acid, and the zaragozic acids, which most frequently are competitive with respect to the farnesyl diphosphate substrate (16–20). SCH 44342 (Fig. 1) was identified by screening chemical analogs of the antihistamine loratadine. SCH 44342 is one of the earliest examples of a nonpeptidic nonthiol FTase

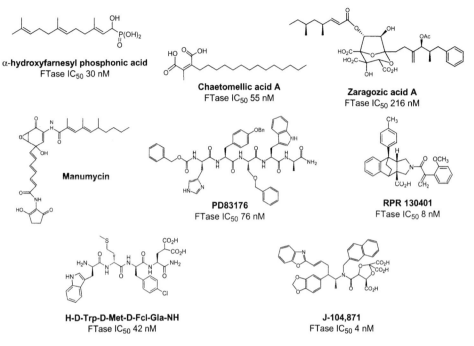

α-hydroxyfarnesyl phosphonic acid
FTase IC$_{50}$ 30 nM

Chaetomellic acid A
FTase IC$_{50}$ 55 nM

Zaragozic acid A
FTase IC$_{50}$ 216 nM

Manumycin

PD83176
FTase IC$_{50}$ 76 nM

RPR 130401
FTase IC$_{50}$ 8 nM

H-D-Trp-D-Met-D-Fcl-Gla-NH
FTase IC$_{50}$ 42 nM

J-104,871
FTase IC$_{50}$ 4 nM

FIG. 2. Structures of FTIs competitive with farnesyl diphosphate.

inhibitor (FTI) competitive with respect to the CaaX substrate (21). Many of the other CaaX mimetics required a prodrug strategy (see Fig. 1), i.e., esterification of the C-terminal carboxylate, to temporarily mask the negative charge and achieve cell activity.

With these compounds, we have learned that FTIs block the farnesylation of the Harvey (Ha)-Ras protein in addition to at least 20 other proteins in cells, demonstrating that FTIs are not specific inhibitors of Ras function (18, 22–26). All the cell-active FTIs also possess antiproliferative and antitumorigenic activity against a variety of *ras*-transformed cell lines and human tumor cell lines in cell culture, and many FTIs block tumor growth in rodent models. Interestingly, in the preclinical models, FTIs were not toxic to nontransformed cells or to mice at doses that were efficacious in tumor assays. Given the structural diversity of these compounds (Figs. 1 and 2) and the fact that they all exhibit similar biological phenotypes, there is strong pharmacological evidence that the biological action of FTIs is mechanism based, i.e., inhibition of FTase. However, these compounds were active against some tumor cell lines lacking mutant *ras* (27), raising the possibility that the biological action of FTIs might be directed at farnesylated proteins in addition to the Ras proteins. The development of these early FTIs also revealed that the Kirsten (Ki)-Ras and N-Ras proteins, the forms of Ras found mutated most frequently in human clinical cancer, become geranyl-geranylated in cells suppressed for FTase activity (28, 29). The reason for this cross-prenylation is the observation that Ki-Ras and N-Ras proteins can serve as substrates for geranylgeranyl protein transferase-I (GGTase-I), although these proteins are not as efficient substrates for GGTase-I as they are for FTase (30, 31). It was speculated that this cross-prenylation might confer resistance to FTIs; however, cells transformed by Ki-*ras* or N-*ras* are sensitive to FTIs alone, although sometimes requiring higher FTI concentrations than those necessary to inhibit Ha-*ras*-transformed cells. Thus, these compounds revealed proof-of-concept efficacy and also illumi-nated some of the complexities of protein prenylation in cells.

III. Newer Compounds

Although the earlier generations of FTIs afforded effective pharmacolog-ical agents for preclinical studies, they were not final chemical entities that could serve as drugs appropriate for clinical evaluation in humans. For example, there was interest in eliminating metabolic liabilities such as the thiol or ester prodrug strategies of CaaX mimetics. The strategies to identify clinical candidates involved further screening of natural products, chemical libraries, and modifications of lead templates. Several excellent reviews have described the chemical development of FTIs (16, 17, 32, 33).

An extensive characterization of a single chemical series has been pub-lished for the Schering-Plough group. The original lead, SCH 44342 (Fig. 1), was found to be competitive with respect to CaaX substrate and as such was the first disclosed CaaX mimetic devoid of a thiol and also not requiring a prodrug strategy for cell activity (21). The Schering-Plough group has described the systematic use of metabolism data to develop a compound with sufficient potency and pharmacokinetics to be evaluated as a drug candidate (34–41). Whereas intermediates in this development process such as SCH 59228 demonstrated good biological activity in animals after oral dosing (42), further modifications led to the clinical development compound SCH 66336 (43). These compounds are potent inhibitors of FTase and poor inhibitors of GGTase-I, and demonstrate good potency and pharmacoki-netic features for biological evaluation.

For medicinal chemistry efforts that evolved from CaaX peptidomimetics as the original lead structures, the goal has been to identify surrogates for the thiol of cysteine. Thiols are metabolically reactive and also can induce hypersensitivity reactions in humans. Of the nonthiol FTI compounds disclosed in the scientific literature, many of the successful efforts have incorporated imidazole groups. The rationale for this approach stems from the discovery of Zn^{2+} ion as being critical for the catalytic activity of FTase (44, 45) and the observation that the thiol of cysteine interacts with zinc in the cystal structure of FTase (46–48). Specific examples of these compounds (see Fig. 1) include BMS-193269 (49), Merck substituted imidazole compounds 21 and 4a (50–52), FTI-2148/2153 from the Sebti and Hamilton laboratories (53), and α-cyanocinnamide derivatives from the Levitzki laboratory (54). The Janssen clinical candidate R115777, which was developed from a lead identified in a screen of a chemical library, also has an imidazole group (55). These compounds are competitive with respect to CaaX substrate and not competitive with respect to FPP. Additional nonthiol compounds have been reported using substituted phenol, furan, or pyridyl moieties and examples include Merck compounds 42 and 43 (56) and Abbott A-170634 (57–60). For compounds lacking an imidazole group, it has been speculated that there is an interaction with a hydrophobic pocket of the FTase active site (51, 56). Several of these compounds also do not require a prodrug, because the negatively charged carboxylate was eliminated or other properties of the compound compensate for this apparent liability. Importantly, the nonthiol FTIs exhibit potencies against FTase and biological activities in cell and animal tumor models that are comparable to if not better than those of thiol-containing FTIs (51–53, 59). The Abbott compound A-170634 is also inhibitory in animal models of angiogenesis (61); the results indicate that the antitumorigenic activity of FTIs may be due to both direct effects on the tumor cells and indirect effects on the tumor vasculature. Imidazoles have also been incorporated into inhibitors of GGTase-I to yield nonthiol GGTIs (53, 62).

Most natural product inhibitors of FTase interact at the isoprenoid-binding site or by an undefined mechanism; however, two have been described as being competitive with respect to the CaaX substrate. Cembranolide (Fig. 1), a diterpene, was the first natural product identified as an inhibitor of the CaaX substrate site of FTase (63). More recently, a steroid natural product called clavaric acid (Fig. 1) was identified that is competitive with respect to the CaaX substrate and uncompetitive with respect to farnesyl diphosphate (64). Clavaric acid also inhibited Ha-Ras processing in cells at micromolar concentrations without inhibiting hydroxymethylglutaryl (HMG)-CoA reductase. Given the similarity of clavaric acid to cholesterol and the fact that cholesterol is a metabolite of the isoprenoid biosynthetic pathway, the identification of a sterol-based compound competitive with respect to CaaX substrate was highly unexpected; efforts to explore the structure–activity relationship of this compound identified several other sterols, but these compounds were competitive with respect to farnesyl diphosphate (*64*).

Polycarboxylate compounds have been identified with a fairly high frequency as inhibitors of FTase that are competitive with respect to farnesyl diphosphate. Unfortunately, the highly charged nature of some of these compounds, such as chaetomellic acid or the zaragozic acids, impairs the ability of these compounds to penetrate cells and these FTIs do not have biological activity (*65*). The polycarboxylate J-104,871 (Fig. 2), however, appears to be an exception to this trend (*66, 67*). J-104,871 emerged from a series of compounds originally identified in a screen of squalene synthase inhibitors. J-104,871 is a highly potent FTI with 300-fold selectivity over GGTase I; the FTI is also devoid of activity against squalene synthase even though it is competitive with respect to FPP in the FTase reaction. Micromolar concentrations of J-104,871 inhibited the growth properties of *ras*-transformed cells in culture. J-104,871 was also active in a mouse tumor xenograft model at doses that inhibited intratumoral Ha-Ras processing.

Because FTIs are specific for FTase and are not necessarily selective for a given substrate, Dong and colleagues have explored the idea of whether it is possible to specifically inhibit the farnesylation of a single substrate (*68*). Using a peptide library, these investigators have identified what they have called "molecular forceps," which recognize the CaaX sequence of Ras proteins and selectively block farnesylation.

IV. Transgenic Mouse Models in Development of Farnesyltransferase Inhibitors

Previous reviews have discussed the efficacy of FTIs in mouse xenograft tumor models and in the mouse mammary tumor virus (MMTV)–v-Ha-*ras* transgenic mouse tumor model (*18, 23, 24, 26, 69*). A variety of inhibitors, such as L-744,832 (*70*), derivatives of PD-083176 (*71*), manumycin (*72*), B-956 (*73*), and SCH 59228 (*42*), blocked tumor growth in the xenograft models. L-744,832 was observed to induce tumor regression in the MMTV–Ha-Ras model (*74*). Although these results were encouraging, several mechanistic questions needed to be addressed. What was the mechanism of tumor regression? Would tumor regression be observed in mice having more than one defined genetic alteration? What would be the effect of an FTI in mice having mutations in the Ki-*ras* or N-*ras* genes? To evaluate these and other issues, a number of studies have reported results using transgenic mouse tumor models (Table I).

TABLE I

ANALYSIS OF FARNESYLTRANFERASE INHIBITORS IN TRANSGENIC TUMOR MODELS

Genotype	Tumor inhibition	Apoptosis induction	G_1/S block
MMTV–v-Ha-*ras*	+	+	−
Wap–v-Ha-*ras*	+	+	+
MMTV–v-Ha-*ras* × p53$^{-/-}$	+	+	+
MMTV–v-Ha-*ras*/*myc*	+	−	+
MMTV–N-*ras*	+	+	−
MMTV–[V12]Ki4B-*ras*	+	ND^a	ND
MMTV–TGF-α	+	+	+
MMTV–TGF-α + DMBA	+	ND	ND
MMTV–TGF-α/*neu*	+	ND	ND
MMTV–*neu*	−	ND	ND

^{a} ND, Not determined.

For example, L-744,832 was evaluated in mice having the genotypes MMTV–v-Ha-*ras* × p53$^{-/-}$ or MMTV–v-Ha-*ras* × c-*myc* (*75*). Tumor inhibition was observed in each of these models; however, the nature and mechanism of the responses varied somewhat with the genotype of the tumor. In the v-Ha-*ras* model, L-744,832 was able to induce tumor regression in mice having either a v-Ha-*ras* × p53$^{-/-}$ or v-Ha-*ras* × *myc* genotype (*75*). The mechanism of this regression appeared to be dependent on the genotype. In v-Ha-*ras* × p53$^{-/-}$, regression was due to both an induction of apoptosis as well as restoration of the G_1 checkpoint. In contrast, little apoptosis was seen in the v-Ha-*ras* × *myc* mice, and tumor regression appeared to be due predominantly to effects on the cell cycle of tumor cells.

The inhibition of tumor formation observed with L-744,832 is clearly not related to transgenes having the MMTV promoter used in these models or due to the thiol properties of this compound as evidenced by results from the Schering-Plough group. They studied transgenic mice expressing v-Ha-*ras* under control of the whey acidic protein (wap) promoter (*43*). Treatment of these mice orally with the nonthiol FTI SCH 66336 prevented mammary tumor formation as well as induced tumor regression in mice having established tumors. The mechanism of tumor regression in the wap–v-Ha-*ras* model was due to the induction of apoptosis accompanied by a decrease in DNA synthesis as measured by *in situ* TUNEL [terminal deoxynucleotidyltransferase (TdT)-mediated dUTP nick end labeling] and bromodeoxyuridine (BrdU) staining, respectively.

Because N-Ras and Ki-Ras become geranylgeranylated in cells treated with an FTI (28, 29), there was speculation that tumors bearing these forms of mutant *ras*, the most common in clinical cancer, would be resistant to FTIs. In cell culture, this does not appear to be the case as antitumor effects are observed independent of whether mutant *ras* alleles are present (27). To extend this observation further, L-744,832 was administered to transgenic mice having the genotypes MMTV–N-*ras* (wild type) or MMTV–[V12]Ki4B-*ras* (76, 77). Inhibition of mammary tumors in the N-*ras* and Ki-*ras* models was observed, and this antitumor effect occurred without inhibition of the processing of the N-Ras or Ki-Ras proteins as monitored in a sodium dodecye sulfate–polyacrylamide gel electrophoresis (SDS–PAGE) gel shift assay. A reduction in lymphoma numbers was also noted in the N-*ras* model, using a limited number of animals. These results support the hypothesis raised in cell culture experiments that proteins in addition to Ras contribute to the biological action of FTIs.

One of the criticisms of the *ras* transgenic mouse mammary tumor models is the fact that the mutant Ras proteins are overexpressed in this tissue; it has also been observed that mutant *ras* genes are rarely found in human breast cancer. In contrast, transforming growth factor α (TGF-α) and the *neu* oncogene contribute to human breast cancer pathophysiology. For these reasons, L-744,832 was tested in transgenic mice that develop breast tumors due to these genes. Three models were evaluated: mice having MMTV–TGF-α, mice having MMTV–TGF-α and treated with the carcinogen DMBA (9,10-dimethyl-1,2-benzanthracene), and mice having both TGF-α and *neu* under the control of the MMTV promoter (78). Tumor regression was observed in all three models; this was associated with an increase in apoptosis and a G_1/S cell cycle arrest. Phospho-MAP (mitogen-associated protein) kinase was also decreased in tumors of FTI-treated mice, indicating that a decrease in DNA synthesis was due at least in part to inhibition of the MAP kinase cascade. The decrease in tumor size was also accompanied by a decrease in circulating plasma concentrations of TGF-α. The ability of L-744,832 to inhibit tumors in the MMTV–TGF-α/*neu* model is interesting because this same FTI does not inhibit tumors in MMTV–c-*neu* mice (75). This difference may be due to effects of L-744,832 on TGF-α regulation. In cell culture, rat intestinal epithelial cells can be transformed by mutant Ha-*ras* or Ki-*ras* and L-744,832 effectively blocks the proliferation of the Ha-*ras* cells but not of the Ki-*ras* cells (79). The inhibition of the Ha-*ras*-transformed cells correlates with inhibition of Ha-Ras processing and a decrease in expression of TGF-α and amphiregulin. These epidermal growth factor receptor ligands are an important autocrine pathway in the transformation of rat intestinal epithelial cells by *ras*. Thus, in this cell culture model, L-744,832 interferes with Ras signaling pathways and autocrine pathways essential to cell survival. Whether an analogous scenario occurs in the transgenic mouse models is not known.

To understand in greater molecular detail the mechanism for the observed antitumor effect of L-744,832 in the TGF-α/neu mice, tumors from control and FTI-treated mice were excised and also tested to determine the phosphorylation state of p70 S6 kinase (80). Doses of L-744,832 that inhibited tumor growth also completely inhibited p70 S6 kinase phosphorylation, indicating that this path leading to DNA synthesis was impaired. Inhibition of this pathway was also observed in various tumor cell lines in culture. The inhibition of S6 kinase is associated with decreased phosphorylation of Phas-1, similar to what is seen in cells treated with rapamycin (80). Phas-1 regulates cap-dependent protein translation by binding to elongation initiation factor 4B in a phosphorylation-dependent manner. Interestingly, the kinetics of this inhibition are rapid, with a 50–60% reduction in S6 kinase activity observed within 30 min of FTI treatment. As protein farnesylation is essentially an irreversible modification (at least prior to protein degradation) (81) and Ras turnover occurs in roughly 24 hr (82), this result suggests that a farnesylated protein other than Ras with a rapid rate of turnover is responsible for the regulation of this DNA synthetic pathway.

V. Mechanisms of Apoptosis

FTIs were originally thought to be cytostatic agents. One of the early surprises in research on the biological mechanism of FTIs was the observation that these compounds were capable of inducing apoptosis in tumor cells under some conditions. The mechanisms for this action have been explored in greater detail. In tumor cells treated with either of several structurally distinct FTIs, there is an upregulation of the proapoptotic proteins Bax and Bcl-xs, release of cytochrome c from mitochondria, and activation of caspase 3 (83–85). The levels of the antiapoptotic protein Bcl2 remain unchanged and caspase 1 is not activated. The apoptosis induced by FTIs is independent of p53 function and can be blocked by the caspase inhibitors Z-DEVD-fmk and ZVAD-fmk or by ectopic overexpression of Bcl2. The susceptibility of tumor cells to undergo apoptosis in response to an FTI is also influenced by the Akt pathway (86). Rat1 cells transformed with mutant Ha-ras do not undergo apoptosis in monolayer culture when treated with an FTI; this is in contrast to cells grown in an anchorage-independent manner. Activation of the Akt pathway in cell-adherent cultures appears to provide a survival signal that is absent in cells deprived of substratum attachment. The complexity of the integrin signals may explain why FTIs in general are able to induce tumor apoptosis in transgenic mouse tumor models but not in tumor xenograft mouse models (87).

VI. Non-Ras Targets

A number of proteins in addition to Ras are farnesylated (18, 20, 22, 24). Many these proteins are members of the Ras superfamily of proteins such as RhoB and RhoE, Rheb, and Rap2 and modulate aspects of Ras signaling and the cytoskeleton. Other farnesylated proteins have important functions in cell regulation such the nuclear lamins, protein tyrosine phosphatase PRL-1, and an inositol phosphatase. More specialized proteins function in the retinal visual signal transduction system (transducin γ subunit, rhodopsin kinase, cGMP phosphodiesterase) or as a molecular chaperone (DnaJ homolog). Because FTIs inhibit the enzyme FTase and are not specific for a particular substrate, the role of these proteins in terms of the biological efficacy of FTIs as well as their potential toxicity is an important issue to consider.

One protein in particular, RhoB, has attracted attention as a key substrate involved in the biology of FTIs. RhoB, which regulates cell adhesion parameters, is both farnesylated and geranylgeranylated. RhoB was originally explored by Prendergast and colleagues because it is a rapidly turning over farnesylated protein that appears to correlate with the kinetics of morphological reversion induced by FTIs (87, 88). RhoB function is limiting for Ras transformation as shown with dominant negative alleles. Furthermore, myristoylated and activated forms of RhoB that function independently of farnesylation confer resistance to FTIs. While the farnesylation of RhoB is inhibited in cells treated with an FTI, RhoB remains geranylgeranylated (89). It was observed that geranylgeranylated RhoB has a growth-suppressive activity due to the ability to induce the cell cycle inhibitor p21 WAF, providing evidence that farnesylated and geranylgeranylated forms of RhoB have at least some distinct functions (90). Thus, FTIs appear to inhibit an essential role of farnesylated RhoB involved with Ras signaling and unmask a growth-suppressive function of geranylgeranylated RhoB. The role of cell adhesion in FTI responses was further elaborated by the discovery that L-744,832 induces $\alpha2(I)$ collagen expression in mutant Ha-ras-transformed Rat1 cells (91). This component of the extracellular matrix was previously shown to suppress Ras transformation. As the expression of RhoB shows cell-type and tissue differences (92), it will be of interest to learn more about the role of RhoB and cell adhesion parameters in Ras physiology and FTI responses in tumors of different tissue types.

VII. Combinations of Farnesyltransferase Inhibitors with Other Therapeutics

The treatment of clinical cancer most often combines more than one chemotherapeutic agent or modality. Thus, it is reasonable to expect that FTIs, should they provide some clinical benefit, might also be used in combination with existing treatment regimens. For this reason, several groups have evaluated FTIs in preclinical models with other chemotherapeutic agents or with radiation treatment. The strategies that have been taken include a more global evaluation of assessing a number of therapeutically useful therapeutics with an FTI as well as testing FTIs in combination with the therapy of choice for a particular tumor type.

One example of the latter is the evaluation of FTI L-744,832 in combination with agents typically used to treat breast cancer [doxorubicin, cisplatin, 5-fluorouracil (5-FU), vinblastine, and taxanes]. Using MCF-7 and MDA-469 human breast cancer cells, it was observed that the combination of L-744,832 with these agents resulted in a greater antiproliferative effect (93). In particular, the combination of L-744,832 with paclitaxel appeared to be greater than additive. A similar enhancement was observed with a structurally different compound, epothilone, which has a mechanism similar to paclitaxel, i.e., tubulin stabilization that leads to a G_2/M cell cycle arrest. The mechanism for this enhanced activity appeared to be related to a more pronounced G_2/M arrest. Interestingly, FTIs, in the absence of a microtubule-modulating agent, typically cause a G_1/S cell cycle arrest in tumor cells having wild-type p53 (94). This arrest is associated with a decrease in Rb phosphorylation caused by upregulation of p21 WAF and a subsequent decrease in cyclin E-associated kinase activity. Independent work by Suzuki and co-workers has also demonstrated that an FTI, SCH 56582, in combination with vincristine was able to arrest Ki-*ras*-transformed NRK cells in G_2/M in an apparently synergistic manner (95). Importantly, these effects were not observed in nontransformed NRK cells. A working model for this pharmacological interaction revolves around the observation that both FTIs and microtubule agents disrupt microtubule dynamics that are important for the survival of tumor cells (95).

The advantages of combination therapy have also been observed in mouse tumor models (43, 53). Liu and colleagues demonstrated that the clinical candidate SCH 66336 is more active against human lung tumor cells in mouse xenografts or against mammary carcinomas in the transgenic wap–v-Ha-*ras* mouse when this agent is combined with either cyclophosphamide, 5-FU, or vincristine (43). More recently, Sun et al. showed that the nonthiol FTI-2153 was more active against A-549 human lung tumors in a mouse xenograft model when tested in combination with either cisplatin, gemcitabine, or paclitaxel (53). Interestingly, a nonthiol GGTase-I inhibitor (GGTI) also showed enhanced antitumor activity in this model when used in combination with the same set of cytotoxic agents (53). These results further demonstrate that FTIs cannot necessarily be thought of as cytostatic agents. If this were the sole mechanism, then one might have predicted that the combination of FTI with a compound acting on the cell cycle might be antagonistic. One example of this phenomenon has been reported with a human melanoma cell line tested in cell culture with the FTI BZA-5B in combination with cisplatin. BZA-5B-treated cells showed increased resistance to cisplatin and had a decreased number of cells undergoing apoptosis (96). Nevertheless, the enhanced activity observed with agents such as SCH 66336, FTI-2153, and L-744,832 in several cell culture and animal tumor models supports clinical evaluation of FTIs in combination with clinically effective anticancer drugs. Indeed, a preliminary report has revealed a phase I protocol of the Janssen compound R115777 in combination with 5-FU/leucovoran in patients having advanced colorectal or pancreatic cancer (97).

Muschel, McKenna, and Bernhard at the University of Pennsylvania have explored the effects of various FTIs in combination with radiation treatment (98, 99). Oncogenes such as ras have been associated with radioresistance. These investigators have demonstrated that FTIs restore the radiosensitivity of ras-transformed cells and that the biological effect correlates with inhibition of Ras processing. In particular, FTIs such as FTI-276 or L-744,832 appear to have a synergistic effect with radiation in tumor cells having mutant Ha-ras (100). The increased cell kill is due to increased apoptosis. Cells that lack a mutant Ha-ras gene do not exhibit this synergistic effect. FTIs also do not appear to confer radiosensitivity to cells having mutant Ki-ras, most likely because Ki-Ras protein becomes geranylgeranylated in cells treated with an FTI (99). However, the combination of FTI-276 with GGTI-298 resulted in inhibition of Ki-Ras processing and led to radiosensitization, further extending the mechanistic correlation between inhibition of Ras processing and increased sensitivity to radiation. Interestingly, in the absence of radiation, the combination of FTI-276 and GGTI-298 does not appear to offer any biological advantage over FTI treatment alone (101).

VIII. Noncancer Applications

Although the primary interest in FTIs is within the context of cancer chemotherapy, other potential applications are also being explored. In mammalian organisms, FTIs have been examined in a corneal model of inflammation (*102*). Topical application of manumycin inhibited macrophage infiltration into the cornea, suggesting a possible application in the treatment of corneal opacity.

Several investigators have reported on protein prenylation in parasites and speculated that inhibition of farnesylation may be an approach to combat this infection (*103–105*). Yokoyama and co-workers purified FTase from *Trypanosoma brucei* and showed that L-745,631, FTI-277, GGTI-298, and SCH-44342 were inhibitory toward this enzyme (*106*). Furthermore, these compounds inhibited the growth of *T. brucei* and *Trypanosoma cruzi* with median inhibitory concentration (IC_{50}) values in the low micromolar range. Because these concentrations of FTIs have not shown toxicity to nontransformed mammalian cells in culture, the authors suggest that FTIs might have utility in parasite infections.

While most studies of protein farnesylation and the evaluation of FTIs involve therapeutic uses in mammalian cells and potentially trypanosomes, it is interesting that the biology of protein farnesylation has also emerged in studies of plants (*107–110*). Plants have all three protein prenyltransferases: FTase, GGTase-I, and Rab-GGTase-II. The FTI manumycin blocks cell cycle progression of cultured tobacco plant cells. Manumycin and α-hydroxyfarnesylphosphonic acid also modulate stomatal closure. The role of protein farnesylation in plants is still emerging and it will be interesting to monitor the potential applications of FTIs in this area of research.

IX. Clinical Development of Farnesyltransferase Inhibitors

Four different FTIs have now been reported to be in clinical trials with cancer patients (Table II). Janssen R115777 (*97, 111, 112*) and Schering-Plough SCH 66336 (*113–115*) are currently in phase II trials as orally administered agents, given twice daily. The dose-limiting toxicities of these

TABLE II

FARNESYLTRANSFERASE INHIBITORS IN CLINICAL TRIALS

FTI	Stage of development	Route of administration
Janssen, R115777	Phase II	Oral
Schering-Plough, SCH 66336	Phase II	Oral
Merck, L-778,123	Phase I	Continuous infusion
Bristol-Myers Squibb, BMS-214662	Phase I	Intravenous

agents were typical of other cancer chemotherapeutics, that is, myelosup-pression and gastrointestinal (GI) toxicities. Earlier phase I trials have been initiated with intravenous agents from Merck (L-778,123), given by a continuous infusion protocol (*116*), and Bristol-Myers Squibb (BMS-214662). To date, only one partial response has been reported in a patient with non-small-cell lung carcinoma treated with SCH 66336. With the dose-limiting toxicities defined, efforts in phase II are to evaluate efficacy. These studies are including combination therapies. For example, a phase I trial was reported for the combination of R115777 with 5-FU/leucovoran in patients with advanced colon or pancreatic cancer (*97*). Both Merck and Schering-Plough are also using pharmacodynamic end points in the clinical evaluation of their FTIs. Doses of L-778,123 have been identified that inhibit the processing of the farnesylated molecular chaperone hDJ in circulating human white blood cells (*116, 117*). Likewise, doses of SCH 66336 were found that inhibit the farnesylation of prelamin A in buccal mucosa cells of treated patients (*113, 114*).

The greatest amount of published information for an FTI currently in the clinic has appeared from the Schering-Plough group. SCH 66336 is a highly potent inhibitor of FTase and is essentially devoid of activity against GGTase-I. It has good potency (75–400 n*M*) against a panel of transformed cells in soft agar growth assays (*37, 43*). Furthermore, SCH 66336 shows activity against primary breast, ovarian, and non-small-cell lung tumor explants in the human tumor colony-forming assay (*118*). Antitumor activity in mouse xenograft and wap–v-Ha-*ras* transgenic models is noted after oral dosing (four times a day) at a dose of 40 mg/kg (*43*). The crystal structure of SCH 66336 bound in the active site of FTase has also been reported (*119*). SCH 66336 spans the active site, interacts directly with atoms of the enzyme, and packs against the isoprenoid chain of farnesyl diphosphate. Although Zn^{2+} ion is bound in the active site of FTase and interacts with the thiol residue of CaaX peptides, there was no indication of an interaction between zinc and SCH 66336.

A summary of the properties of BMS-214662 indicates that this com-pound is highly effective in mouse xenograft and carcinogen-induced tumor models (*120*). The mechanism of tumor inhibition and in some cases regres-sion appears to be due to an induction of apoptosis. A tumor regression and cure was also seen in animals bearing HCT-116/VM-26 human colon tumor cells; this line is known to be insensitive to paclitaxel. BMS-214662 was effective in these preclinical models when administered by either the oral, intravenous, or intraperitoneal route.

Information about R115777 to date has appeared in abstract form (*55, 121–124*). R115777 has subnanomolar potency against FTase, has excellent antiproliferative activity in cell culture, and is active in tumor xenografts using human tumor cell lines. The properties of the Merck FTI L-778,123 have not been disclosed.

X. Conclusions

This is an exciting time in the study of protein prenylation. The FTIs have been developed to the point that the fundamental hypotheses generated in preclinical studies can be directly tested in people. Since the time when efforts to develop inhibitors of FTase for cancer began, two fundamental pieces of information have been learned. First, the number of proteins known to be farnesylated is much larger than originally recognized, with at least 20 proteins identified so far. Some of these proteins may contribute to the biology of FTIs. Second, the cross-prenylation of N-Ras and Ki-Ras proteins by GGTase-I in cells treated with FTIs showed that inhibiting the function of these forms of Ras, which are most closely associated with the clinical disease in humans, was more complicated than originally envisioned. Nevertheless, the efficacy of FTIs in the preclinical models warranted evaluation in the clinic. Given that FTIs can no longer be viewed as "laser beam" therapy for Ras-mediated malignancy, it is not surprising that the dose-limiting toxicities noted in the phase I clinical trials are similar to those found with currently used antiproliferative drugs. The question will be whether inhibition of a mechanistically new target, FTase, will afford advantages in the clinic either as a single agent or in combination therapies. The results of current FTI trials are much anticipated.

ACKNOWLEDGMENTS

I thank Sam Graham for many helpful comments to the manuscript. This review was written based on the literature available up to September 1999.

REFERENCES

1. Hancock, J. F., Magee, A. I., Childs, J. E., and Marshall, C. J. (1989). *Cell* **57,** 1167.
2. Schafer, W. R., Kim, R., Sterne, R., Thorner, J., Kim, S. H., and Rine, J. (1989). *Science* **245,** 379.
3. Casey, P. J., Solski, P. A., Der, C. J., and Buss, J. E. (1989). *Proc. Natl. Acad. Sci. U.S.A.* **86,** 8323.
4. Reiss, Y., Goldstein, J. L., Seabra, M. C., Casey, P. J., and Brown, M. S. (1990). *Cell* **62,** 81.
5. Schaber, M. D., O'Hara, M. B., Garsky, V. M., Mosser, S. D., Bergstrom, J. D., Moores,

S. L., Marshall, M. S., Friedman, P. A., Dixon, R. A. F., and Gibbs, J. B. (1990). *J. Biol. Chem.* **265**, 14701.

6. Manne, V., Roberts, D., Tobin, A., O'Rourke, E., De Virgilio, M., Meyers, C., Ahmed, N., Kurz, B., Resh, M., Kung, H. F., and Barbacid, M. (1990). *Proc. Natl. Acad. Sci. U.S.A.* **87**, 7541.

7. Schafer, W. R., Trueblood, C. E., Yang, C. C., Mayer, M. P., Rosenberg, S., Poulter, C. D., Kim, S. H., and Rine, J. (1990). *Science* **249**, 1133.

8. Barbacid, M. (1987). *Annu. Rev. Biochem.* **56**, 779.

9. Bos, J. L. (1989). *Cancer Res.* **49**, 4682.

10. Lowy, D. R., and Willumsen, B. M. (1993). *Annu. Rev. Biochem.* **62**, 851.

11. Gibbs, J. B. (1992). *Semin. Cancer Biol.* **3**, 383.

12. Ahmadian, M. R., Zor, T., Vogt, D., Kabsch, W., Selinger, Z., Wittinghofer, A., and Scheffzek, K. (1999). *Proc. Natl. Acad. Sci. U.S.A.* **96**, 7065.

13. Wigler, M., Field, J., Powers, S., Brock, D., Toda, T., Cameron, S., Nikawa, J., Michaeli, T., Colicelli, J., and Ferguson, K. (1988). *Cold Spring Harbor Symp. Quant. Biol.* **53**, 649.

14. Glomset, J. A., Gelb, M. H., and Farnsworth, C. C. (1990). *Trends Biochem. Sci.* **15**, 139.

14a. Spence, R. A. and Casey, P. J. (Chapter 1); Terry, K. L., Long, S. B., and Beese, L. S. (Chapter 2); and Urano, J., Yang, W., and Tamanoi, F. (2000). *In* "The Enzymes, Vol. XXI. Academic Press, San Diego, California. [This volume]

15. Schafer, W. R., and Rine, J. (1992). *Annu. Rev. Genet.* **30**, 209.

16. Graham, S. L., and Williams, T. M. (1996). *Exp. Opin. Ther. Patents* **6**, 1295.

17. Leonard, D. M. (1997). *J. Med. Chem.* **40**, 2971.

18. Gibbs, J. B., and Oliff, A. (1997). *Annu. Rev. Pharmacol. Toxicol.* **37**, 143.

19. Sebti, S., and Hamilton, A. D. (1997). *Curr. Opin. Oncol.* **9**, 557.

20. Gibbs, J. B., Graham, S. L., Hartman, G. D., Koblan, K. S., Kohl, N. E., Omer, C. A., and Oliff, A. (1997). *Curr. Opin. Chem. Biol.* **1**, 197.

21. Bishop, W. R., Bond, R., Petrin, J., Wang, L., Patton, R., Doll, R., Njoroge, G., Catino, J., Schwartz, J., Windsor, W., Syto, R., Schwartz, J., Carr, D., James, L., and Kirschmeier, P. (1995). *J. Biol. Chem.* **270**, 30611.

22. Cox, A. D., and Der, C. J. (1997). *Biochim. Biophys. Acta* **1333**, F51.

23. Sebti, S. M., and Hamilton, A. D. (1997). *Pharmacol. Ther.* **74**, 103.

24. Oliff, A. (1999). *Biochim. Biophys. Acta* **1423**, C19.

25. Omer, C. A., and Kohl, N. E. (1997). *Trends Pharmacol. Sci.* **18**, 437.

26. Lobell, R. B., and Kohl, N. E. (1998). *Cancer Metastasis Rev.* **17**, 203.

27. Sepp-Lorenzino, L., Ma, Z., Rands, E., Kohl, N. E., Gibbs, J. B., Oliff, A., and Rosen, N. (1995). *Cancer Res.* **55**, 5302.

28. Whyte, D. B., Kirschmeier, P., Hockenberry, T. N., Nunez Oliva, I., James, L., Catino, J. J., Bishop, W. R., and Pai, J. K. (1997). *J. Biol. Chem.* **272**, 14459.

29. Rowell, C. A., Kowalczyk, J. J., Lewis, M. D., and Garcia, A. M. (1997). *J. Biol. Chem.* **272**, 14093.

30. James, G. L., Goldstein, J. L., and Brown, M. S. (1995). *J. Biol. Chem.* **270**, 6221.

31. Zhang, F. L., Kirschmeier, P., Carr, D., James, L., Bond, R. W., Wang, L., Patton, R., Windsor, W. T., Syto, R., Zhang, R., and Bishop, W. R. (1997). *J. Biol. Chem.* **272**, 10232.

32. Williams, T. M. (1998). *Exp. Opin. Ther. Patents* **8**, 553.

33. Williams, T. M. (1999). *Exp. Opin. Ther. Patents* **8**, 1263.

34. Njoroge, F. G., Vibulbhan, B., Rane, D. F., Bishop, W. R., Petrin, J., Patton, R., Bryant, M. S., Chen, K. J., Nomeir, A. A., Lin, C. C., Liu, M., King, I., Chen, J., Lee, S., Yaremko, B., Dell, J., Lipari, P., Malkowski, M., Li, Z., Catino, J., Doll, R. J., Girijavallabhan, V., and Ganguly, A. K. (1997). *J. Med. Chem.* **40**, 4290.

35. Njoroge, F. G., Doll, R. J., Vibulbhan, B., Alvarez, C. S., Bishop, W. R., Petrin, J.,

Kirschmeier, P., Carruthers, N. I., Wong, J. K., Albanese, M. M., Piwinski, J. J., Catino, J., Girijavallabhan, V., and Ganguly, A. K. (1997). *Bioorg. Med. Chem.* **5,** 101.

36. Njoroge, F. G., Vibulbhan, B., Pinto, P., Bishop, W. R., Bryant, M. S., Nomeir, A. A., Lin, C., Liu, M., Doll, R. J., Girijavallabhan, V., and Ganguly, A. K. (1998). *J. Med. Chem.* **41,** 1561.

37. Njoroge, F. G., Taveras, A. G., Kelly, J., Remiszewski, S., Mallams, A. K., Wolin, R., Afonso, A., Cooper, A. B., Rane, D. F., Liu, Y. T., Wong, J., Vibulbhan, B., Pinto, P., Deskus, J., Alvarez, C. S., del Rosario, J., Connolly, M., Wang, J., Desai, J., Rossman, R. R., Bishop, W. R., Patton, R., Wang, L., Kirschmeier, P., Bryant, M. S., Nomeir, A. A., Lin, C. C., Liu, M., McPhail, A. T., Doll, R. J., Girijavallabhan, V. M., and Ganguly, A. K. (1998). *J. Med. Chem.* **41,** 4890.

38. Mallams, A. K., Njoroge, F. G., Doll, R. J., Snow, M. E., Kaminski, J. J., Rossman, R. R., Vibulbhan, B., Bishop, W. R., Kirschmeier, P., Liu, M., Bryant, M. S., Alvarez, C., Carr, D., James, L., King, I., Li, Z., Lin, C. C., Nardo, C., Petrin, J., Remiszewski, S. W., Taveras, A. G., Wang, S., Wong, J., Catino, J., Ganguly, A. K., *et al.* (1997). *Bioorg. Med. Chem.* **5,** 93.

39. Mallams, A. K., Rossman, R. R., Doll, R. J., Girijavallabhan, V. M., Ganguly, A. K., Petrin, J., Wang, L., Patton, R., Bishop, W. R., Carr, D. M., Kirschmeier, P., Catino, J. J., Bryant, M. S., Chen, K. J., Korfmacher, W. A., Nardo, C., Wang, S., Nomeir, A. A., Lin, C. C., Li, Z., Chen, J., Lee, S., Dell, J., Lipari, P., Malkowski, M., Yaremko, B., King, I., and Liu, M. (1998). *J. Med. Chem.* **41,** 877.

40. Taveras, A. G., Deskus, J., Chao, J. P., Vaccaro, C. J., Njoroge, F. G., Vibulbhan, B., Pinto, P., Remiszewski, S., del Rosario, J., Doll, R. J., Alvarez, C., Lalwani, T., Mallams, A. K., Rossman, R. R., Afonso, A., Girijavallabhan, V. M., Ganguly, A. K., Pramanik, B., Heimark, L., Bishop, W. R., Wang, L., Kirschmeier, P., James, L., Carr, D., Patton, R., Bryant, M. S., Nomeir, A. A., and Liu, M. (1999). *J. Med. Chem.* **42,** 2651.

41. Bryant, M. S., Korfmacher, W. A., Wang, S., Nardo, C., Nomeir, A. A., and Lin, C. C. (1997). *J. Chromatogr. A* **777,** 61.

42. Liu, M., Bryant, M. S., Chen, J. P., Lee, S. N., Yaremko, B., Li, Z. J., Dell, J., Lipari, P., Malkowski, M., Prioli, N., Rossman, R. R., Korfmacher, W. A., Nomeir, A. A., Lin, C. C., Mallams, A. K., Doll, R. J., Catino, J. J., Girijavallabhan, V. M., Kirschmeier, P., and Bishop, W. R. (1999). *Cancer Chemother. Pharmacol.* **43,** 50.

43. Liu, M., Bryant, M. S., Chen, J. P., Lee, S. N., Yaremko, B., Lipari, P., Malkowski, M., Ferrari, E., Nielsen, L., Prioli, N., Dell, J., Sinha, D., Syed, J., Korfmacher, W. A., Nomeir, A. A., Lin, C. C., Wang, L., Taveras, A. G., Doll, R. J., Njoroge, F. G., Mallams, A. K., Remiszewski, S., Catino, J. J., Girijavallabhan, V. M., Kirschmeier, P., and Bishop, W. R. (1998). *Cancer Res.* **58,** 4947.

44. Huang, C. C., Casey, P. J., and Fierke, C. A. (1997). *J. Biol. Chem.* **272,** 20.

45. Fu, H. W., Beese, L. S., and Casey, P. J. (1998). *Biochemistry* **37,** 4465.

46. Park, H. W., Boduluri, S. R., Moomaw, J. F., Casey, P. J., and Beese, L. S. (1997). *Science* **275,** 1800.

47. Long, S. B., Casey, P. J., and Beese, L. S. (1998). *Biochemistry* **37,** 9612.

48. Strickland, C. L., Windsor, W. T., Syto, R., Wang, L., Bond, R., Wu, Z., Schwartz, J., Le, H. V., Beese, L. S., and Weber, P. C. (1998). *Biochemistry* **37,** 16601.

49. Hunt, J. T., Lee, V. G., Leftheris, K., Seizinger, B., Carboni, J., Mabus, J., Ricca, C., Yan, N., and Manne, V. (1996). *J. Med. Chem.* **39,** 353.

50. Ciccarone, T. M., MacTough, S. C., Williams, T. M., Dinsmore, C. J., O'Neill, T. J., Shah, D., Culberson, J. C., Koblan, K. S., Kohl, N. E., Gibbs, J. B., Oliff, A. I., Graham, S. L., and Hartman, G. D. (1999). *Bioorg. Med. Chem. Lett.* **9,** 1991.

51. Anthony, N. J., Gomez, R. P., Schaber, M. D., Mosser, S. D., Hamilton, K. A., TJ, O. N., Koblan, K. S., Graham, S. L., Hartman, G. D., Shah, D., Rands, E., Kohl, N. E., Gibbs, J. B., and Oliff, A. I. (1999). *J. Med. Chem.* **42,** 3356.
52. Williams, T. M., Bergman, J. M., Brashear, K., Breslin, M. J., Dinsmore, C. J., Hutchinson, J. H., MacTough, S. C., Stump, C. A., Wei, D. D., Zartman, C. B., Bogusky, M. J., Culberson, J. C., Buser Doepner, C., Davide, J., Greenberg, I. B., Hamilton, K. A., Koblan, K. S., Kohl, N. E., Liu, D. M., Lobell, R. B., Mosser, S. D., O'Neill, T. J., Rands, E., Schaber, M. D., Wilson, F., Senderak, E., Motzel, S. L., Gibbs, J. B., Graham, S. L., Heimbrook, D. C., Hartman, G. D., Oliff, A. I., and Huff, J. R. (1999). *J. Med. Chem.* **42,** 3779.
53. Sun, J. Z., Blaskovich, M. A., Knowles, D., Qian, Y. M., Ohkanda, J., Bailey, R. D., Hamilton, A. D., and Sebti, S. M. (1999). *Cancer Res.* **59,** 4919.
54. Poradosu, E., Gazit, A., Reuveni, H., and Levitzki, A. (1999). *Bioorg. Med. Chem.* **7,** 1727.
55. Venet, M., Angibaud, P., Sanz, G., Poignet, H., End, D., and Bowden, C. (1998). *Proc. Am. Assoc. Cancer Res. Annu. Meet.* **39,** 318.
56. Breslin, M. J., deSolms, S. J., Giuliani, E. A., Stokker, G. E., Graham, S. L., Pompliano, D. L., Mosser, S. D., Hamilton, K. A., and Hutchinson, J. H. (1998). *Bioorg. Med. Chem. Lett.* **8,** 3311.
57. Augeri, D. J., O'Connor, S. J., Janowick, D., Szczepankiewicz, B., Sullivan, G., Larsen, J., Kalvin, D., Cohen, J., Devine, E., Zhang, H. C., Cherian, S., Saeed, B., Ng, S. C., and Rosenberg, S. (1998). *J. Med. Chem.* **41,** 4288.
58. Augeri, D. J., Janowick, D., Kalvin, D., Sullivan, G., Larsen, J., Dickman, D., Ding, H., Cohen, J., Lee, J., Warner, R., Kovar, P., Cherian, S., Saeed, B., Zhang, H., Tahir, S., Ng, S. C., Sham, H., and Rosenberg, S. H. (1999). *Bioorg. Med. Chem. Lett.* **9,** 1069.
59. O'Connor, S. J., Barr, K. J., Wang, L., Sorensen, B. K., Tasker, A. S., Sham, H., Ng, S. C., Cohen, J., Devine, E., Cherian, S., Saeed, B., Zhang, H., Lee, J. Y., Warner, R., Tahir, S., Kovar, P., Ewing, P., Alder, J., Mitten, M., Leal, J., Marsh, K., Bauch, J., Hoffman, D. J., Sebti, S. M., and Rosenberg, S. H. (1999). *J. Med. Chem.* **42,** 3701.
60. Shen, W., Fakhoury, S., Donner, G., Henry, K., Lee, J., Zhang, H. C., Cohen, J., Warner, R., Saeed, B., Cherian, S., Tahir, S., Kovar, P., Bauch, J., Ng, S. C., Marsh, K., Sham, H., and Rosenberg, S. (1999). *Bioorg. Med. Chem. Lett.* **9,** 703.
61. Gu, W. Z., Tahir, S. K., Wang, Y. C., Zhang, H. C., Cherian, S. P., O'Connor, S. J., Leal, J. A., Rosenberg, S. H., and Ng, S. C. (1999). *Eur. J. Cancer* **35,** 1394.
62. Vasudevan, A., Qian, Y. M., Vogt, A., Blaskovich, M. A., Ohkanda, J., Sebti, S. M., and Hamilton, A. D. (1999). *J. Med. Chem.* **42,** 1333.
63. Coval, S. J., Patton, R. W., Petrin, J. M., James, L., Rothofsky, M. L., Lin, S. L., Patel, M., Reed, J. K., McPhail, A. T., and Bishop, W. R. (1996). *Bioorg. Med. Chem. Lett.* **6,** 909.
64. Lingham, R. B., Silverman, K. C., Jayasuriya, H., Kim, B. M., Amo, S. E., Wilson, F. R., Rew, D. J., Schaber, M. D., Bergstrom, J. D., Koblan, K. S., Graham, S. L., Kohl, N. E., Gibbs, J. B., and Singh, S. B. (1998). *J. Med. Chem.* **41,** 4492.
65. Gibbs, J. B., Pompliano, D. L., Mosser, S. D., Rands, E., Lingham, R. B., Singh, S. B., Scolnick, E. M., Kohl, N. E., and Oliff, A. (1993). *J. Biol. Chem.* **268,** 7617.
66. Aoyama, T., Satoh, T., Yonemoto, M., Shibata,, J., Nonoshita, K., Arai, S., Kawakami, K., Iwasawa, Y., Sano, H., Tanaka, K., Monden, Y., Kodera, T., Arakawa, H., Suzuki Takahashi, I., Kamei, T., and Tomimoto, K. (1998). *J. Med. Chem.* **41,** 143.
67. Yonemoto, M., Satoh, T., Arakawa, H., Suzuki Takahashi, I., Monden, Y., Kodera, T., Tanaka, K., Aoyama, T., Iwasawa, Y., Kamei, T., Nishimura, S., and Tomimoto, K. (1998). *Mol. Pharmacol.* **54,** 1.

68. Dong, D. L., Liu, R. P., Sherlock, R., Wigler, M. H., and Nestler, H. P. (1999). *Chem. Biol.* **6**, 133.
69. Omer, C. A., Anthony, N. J., Buser Doepner, C. A., Burkhardt, A. L., deSolms, S. J., Dinsmore, C. J., Gibbs, J. B., Hartman, G. D., Koblan, K. S., Lobell, R. B., Oliff, A., Williams, T. M., and Kohl, N. E. (1997). *Biofactors* **6**, 359.
70. Kohl, N. E., Wilson, F. R., Mosser, S. D., Giuliani, E. A., deSolms, S. J., Conner, M. W., Anthony, N. J., Holtz, W. J., Gomez, R. P., Lee, T.-J., Smith, R. L., Graham, S. L., Hartmen, G. D., Gibbs, J. B., and Oliff, A. (1994). *Proc. Natl. Acad. Sci. U.S.A.* **91**, 9141.
71. McNamara, D. J., Dobrusin, E., Leonard, D. M., Shuler, K. R., Kaltenbronn, I S., Quin, J., III, Bur, S., Thomas, C. E., Doherty, A. M., Scholten, J. D., Zimmerman, K. K., Gibbs, B. S., Gowan, R. C., Latash, M. P., Leopold, W. R., Przybranowski, S. A., and Sebolt Leopold, J. S. (1997). *J. Med. Chem.* **40**, 3319.
72. Ito, T., Kawata, S., Tamura, S., Igura, T., Nagase, T., Miyagawa, J.-I., Yamazaki, E., Ishiguro, H., and Matsuzawa, Y. (1996). *Jpn. J. Cancer Res.* **87**, 113.
73. Nagasu, T., Yoshimatsu, K., Rowell, C., Lewis, M. D., and Garcia, A. M. (1995). *Cancer Res.* **55**, 5310.
74. Kohl, N. E., Omer, C. A., Conner, M. W., Anthony, N. J., Davide, J. P., deSolms, S. J., Giuliani, E. A., Gomez, R. P., Graham, S. L., Hamilton, K., Handt, L. K., Hartmen, G. D., Koblan, K. S., Kral, A. M., Miller, P. J., Mosser, S. D., O'Neil, T. J., Rands, E., Schaber, M. D., Gibbs, J. B., and Oliff, A. (1995). *Nature Med.* **1**, 792.
75. Barrington, R. E., Subler, M. A., Rands, E., Omer, C. A., Miller, P. J., Hundley, J. E., Koester, S. K., Troyer, D. A., Bearss, D. J., Conner, M. W., Gibbs, J. B., Hamilton, K., Koblan, K. S., Mosser, S. D., O'Neill, T. J., Schaber, M. D., Senderak, E. T., Windle, J. J., Oliff, A., and Kohl, N. E. (1998). *Mol. Cell. Biol.* **18**, 85.
76. Mangues, R., Corral, T., Kohl, N. E., Symmans, W. F., Lu, S., Malumbres, M., Gibbs, J. B., Oliff, A., and Pellicer, A. (1998). *Cancer Res.* **58**, 1253.
77. Omer, C. A., Chen, H. Y., Conner, M. W., deSolms, S. J., Dinsmore, C. J., Graham, S. L., Hartman, G. D., Williams, T. M., and Oliff, A. (1999). *Proc. Am. Assoc. Cancer Res. Annu. Meet.* **40**, 523.
78. Norgaard, P., Law, B., Joseph, H., Page, D. L., Shyr, Y., Mays, D., Pietenpol, J. A., Kohl, N. E., Oliff, A., Coffey, R. J., Poulsen, H. S., and Moses, H. L. (1999). *Clin. Cancer Res.* **5**, 35.
79. Sizemore, N., Cox, A. D., Barnard, J. A., Oldham, S. M., Reynolds, E. R., Der, C. J., and Coffey, R. J. (1999). *Gastroenterology* **117**, 567.
80. Law, B. K., Norgaard, P., Gnudi, L., Kahn, B. B., Poulson H. S., and Moses, H. L. (1999). *J. Biol. Chem.* **274**, 4743.
81. Zhang, L., Tschantz, W. R., and Casey, P. J. (1997). *J. Biol. Chem.* **272**, 23354.
82. Ulsh, L. S., and Shih, T. Y. (1984). *Mol. Cell. Biol.* **4**, 1647.
83. Hung, W. C., and Chuang, L. Y. (1998). *Int. J. Oncol.* **12**, 1339.
84. Suzuki, N., Urano, J., and Tamanoi, F. (1998). *Proc. Natl. Acad. Sci. U.S.A.* **95**, 15356.
85. Vitale, M., Di Matola, T., Rossi, G., Laezza, C., Fenzi, G., and Bifulco, M. (1999). *Endocrinology* **140**, 698.
86. Du, W., Liu, A., and Prendergast, G. C. (1999). *Cancer Res.* **59**, 4208.
87. Prendergast, G. C., and Du, W. (1999). *Drug Resist. Updates* **2**, 81.
88. Lebowitz, P. F., and Prendergast, G. C. (1998). *Oncogene* **17**, 1439.
89. Lebowitz, P. F., Casey, P. J., Prendergast, G. C., and Thissen, J. A. (1997). *J. Biol. Chem.* **272**, 15591.
90. Du, W., Lebowitz, P. F., and Prendergast, G. C. (1999). *Mol. Cell. Biol.* **19**, 1831.
91. Du, W., Lebowitz, P. F., and Prendergast, G. C. (1999). *Cancer Res.* **59**, 2059.

92. Fritz, G., Gnad, R., and Kaina, B. (1999). *Anticancer Res.* **19,** 1681.
93. Moasser, M. M., Sepp Lorenzino, L., Kohl, N. E., Oliff, A., Balog, A., Su, D. S., Danishefsky, S. J., and Rosen N. (1998). *Proc. Natl. Acad. Sci. U.S.A.* **95,** 1369.
94. Sepp Lorenzino, L., and Rosen, N. (1998). *J. Biol. Chem.* **273,** 20243.
95. Suzuki, N., Del Villar, K., and Tamanoi, F. (1998). *Proc. Natl. Acad. Sci. U.S.A.* **95,** 10499.
96. Fokstuen, T., Rabo, Y. B., Zhou, J. N., Karlson, J., Platz, A., Shoshan, M. C., Hansson, J., and Linder, S. (1997). *Anticancer Res.* **17,** 2347.
97. Peeters, M., Cutsem, E. V., Marsé, H., Palmer, P., Walraven, V., and Willems, L. (1999). *Proc. Am. Soc. Clin. Oncol.* **17,** 859.
98. Muschel, R. J., Soto, D. E., McKenna, W., and Bernhard, E. J. (1998). *Oncogene* **17,** 3359.
99. Bernhard, E. J., McKenna, W. G., Hamilton, A. D., Sebti, S. M., Qian, Y., Wu, J. M., and Muschel, R. J. (1998). *Cancer Res.* **58,** 1754.
100. Bernhard, E. J., Kao, G., Cox, A. D., Sebti, S. M., Hamilton, A. D., Muschel, R. J., and McKenna, W. G. (1996). *Cancer Res.* **56,** 1727.
101. Sun, J., Qian, Y., Hamilton, A. D., and Sebti, S. M. (1998). *Oncogene* **16,** 1467.
102. Sonoda, K., Sakamoto, T., Yoshikawa, H., Ashizuka, S., Ohshima, Y., Kishihara, K., Nomoto, K., Ishibashi, T., and Inomata, H. (1998). *Invest. Ophthalmol. Vis. Sci.* **39,** 2245.
103. Chakrabarti, D., Azam, T., DelVecchio, C., Qiu, L., Park, Y. I., and Allen, C. M. (1998). *Mol. Biochem. Parasitol.* **94,** 175.
104. Osman, A., Niles, E. G., and LoVerde, P. T. (1999). *Mol. Biochem. Parasitol.* **100,** 27.
105. Yokoyama, K., Trobridge, P., Buckner, F. S., Scholten, J., Stuart, K. D., Van Voorhis, W. C., and Gelb, M. H. (1998). *Mol. Biochem. Parasitol.* **94,** 87.
106. Yokoyama, K., Trobridge, P., Buckner, F. S., VanVoorhis, W. C., Stuart, K. D., and Gelb, M. H. (1998). *J. Biol. Chem.* **273,** 26497.
107. Nambara, E., and McCourt, P. (1999). *Curr. Opin. Plant Biol.* **2,** 388.
108. Pei, Z. M., Ghassemian, M., Kwak, C. M., McCourt, P., and Schroeder, J. I. (1998). *Science* **282,** 287.
109. Rodriguez Concepcion, M., Yalovsky, S., and Gruissem, S. (1999). *Plant Mol. Biol.* **39,** 865.
110. Zhou, D., Qian, D., Cramer, C. L., and Yang, Z. (1997). *Plant J.* **12,** 921.
111. Hudes, G. R., Schol, J., Baab, J., Rogatko, A., Bol, C., I. Horak, Langer, C., Goldstein, L. J., Szarka, C., Meropol, N. J., and Weiner, L. (1999). *Proc. Am. Soc. Clin. Oncol.* **17,** 601.
112. Zujewski, J., Horak, I. D., Bol, C. J. J. G., Woestenborghs, R., End, D., Chiao, J., Belly, R. T., Kohler, D., Chow, C., Noone, M., Hakim, F. T., Larkin, G., Gress, R. E., Nussenblatt, R. B., Kremer, A. B., and Cowan, K. H. (1999). *Proc. Am. Soc. Clin. Oncol.* **17,** 739.
113. Adjei, A. A., Erlichman, C., Davis, J. N., Reid, J., Sloan, J., Statkevich, P. Zhu, Y., Marks, R. S., Pito, H. C., Goldberg, R., Hanson, L., Alberts, S., Cutler, D., and Kaufmann, S. H. (1999). *Proc. Am. Soc. Clin. Oncol.* **17,** 598.
114. Hurwitz, H. I., Colvin, O. M., Petros, W. P., Williams, R., Conway, D., Adams, D. J., Casey, P. J., Calzetta, A., Mastorides, P., Statkevich, P., and Cutler, D. (1999). *Proc. Am. Soc. Clin. Oncol.* **17,** 599.
115. Eskens, F., Awada, A., Verweij, J., Cutler, D. L., Hanauske, A., and Piccart, M. (1999). *Proc. Am. Soc. Clin. Oncol.* **17,** 600.
116. Britten, C. D., Rowinsky, E., Yao, S.-L., Soignet, S., Rosen, N., Eckhardt, S. G., Drengler, R., Hammond, L., Siu, L. L., Smith, L., McCreery, H., Pezzulli, S., Lee, Y., Lobell, R., Deutsch, P., Hoff, D. V., and Spriggs, D. (1999). *Proc. Am. Soc. Clin. Oncol.* **17,** 597.
117. Soignet, S., Yao, S. L., Britten, C., Spriggs, D., Pezzulli, S., McCreery, H., Mazina, K.,

Deutsch, P., Lee, Y., Lobell, R., Rosen, N., and Rowinsky, E. (1999). *Proc. Am. Assoc. Cancer Res. Annu. Meet.* **40,** 517.

118. Petit, T., Izbicka, E., Lawrence, R. A., Bishop, W. R., Weitman, S., and Von Hoff, D. D. (1999). *Ann. Oncol.* **10,** 449.

119. Strickland, C. L., Weber, P. C., Windsor, W. T., Wu, Z., Le, H. V., Albanese, M. M., Alvarez, C. S., Cesarz, D., del Rosario, J., Deskus, J., Mallams, A. K., Njoroge, F. G., Piwinski, J. J., Remiszewski, S., Rossman, R. R., Taveras, A. G., Vibulbhan, B., Doll, R. J., Girijavallabhan, V. M., and Ganguly, A. K. (1999). *J. Med. Chem.* **42,** 2125.

120. Ferrante, K., Winograd, B., and Canetta, R. (1999). *Cancer Chemother. Pharmacol.* **43**(Suppl. S), S61.

121. Smets, G., Xhonneux, B., Cornelissen, F., End, D., Bowden, C., and Wouters, W. (1998). *Proc. Am. Assoc. Cancer Res. Annu. Meet.* **39,** 318.

122. Skrzat, S., Angibaud, P., Venet, M., Sanz, G., Bowden, C., and End, D. (1998). *Proc. Am. Assoc. Cancer Res. Annu. Meet.* **39,** 317.

123. End, D., Skrzat, S., Devine, A., Angibaud, P., Venet, M., Sanz, G., and Bowden, C. (1998). *Proc. Am. Assoc. Cancer Res. Annu. Meet.* **39,** 270.

124. Heliez, C., Delmas, C., Bonnet, J., Moyal, E., End, D., Daly Schveitzer, N., Favre, G., and Toulas, C. (1999). *Proc. Am. Assoc. Cancer Res. Annu. Meet.* **40,** 640.

5

Protein Geranylgeranyltransferase Type I

KOHEI YOKOYAMA · MICHAEL H. GELB
Departments of Chemistry and Biochemistry
University of Washington
Seattle, Washington 98195

THE ENZYMES, Vol. XXI

I. Introduction

A. PROTEIN PRENYLATION

Protein prenylation, the attachment of 15-carbon farnesyl and 20-carbon geranylgeranyl groups to proteins, is a posttranslational protein modification that occurs in most, if not all, eukaryotic cells (*1–6*). The first farnesylated polypeptides to be identified are a class of sex phermones from jelly fungi (*7*). In the 1980s, studies of the effects of hydroxymethylglutaryl-coenzyme A reductase inhibitors (the enzyme that produces mevalonate, the first committed precursor of prenyl groups) led to the discovery that exogenously added radiolabeled mevalonate is incorporated into a specific set of mammalian cell proteins (*1*). The first protein of this type to be identified was human lamin B, a structural protein bound to the nuclear envelope (*8, 9*). Rigorous structural analysis using mass spectrometry with authentic standards established that the mevalonate-derived protein ligand is a farnesyl group (*10*). This lipid is attached to the C-terminal cysteine residue through a thioether linkage between carbon-1 of farnesyl and the sulfur of cysteine. At the same time, it was also shown that Ras proteins are labeled with mevalonate (*11–13*) and that the lipid released from the protein after methyl iodide treatment comigrated with farnesol on a high-performance liquid chromatography (HPLC) column (*13*), suggesting that it is farnesylated.

Early on in our studies of radiolabeling of human cell proteins by mevalonate, we found that exhaustive proteolysis of cell protein yielded two peaks of radioactive material on a Sephadex LH-20 gel-filtration column. The lower molecular weight fraction contained farnesylated amino acids/short peptides. When the higher molecular weight fraction was treated with Raney nickel to cleave the putative thioether linkage, radioactivity became pentane soluble. Radiometric gas chromatography analysis of the organic extract showed that the labeled component comigrated with the 20-carbon hydrocarbon 3,7,11,15-tetramethyl-2,6,10,14-hexadecatetraene, consistent with a thioether-linked geranylgeranyl group (*14*). The first geranylgeranylated protein discovered was the γ subunit of brain heterotrimeric G proteins as shown by mass spectrometric (*15*) and HPLC (*16*) analyses of the released lipid.

Many members of the Rab family of GTP-binding proteins contain a pair of cysteines near the C terminus, and both are geranylgeranylated (*17*). Listings of proteins known to be prenylated are given in the review articles cited above.

To date, all known monofarnesylated and monogeranylgeranylated proteins contain the lipid on the cysteine of the C-terminal sequence CaaX, where *a* is usually but not necessarily an aliphatic residue, and

X is a variety of amino acids. After prenylation, aaX is removed by a membrane-bound endoprotease *(18–21)*, and the newly formed *S*-prenylcysteine C terminus is methylated on the α-carboxyl group by a membrane-bound methyltransferase *(22–24)*. Doubly geranylgeranylated Rab proteins are not proteolyzed, and C-terminal methylation occurs if the motif is C(*S*-geranylgeranyl)-X-C(*S*-geranylgeranyl) but not if it is C(*S*-geranylgeranyl)-C(*S*-geranylgeranyl) or C(*S*-geranylgeranyl)-C(*S*-geranylgeranyl)XX *(25–27)*.

B. PROTEIN PRENYLTRANSFERASES

The first protein prenyltransferase to be characterized was rat protein farnesyltransferase (PFT) *(28)*, and a review of this enzyme is presented in [1] in this volume *(28a)*. Next came protein geranylgeranyltransferase-I (PGGT-I), which was detected in 1991 in the soluble fraction of eukaryotic cells *(29–31)* and is the topic of the current chapter. The final protein prenyltransferase to be identified was protein geranylgeranyltransferase-II, also known as Rab geranylgeranyltransferase, which is presented in [6] in this volume *(31a)*. No additional protein prenyltransferases have been described to date, and the three known protein prenyltransferases carry out the set of prenylation reactions that seem to account for all known types of C-terminal prenyl motifs found in eukaryotic proteins. Inspection of the complete genome of *Saccharomyces cerevisiae* suggests that it contains single genes for each of the subunits of the three different protein prenyltransferases.

The prenyl donor for the PFT-catalyzed reaction is farnesyl pyrophosphate (FPP), whereas geranylgeranyl pyrophosphate (GGPP) is the prenyl donor for PGGT-I and for Rab geranylgeranyltransferase. PFT and PGGT-I transfer the prenyl group to proteins and peptides with a C-terminal CaaX motif. Rab geranylgeranyltransferase operates on Rab proteins with C-terminal CXC, CC, and CCXX motifs but does not geranylgeranylate peptides with these motifs.

II. Protein Geranylgeranyltransferase Type I

A. DETECTION OF PROTEIN GERANYLGERANYLTRANSFERASE TYPE I AND LEUCINE RULE

Bovine brain cytosol was shown to contain a PGGT-I that transfers the geranylgeranyl group from GGPP to the cysteine SH in the CaaX motif of the GTP-binding protein Rap 2B (CaaX sequence of CVIL) *(29)*, which is known to be geranylgeranylated *in vivo*. The brain enzyme also geranyl-geranylated a number of synthetic peptides that were patterned after the C termini of known geranylgeranylated proteins including the G protein γ_6 subunit (CaaX sequence of CAIL), and GTP-binding proteins smg-p21B and G25K (CaaX sequences of CQLL and CVLL, respectively) (*29*). H-Ras has C terminus CVLS and is farnesylated. Replacement of CVLS with CVLL leads to an H-Ras variant that is selectively geranylgeranylated by bovine brain PGGT-I (*30*). In both studies, it was shown that PGGT-I activity could be resolved from PFT activity by anion-exchange chromatography (*29, 30*).

All these geranylgeranyl acceptors contain a CaaX motif with leucine in the X position. When the C-terminal leucine of the G protein γ_6 subunit C-terminal peptide NPFREKKFFCAIL was replaced with serine (serine is a common X residue in farnesylated proteins), the peptide is transformed from a geranylgeranyl acceptor to a farnesyl acceptor (*29*). Likewise, when the C-terminal serine of the lamin B-derived peptide, GTPRASNRSCAIS, was replaced with leucine, the farnesyl-accepting peptide became a good substrate for PGGT-I (*29*). The significance of C-terminal leucine for promoting recognition by PGGT-I was also reported in studies with protein substrates and their C-terminal mutants (*32, 33*). It is now generally accepted that most, if not all, CaaX motifs that contain leucine in the X position are good substrates for PGGT-I and are poor substrates for PFT (leucine rule). It has been possible to switch the prenyl group from farnesyl to geranylgeranyl *in vivo* by expressing a mutant form of a normally farnesylated protein containing the wild-type X residue replaced by leucine (*34*). In addition, proteins that do not contain a CaaX motif can become prenylated in mammalian cells if a CaaX motif is engineered onto the C terminus (*12*). The predictability of the leucine rule is generally good, but additional studies (Section II,F,2) have shown that amino acids in the second and third positions of CaaX as well as non-CaaX residues in protein substrates can moderately influence the substrate specificity of protein prenyltransferases.

B.   Protein Geranylgeranyltransferase Type I Assays

The most common way to assay PGGT-I is to trap protein substrate on a cellulose paper disk after it has been radiogeranylgeranylated in the presence of radiolabeled GGPP that contains ^{3}H or ^{14}C at carbon-1 (*29, 30, 35,*) and to subject the disk to scintillation counting. This assay is sensitive, simple to execute, and generally gives low levels of counts per minute (cpm) in the minus enzyme and minus peptide/protein substrate controls. CaaX-containing peptides with an N-terminal biotin moiety can also be used to assay PGGT-I (29, 31). In this case, radiogeranylgeranylated peptide is captured with solid-phase streptavidin. In our hands, the biotinylated peptide assay gives higher background cpm in control reactions than does the paper disk assay.

Perhaps the most rapid assay, which has been used in large-scale screenings for PGGT-I inhibitors, is the scintillant proximity assay (36). This assay makes use of tritiated GGPP, a biotinylated peptide, and a commercial scintillant bead attached to streptavidin. Binding of the radiogeranylgeranylated and biotinylated peptide to streptavidin places the tritium close to the scintillant bead for high-efficiency capture of the emitted β particle. Large amounts of tritium present as tritiated GGPP are not detected as long as the reaction is quenched with 1.5 M $MgCl_2$, 0.2 M H_3PO_4 to reduce nonspecific binding of radiolabeled substrate to scintillant–streptavidin. After the prenylation reaction, scintillant–streptavidin and quench reagent are added, and the reaction mixture is placed in the scintillation counter without additional steps.

The need for radiolabeled GGPP is eliminated by use of a fluorimetric PGGT-I assay. This assay makes use of the peptide N-dansyl-GCIIL (37) or N-dansyl-GCVLL (38). Attachment of the geranylgeranyl group makes the environment of the fluorophore more hydrophobic, and the spectral shift in the dansyl emission is detected. This assay provides for continuous monitoring of PGGT-I activity. It requires ~100 ng of PGGT-I, and is thus less sensitive than radiometric assays, which require a few nanograms of enzyme.

PGGT-I assays are typically carried out in phosphate or Tris buffer at a pH near neutrality. $MgCl_2$ (~2 mM) and $ZnCl_2$ (~20 μM) are typically added as cofactors (Section II,C). It is a good idea to include 1 mM dithiothreitol (DTT) in the buffer to keep enzyme and prenyl acceptor in the reduced form.

C. Purification of Protein Geranylgeranyltransferase Type I, Subunit Composition, and Metal Ion Requirements

Two groups independently reported the ~50,000-fold purification of bovine brain PGGT-I to near homogeneity (39, 40). The key step in the purification was the use of an affinity column containing a bound CaaX peptide, the G protein γ_2-subunit peptide YREKKFFCAIL in one case and the peptide SSCILL in the other case. The latter peptide was discovered by a systematic approach (40). Two peptide mixtures were made on the basis of the sequence of the G protein γ_6 subunit, and contained a mixture of all 20 amino acids at the a_1 or a_2 position: TPVPGKARKKSSC-a_1-LL and TPVPGKARKKSSCQ-a_2-L. Only the former peptide mixture present at 1–5 μM competed well with biotinylated peptide substrate for geranylgeranylation by PGGT-I. This peptide mixture was fractionated by HPLC, and individual fractions were tested in the PGGT-I competition assay. The peptide with highest affinity for PGGT-I was identified as a_2 equals isoleucine by mass spectrometry and Edman sequencing. This peptide was synthesized in larger quantities, and at 0.7 μM it caused 50% inhibition of the geranylgeranylation of the biotin–γ_6 peptide (with C-terminus CAIL) present at 50 μM. Thus, the discovered peptide binds 70-fold tighter to PGGT-I than biotin–γ_6 peptide. Attachment of this peptide to Sepharose beads afforded resin that retained PGGT-I activity; however, the purification factor was not substantial. Because this peptide contains four positively charged residues and can thus function as an anion-exchange column, the truncated version, SSCILL, was prepared and attached to Sepharose. This affinity gel provided a 370-fold purification of bovine brain PGGT-I and was the key step in the overall 42,000-fold purification (40).

Purified mammalian PGGT-I was found to consist of ~40- and ~48-kDa subunits (39, 40). Peptides isolated from the ~48-kDa subunit were found to be identical in sequence to regions of the α subunit from mammalian PFT (39), strongly suggesting that PGGT-I and PFT share a common α subunit. This confirms earlier suggestions of subunit sharing based on the ability of multiple antibodies against PFT α subunit to recognize the α subunit of PGGT-I (41). In addition, studies in S. cerevisiae show that both PFT and PGGT-I activities are greatly reduced in a yeast strain bearing a mutation in the RAM2 gene (42). RAM2 is now known to encode the α subunit of yeast PFT and PGGT-I.

PGGT-I activity was lost when the enzyme was treated with 1 mM EDTA for 2–3 days (39, 43). Activity could be restored when Mg^{2+} and Zn^{2+} were added (39, 43). The enzyme was half maximally restored when ~0.5 mM $MgCl_2$ was added in the presence of micromolar $ZnCl_2$, or when ~2 μM $ZnCl_2$ was added in the presence of millimolar $MgCl_2$ (39, 43). When large amounts of recombinant PGGT-I became available via overexpression, it could be determined that the enzyme contains 0.8 ± 0.1 mol of Zn^{2+} per heterodimer, showing that this enzyme is a zinc-metalloenzyme (44).

Treatment of mammalian PGGT-I with a radiolabeled analog of GGPP bearing a photoaffinity probe (2-diazo-3,3,3-trifluoropropionyl) in place of the gem-dimethyl group led to specific labeling of the β subunit on exposure to light (43, 45). This result suggests that the GGPP-binding pocket lies within the β subunit, and this is consistent with the modeled structure of PGGT-I based on the X-ray structure of PFT (see Section II,G).

D. PROTEIN GERANYLGERANYLTRANSFERASE TYPE I β-SUBUNIT SEQUENCE AND COMPARISON WITH PROTEIN FARNESYLTRANSFERASE β SUBUNIT

cDNA cloning of rat and human PGGT-I-β yield predicted molecular masses of 42.5 and 42.4 kDa, respectively. The sequence of rat PGGT-I-β is shown aligned with rat PFT-β (Fig. 1) (46). The sequence suggests that

```
             . . . .10 . . . .20 . . . .30 . . . 40 . . . .50 . . . .60
rapftb     1:MASSSSFTYYCPPSSSPVWOEFLISLRPEHARERLQDDSVETVTSIEQAKVEEKIQEVFS: 60
rapggtIb   1:*----------------------------------------------------MAATEDD: 7

                                        *  *  *    #   #
             . . . .70 . . . .80 . . . .90 . . . 100 . . . 110 . . . 120
rapftb    61:SYKFNHLVPRLVLQREKHFHYLKRGLRQLTDAYECLDASRPWLCYWILHSLELLDE....:116
rapggtIb   8:RLAGSGEGERLDFLRDRHVRFFQRCLQVLPERYSSLETSRLTIAFFALSGLDMLDSLDVV: 67

                                                            #  ##
             . . . 130 . . . 140 . . . 150 . . . 160 . . . 170 . . . 180
rapftb   116:.........PIPQIVATDVCQFLELCQSPDGGFGG...GPGQYP........HLAPTYA:155
rapggtIb  68:NKDDIIEWIYSLQVLPTEDRSNLDRCGFRGSSYLGIPFNPSKNPGTAHPYDSGHIAMTYT:127

                                                       !
                                                    #  !
             . . . 190 . . . 200 . . . 210 . . . 220 . . . 230 . . . 240
rapftb   156:AVNALCIIGTEEAYNVINREKLLQYLYSLKQPDGSF.LMHVGGEVDVRSAYCAASVASLT:214
rapggtIb 128:GLSCLIILGDD..LSRVDKEACLAGLRALQLEDGSFCAVPEGSENDMRFVYCASCICYML:185

                                               !  !!  !
             . . . 250 . . . 260 . . . 270 . . . 280 . . . 290 . . . 300
rapftb   215:NIITPDLFEGTAEWIARCQNWEGGIGGVPGMEAHGGYTFCGLAALVIL.KKERSLNLKSL:273
rapggtIb 186:NNWSGMDMKKAISYLRRSMSYDNGLAQGAGLESHGGSTFCGIASLCLMGRLEEVFSEKEL:245

                       !  !   Z  Z#   !
             . . . 310 . . . 320 . . . 330 . . . 340 . . . 350 . . . 360
rapftb   274:LQWVTSRQMRFEGGFQGRCNKLVDGCYSFWQAGLLPLLHRALHAQGDPALSMSHWMFHQQ:333
rapggtIb 246:NRIKRWCINRQQNGYHGRPNKPVDTCYSFWVGATLKLLKIFQYT.............NFE:292

                              #Z
             . . . 370 . . . 380 . . . 390 . . . 400 . . . 410 . . . 420
rapftb   334:ALQEYILMCCQCPAGGLLDKPGKSRDFYHTCYCLSGLSIAQHFGSGAMLHDV.VMGVPEN:392
rapggtIb 293:KNRNYILSTQDRLVGGFAKWPDSHPDALHAYFGICGLSLMEESGICKVHPALNVSTRTSE:352

             . . . 430 . . . 440 . . . 450 . . . 460 . .
rapftb   393:VLQPTHPVYNIGPDKVIQATTHFLQKPVPGFEECEDAVTSDPATD:437
rapggtIb 353:RLRDLHQSWKTKDSKQCSDNVHISS--------------------:377
```

FIG. 1. Alignment of all rat PFT-β (rapftb) with rat PGGT-I-β (rapggtIb), using the PILEUP routine within the Genetics Computer Group (GCG, Madison, WI) software package (Wisconsin Package version 9.1). Identical residues are in black boxes, and similar residues are in gray boxes. Z, Zn^{2+} ligands; *, residues that may contribute to the anionic enzyme patch that interacts with the polybasic region near the CaaX motif of substrates (see Section II,F,2); #, PFT-β residues that are in van der Waals contact with bound N-acetyl-CVIM, based on the X-ray structure (64), and proposed PGGT-I-β residues that serve the same function; !, PFT-β residues that are in van der Waals contact with bound FPP, based on the X-ray structure (52), and proposed PGGT-I-β residues that serve the same function.

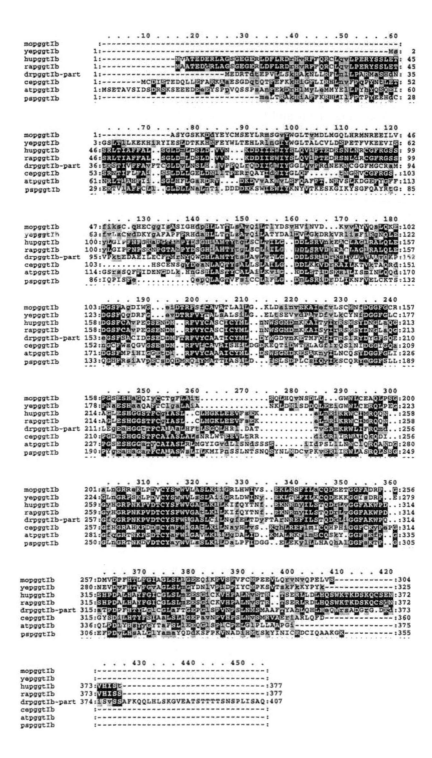

the protein encoded by *S. cerevisiae* gene *CDC43* is the yeast homolog of mammalian PGGT-I. Coexpression of the PGGT-I-β gene with the PFT-α gene gives PGGT-I protein with enzymatic properties indistinguishable from those of native PGGT-I, providing strong evidence that PGGT-I and PFT share a common α subunit (*46*). Coinfection of Sf9 insect cells with baculoviruses expressing both subunits of PGGT-I produces large quantities of recombinant enzyme that can be purified to homogeneity to give active enzyme with a specific activity that is similar to that of native PGGT-I (*44*). However, it is our experience that significant amounts of nonnative enzyme are produced in this system, and purification of fully active recombinant PGGT-I is best carried out by careful column chromatography and analysis of the specific enzymatic activity across the peak of PGGT-I protein. Figure 2 gives the alignment of all PGGT-I-β subunits present in GenBank at the time of writing of this chapter. The analysis of conserved amino acids in the context of the structure and function of the enzyme is given in Section II,G below.

The existence of at least 13 pseudogenes in humans related to PGGT-I-β has been reported (*47*). The functional significance of this finding is unknown, as it appears that these pseudogenes do not encode a functional β subunit.

Recombinant yeast PGGT-I has been overexpressed in *Escherichia coli* and purified to homogeneity (*37*). Like mammalian PGGT-I, the yeast enzyme loses activity when dialyzed against metal chelator, and activity is restored on addition of micromolar Zn^{2+} and millimolar Mg^{2+} (*37*).

E. STEADY STATE KINETIC MECHANISM

Because PGGT-I uses two substrates, the possibility of ordered versus random addition of substrates to the enzyme to form the catalytically active complex needs to be considered. Also, the enzyme may proceed via a ternary complex of enzyme with both intact substrates bound or via a ping–pong mechanism, in which GGPP is chemically converted to an inter-

---

FIG. 2. Alignment of all PGGT-I β subunits present in GenBank. Sequences are as follows: mopggtlb, mouse PGGT-I-β; ye, *Saccharomyces cerevisiae;* hu, human; ra, rat; dr, *Drosophila melanogaster;* ce, *Caenorhabditis elegans;* at, *Arabidopsis thaliana;* ps, *Pisum sativum.* Sequences were aligned with the PILEUP routine within the Genetics Computer Group (GCG, Madison, WI) software package (Wisconsin Package version 9.1). Identical residues present in at least two sequences are in black boxes (upper-case letters), and similar residues are in gray boxes (lower-case letters). Residue numbers for each sequence are given on the left and right side of the sequences, and numbering above the sequence groups is for the alignment cluster.

mediate, such as a geranylgeranylated enzyme, prior to CaaX binding and prenyl transfer to cysteine. Although there are no studies that directly bear on the issue of a covalent geranylgeranylated PGGT-I, stereochemical studies with PFT and isotopically labeled FPP show that farnesylation proceeds with inversion of stereochemistry at carbon-1, suggesting a direct attack of the SH of CaaX on the electrophilic carbon of FPP. Steady state kinetic studies of mammalian PGGT-I using dead end analogs of both substrates show inhibition patterns typical of a random sequential mechanism, in which either substrate can bind first to free enzyme followed by the second substrate to form a ternary complex that undergoes prenyl transfer (43, 44). Circumstantial evidence of this random sequential mechanism comes from direct binding studies showing that PGGT-I binds GGPP tightly in the absence of CaaX to form a binary complex that can be isolated by gel filtration (40), and that enzyme binding to CaaX peptides attached to a solid support is the basis for its purification (39, 40).

Additional studies support the idea that the PGGT-I mechanism proceeds preferentially by an ordered mechanism in which GGPP binds first, followed by CaaX. When the binary complex of PGGT-I bound to radiolabeled GGPP is mixed with an excess of nonlabeled GGPP and CaaX, most of the radiolabeled prenyl group is transferred to peptide; this shows that bound GGPP proceeds toward product without dissociation from the enzyme (40). On the other hand, it has not been possible to trap radiolabeled CaaX when the PGGT-I · CaaX complex is mixed with excess unlabeled CaaX and GGPP (43). Trapping of bound GGPP as product implies that the PGGT-I operates at steady state rather than rapid equilibrium, i.e., enzyme-bound GGPP is not in binding equilibrium with free GGPP. The kinetic equations describing the random sequential steady state mechanism are complex in general (48). For example, under steady state conditions, the flux through one of the substrate-binding pathways is usually faster than the pathway in which the substrates add in the reverse order. In this case, addition of high concentrations of substrate that adds second in the preferred kinetic pathway may inhibit the enzymatic reaction by forcing the enzyme to proceed through the slower pathway. This may be the reason that high concentrations of CaaX substrate but not of GGPP give rise to substrate inhibition (29). In summary, it appears that mammalian PGGT-I operates by a sequential mechanism involving the preferred addition of GGPP prior to CaaX. Further work is needed to determine if the reaction mechanism is strictly ordered. Steady state analysis of yeast PGGT-I using dead end inhibitors shows kinetic patterns typical of an ordered mechanism involving GGPP binding followed by prenyl acceptor binding (49).

Rapid quench experiments have been used to study the rate of prenyl transfer when the binary complex PGGT-I · GGPP is mixed with G protein

γ_6 peptide (36). Under these single turnover conditions, the geranylgeranyl group is transferred to the peptide with a first-order rate constant of 0.49 ± 0.06 sec^{-1}. The rate of prenyl transfer was not limited by peptide binding because the same rate constant was obtained when higher concentrations of peptide were used. When the rapid quench experiment was repeated with PGGT-I · FPP and the same peptide acceptor, the observed first-order rate constant was 37-fold smaller than for geranylgeranyl transfer, 0.015 ± 0.06 sec^{-1} (36). Interestingly, under steady state multiple turnover, PGGT-I geranylgeranylates the γ_6 peptide with a k_{cat} of 0.032 sec^{-1} under conditions in which both GGPP and peptide are saturating (36). The corresponding values for PGGT-I-catalyzed farnesylation of the γ_6 peptide is 0.012 sec^{-1}. These experiments led to the conclusion that PGGT-I-catalyzed geranylgeranylation is limited by a step that occurs after the chemical transfer of the geranylgeranyl group from GGPP to the SH group of the prenyl acceptor; this slow step could be release of geranylgeranylated peptide product from enzyme. On the other hand, the rate of the chemical step for farnesyl transfer by PGGT-I is probably rate limiting for the steady state reaction cycle. These data resolve the paradox that steady state geranylgeranylation and farnesylation of the γ_6 peptide occurs with the same maximal velocity (43) despite the fact that the actual rate of prenyl transfer may be slower for FPP, which may not sit in an optimal position on PGGT-I for reaction with the SH group of the prenyl acceptor. This phenomenon of a rate-limiting step after prenyl transfer was first observed with PFT (50).

F. SUBSTRATE SPECIFICITY

1. *Prenyl Donor Specificity*

GGPP and FPP bind to micromolar concentrations of PGGT-I with apparent similar affinity as seen by addition of increasing amounts of radiolabeled isoprenoids to enzyme and separation of the binary complex by gel filtration (51). This result suggests that the dissociation constants for the two binary complex are the same or that they differ but have less than micromolar values. The latter was shown to be the case. Binding studies with nanomolar amounts of PGGT-I show that the K_d for GGPP is about 3 nM and the K_d for FPP is much higher, 1 μM (36). This differential binding was confirmed by competition experiments in which various amounts of unlabeled GGPP were added to displace radiolabeled FPP from the enzyme and vice versa. Such studies show that PGGT-I binds its normal substrate GGPP ~300-fold tighter than FPP. Using the same approach it was found that PFT binds FPP ~15-fold tighter than GGPP (36). The ability of PGGT-I to bind GGPP much more tightly than FPP may be due to the extra binding energy afforded by the contact of an additional isoprenoid unit with the isoprenoid-binding pocket on PGGT-I, which is presumably deeper than that on PFT.

Although PGGT-I binds GGPP in preference to FPP, the PGGT-I · FPP complex is highly catalytically active. For example, the turnover number for PGGT-I-catalyzed geranylgeranylation of the G protein γ_6 subunit-derived peptide NPFREKKFFCAIL is 0.3 sec^{-1}, and the value for the farnesylation reaction of the same peptide is only threefold smaller (43). This results argues that FPP can occupy the GGPP-binding pocket of PGGT-I in a catalytically productive role; however, the role of the microscopic prenyl transfer step is faster for geranylgeranyl versus farnesyl (see Section II,E). In contrast, PFT transfers GGPP poorly relative to FPP, presumably because carbon-1 of GGPP is positioned too far from the attacking SH of the CaaX acceptor (52). In summary, although PGGT-I can bind FPP in a catalytically productive manner, displacement of FPP by tighter binding GGPP *in vivo* probably ensures that little, if any, PGGT-I · FPP is present in cells.

2. Prenyl Acceptor Specificity

The importance of the CaaX motif for recognition by PGGT-I is evident from the fact that truncated G protein γ_6-subunit peptide NPFREKKFFC (missing aaX) is not detectably geranylgeranylated. The peptide derived from the C terminus of the Rab geranylgeranyltransferase substrate p25a GTP-binding protein, LTDQQAPPHQDCAC, is also not detectably modified by PGGT-I (29).

The rate of geranylgeranylation of 1 μM H-Ras-CAIL (H-Ras with its C-terminal CVLS replaced by the G protein γ_6 subunit CaaX box CAIL) by human PGGT-I · GGPP is >1000-fold faster than the rate with 1 μM H-Ras-CVLS, showing that PGGT-I can greatly discriminate between geranylgeranyl versus farnesyl acceptors (46). With bovine PGGT-I, the G protein γ_6-subunit peptide was geranylgeranylated 20-fold faster than the lamin B peptide (40), and the same enzyme geranylgeranylated H-Ras with C-terminus CVLL ~20-fold faster than H-Ras-CVLS (30). These studies with the bovine enzyme were carried out with 4–5 μM prenyl acceptor, which is higher than the 1 μM concentration used in the human enzyme study. The difference in prenyl transfer rates measured with geranylgeranyl versus farnesyl acceptors may be due in part to relatively higher values of K_m for the latter, and this would lead to a compression of the rate differences at higher prenyl acceptor concentrations. As described next, prenyl acceptor specificity is best gauged by measuring the specificity constant, k_{cat}/K_m, for the desired peptide or protein.

In vivo, the various CaaX-containing proteins will compete with each

TABLE I

MAMMALIAN PROTEIN GERANYLGERANYLTRANSFERASE TYPE I SPECIFICITY CONSTANTS FOR
CaaX-CONTAINING PROTEINS AND PEPTIDES[a]

Prenyl acceptor	C-Terminal sequence	k_{cat}/K_m $(M^{-1} sec^{-1})$
G protein γ_6-subunit peptide	NPFREKKFFCAIL	4.5×10^5
G protein γ_6-subunit peptide (L-to-S mutant)	NPFREKKFFCAIS	3.7×10^3
SSCAIL peptide	SSCAIL	1×10^5
Lamin B peptide (S-to-L mutant)	GTPRASNRSCAIL	4.9×10^5
Lamin B peptide	GTPRASNRSCAIS	1.7×10^3
Human placenta G25K peptide	ALEPPETEPKRKCCIF	2.4×10^3
Human placenta G25K protein (GST fusion protein)	ALEPPETEPKRKCCIF	0.36×10^3
K-Ras4B protein	KKKKKKSKTKCVIM	4.9×10^4
TC21 protein	TRKEKDKKGCHCVIF	5.9×10^4

[a] From Refs. *36* and *43*.

other for geranylgeranylation by PGGT-I · GGPP (PGGT-I · FPP should be a minor component *in vivo;* see Section II,F,1). This competition is described by Eq. (1).

$$v_1/v_2 = [(k_{cat}/K_m)_1/(k_{cat}/K_m)_2][S_1]/[S_2] \qquad (1)$$

Here, the ratio of velocities of the geranylgeranylation of two substrates S_1 and S_2 by PGGT-I · GGPP when both substrates are present at the same concentration is given by the ratio of k_{cat}/K_m values for the two substrates (K_m is for the prenyl acceptor). The parameter k_{cat}/K_m is known as the specificity constant (*48*). Note that it is this ratio of kinetic parameters and not the values of k_{cat} and K_m individually that dictates prenyl acceptor specificity. Relative values of k_{cat}/K_m for pairs of prenyl acceptors have been obtained by incubating the binary complex of tritiated GGPP bound to PGGT-I with competing peptides and analyzing the ratio of prenylated peptide products (*36*). Table I lists the specificity constants for CaaX-containing proteins and peptides measured by this competitive method or computed by using values of k_{cat} and K_m that were measured from plots of reaction velocity versus concentration of prenyl acceptor. Peptides with CaaL (G protein γ_6 subunit, lamin B mutant, and SSCAIL) have similar specificity constants, which are ~100-fold larger than the corresponding peptides with CaaS. The CaaX specificity of PGGT-I from *S. cerevisiae* appears to be similar to that of the mammalian enzyme (CAIL/CVVL/ CIIL/CTIL ≫ CVLF/CVIM/CAIM/CAIV ≫ CVLS/CIIC/CIIS) (*53*).

Interestingly, there are two variants of the GTP-binding protein G25K in humans, one with C-terminus CVLL (brain form) and one with CCIF (placenta form). The former is a good substrate for bovine brain PGGT-I (29) as expected on the basis of the leucine rule, whereas the latter is 150-fold less preferred than G protein γ_6 subunit (Table I). The prenyl group attached to native placenta G25K is not known. The CCIF-containing G25K peptide is a much poorer substrate for PFT than for PGGT-I (29), which suggests that it is not farnesylated *in vivo*. Placenta G25K may be a good substrate for Rab geranylgeranyltransferase because this enzyme is known to doubly geranylgeranylate proteins with C-terminal motif CCXX (54).

A systematic study of the influence of the X residue of CaaX on the recognition by PGGT-I has been carried out for the peptide series KKSSCVIX (55). Values of k_{cat}/K_m for peptides with X = G, S, A, W, and E are the lowest in the series, being 1000- to 6000-fold less preferred than X = L. When X = C, Y, H, N, T, aminobutyric acid, L-penicillamine, and homoserine, k_{cat}/K_m values are 150- to 300-fold smaller than when X = L. The last group is X = M, V, I, F, homocysteine, aminopentanoic acid, and aminohexanoic acid, with k_{cat}/K_m values 11- to 37-fold smaller than when X = L. Such studies not only show that a hydrophobic residue is preferred at the X position but that γ branching that occurs in leucine promotes interaction with PGGT-I more than β branching (i.e., V and I) or no branching (i.e., M). Tryptophan is hydrophobic but bulky and impedes binding of peptide to PGGT-I. Branching at the β position by addition of an aromatic ring is less preferred than aliphatic β branching (i.e., Y and F).

The GTP-binding protein RhoB with C-terminal sequence CCKVL is interesting because previous studies have shown that is contains roughly equal amounts of farnesyl and geranylgeranyl when expressed in COS cells grown in the presence of radiolabeled mevalonate (56), despite the fact that this protein is predicted to be geranylgeranylated (leucine rule). RhoB is a poor substrate for PFT in the presence of FPP or GGPP but is a good substrate for PGGT-I in the presence of GGPP and FPP (36, 51). Values of k_{cat} for PGGT-I-catalyzed geranylgeranylation and farnesylation of RhoB are similar, as are the values of K_m for RhoB for the two reactions (51). Farnesylation of RhoB by PGGT-I may not be possible *in vivo* because of GGPP/FPP competition for binding to PGGT-I (see Section II,F,1). In the presence of RhoB, PGGT-I maintains its relatively high affinity for GGPP compared with FPP (36). If RhoB is a poor substrate for PFT and if PGGT-I · FPP is present only at low levels in cells, how does RhoB become farnesylated in COS cells? One possibility is that the farnesylation of RhoB in COS cells labeled with tritiated mevalonate may be an artifact of the radiolabeling procedure. Cell studies with tritiated mevalonate are usually carried out in the presence of a hydroxymethylglutaryl-CoA reductase inhibitor (statin) to block endogenous mevalonate production so that the specific radioactivity of the intracellular mevalonate pool is increased. We have shown that treatment of cells with statins can lead to an increase in the FPP/GGPP ratio, and this in turn can distort the ratio of prenyl groups attached to proteins (57). Thus, before it is concluded that RhoB contains a mixture of C_{15} and C_{20} prenyl groups, the structure of the protein-bound prenyl group must be established by mass spectrometry when RhoB is isolated from cells grown in the absence of statins. Because RhoB is a better substrate for PGGT-I-catalyzed farnesylation than are most geranylgeranyl acceptors, the FPP/GGPP distortion induced by statins may be most apparent with this GTP-binding protein.

The GTP-binding protein TC21 is a Ras-related protein with oncogenic potential. It contains a C-terminal CVIF motif and can be farnesylated *in vitro* by PFT and geranylgeranylated *in vitro* by PGGT-I (58). Cells transformed by oncogenic TC21 are resistant to transformation reversal by PFT inhibitors, suggesting that this protein could be geranylgeranylated in cells, especially when PFT is inhibited (58). *In vitro*, PGGT-I maintains its high affinity for GGPP versus FPP in the presence of TC21 (36), suggesting that PGGT-I does not farnesylate this protein in cells. Prenyl acceptor competition studies show that the k_{cat}/K_m for TC21 is eightfold smaller than that for the G protein γ_6-subunit peptide (Table I). Interestingly, the k_{cat}/K_m for farnesylation of TC21 by PFT · FPP is only 10-fold smaller than the specificity constant for farnesylation of the lamin B peptide (CAIS) by PFT (36). Thus, TC21 seems to compete equally well with typical PFT and PGGT-I substrates and could be geranylgeranylated by PGGT-I in cells treated with PFT-specific inhibitors.

The prenylation specificity of K-Ras4B with C-terminus CVIM is an important issue for cancer therapy based on PFT inhibition. Studies have shown that the farnesylation of this protein is more difficult to block with PFT inhibitors such as benzodiazepine peptidomimetics than is the farnesylation of many other proteins including H-Ras (59). Whereas values of K_m for many farnesyl acceptors are about 5–10 μM, the K_m for K-Ras4B is about 50-fold lower (59). Competition experiments show that the k_{cat}/K_m for K-Ras4B is 37-fold larger than that for the lamin B peptide (36). Thus, higher concentrations of CaaX-mimetic PFT inhibitors are needed to competitively block the binding of K-Ras4B to PFT. Presumably an inhibitor that is competitive with respect to FPP but not with CaaX mimetic will more readily block K-Ras4B farnesylation *in vivo*. K-Ras4B is also a respectable substrate for PGGT-I-catalyzed geranylgeranylation, having a value of k_{cat}/K_m that is only 10-fold smaller than that for the G protein γ_6-subunit peptide (36). These results probably explain why K-Ras4B is strictly farnesylated in cells (13) but becomes geranylgeranylated, presumably by PGGT-I, in cells treated with PFT inhibitors (60–63). As a cautionary note, these studies were carried out with mevalonate radiolabeling in the presence of statins, which may alter the type of prenyl group attached to K-Ras4B (see Section II,F,1).

When the CaaX motif of H-Ras is changed from CVLS to that of K-Ras4B (CVIM), the affinity of H-Ras for PFT was not increased significantly (59). An unusual feature of K-Ras4B is a highly basic C terminus (eight lysines near the CVIM sequence) that is not present in H-Ras. The affinity of H-Ras for PFT was not substantially increased when the polylysine region was incorporated into H-Ras (59). However, the affinity increased to that of K-Ras4B when both the polylysine and CVIM sequences were incorporated into H-Ras (59). Examination of the X-ray structure of rat PFT reveals an anionic patch on the surface of the enzyme coming from α-subunit residues D59, E114, E125, E150, E151, and E161 and β-subunit residues E94 and D97 (64). Because the X-ray structure of this enzyme with bound CVIM is available, it was possible to model the C terminus of K-Ras4B into the active site, and this suggests that the polybasic region can dock against this anionic enzyme patch (64). PGGT-I also contains this anionic patch, because it has an identical α subunit and two of the three anionic β-subunit residues are anionic (D91 replaced by E, E94 replaced by S, and D97 replaced by E). A number of geranylgeranyl acceptors have polybasic C termini, but a systematic study of the importance of these cationic residues for promoting binding to the enzyme has not been presented.

It is clear from the studies with Ras proteins that protein sequence outside of the CaaX box can influence the interaction with PFT and PGGT-I. A second example of this phenomenon is the prenylation of G protein γ subunits. Expression of the γ_2 subunit in insect cells with the baculovirus system gives rise to a protein that is exclusively geranylgeranylated (65). When the C-terminal CAIL motif is changed to CAIS, the protein now receives a roughly equal mixture of C_{15} and C_{20} prenyl groups (65). This seems to violate the leucine rule (however, see below). Because the insect cells were cultured in the presence of radiolabeled mevalonate but in the absence of statins, the intracellular concentrations of isoprenoid pyrophosphates may not be distorted. Importantly, the γ_1 subunit (CVIS motif) is modified exclusively by farnesyl in mammalian cells but was found to contain a mixture of C_{15} and C_{20} prenyl groups when overexpressed in insect cells (65). Farnesylation of γ_2 is catalyzed by PFT and not PGGT-I in insect cells because it is inhibited by a PFT-specific inhibitor (65). This result supports the prediction that PGGT-I does not transfer farnesyl groups *in vivo*. Also, farnesylation of γ_1 is higher relative to geranylgeranylation when mammalian PFT is co-overexpressed in insect cells along with the G protein subunit (65). Studies with G protein γ-subunit chimeras in insect cells reveal that a γ_1 chimera with native C and N termini but with a middle section grafted from the corresponding section of γ_2 was predominantly geranylgeranylated, suggesting that regions of the protein outside of the CaaX motif can alter prenylation specificity (65). Whether such trends hold for γ subunits and other prenylated proteins in mammalian cells remains to be seen.

However, the following explanation for the anomalous result of PGGT-I-catalyzed geranylation of γ_2 mutant with CAIS and γ_1 with CVIS in insect cells must be considered. Studies of bovine brain PGGT-I showed the following specificity behavior. The k_{cat} for geranylgeranylation of mutant γ_6 peptide NPFREKKFFCAIS by PGGT-I is 10-fold smaller than the k_{cat} with the CAIL peptide (43). The K_m for the serine peptide is 10-fold larger than the value for the peptide with leucine (43). Thus, the sluggishness to geranylgeranylate the CAIS peptide may be at least partly overcome by overexpressing γ_1 and γ_2 subunits in insect cells as pointed out previously (65). In summary, the CaaX motif seems to be the major dictator of prenylation specificity, but other regions in the protein, especially polybasic C termini and perhaps the middle segment of G protein γ subunits, may influence the specificity moderately and especially under conditions in which PFT and/or PGGT-I is inhibited.

One member of the Rab family of GTP-binding proteins, Rab 8, contains C-terminal sequence CVLL and is a surpassingly good substrate for PGGT-I and Rab geranylgeranyltransferase in a cell-free assay (66). Prenylation of Rab 8 when overexpressed in human embryonic kidney 293 cells was not prevented by adding a PGGT-I-specific inhibitor, whereas the geranylgeranylation of the known PGGT-I substrate CDC42 was blocked (66). The Y87D mutant of Rab 8 does not bind to the Rep component of Rab geranylgeranyltransferase (see [6] in this volume (31a)). The prenylation of this mutant was reduced by 60–70% when expressed in cells (66). These results suggest that Rab 8 may be prenylated by both geranylgeranyltransferase *in vivo*.

G. COMPARISON OF PROTEIN GERANYLGERANYLTRANSFERASE TYPE I
    WITH PROTEIN FARNESYLTRANSFERASE X-RAY STRUCTURE

As already mentioned, PFT and PGGT-I share an identical α subunit. The alignment of the PFT and PGGT-I β-subunit sequences is shown in Fig. 1. The sequences are 46% similar. Relative to PGGT-I-β, PFT-β contains an ~50-amino acid N-terminal extension and an ~20-amino acid C-terminal extension. This has no obvious consequences on the functioning of PFT and PGGT-I as the termini of the β subunits are far from the catalytic site. The X-ray structure of rat PFT shows that the active site Zn^{2+} is coordi-

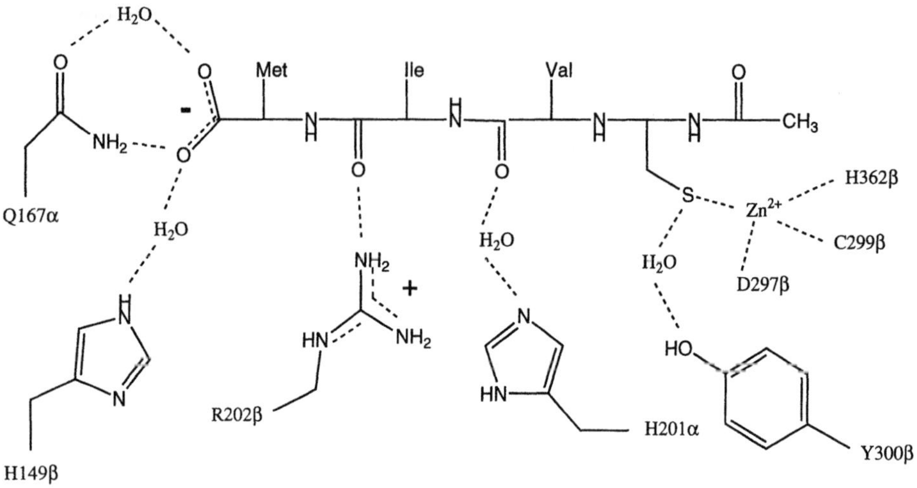

FIG. 3. Residues of rat PFT that interact with the backbone of bound N-acetyl-CVIM (64).
All residues are conserved in PGGT-I.

nated to D292β, C299β, H362β, and a water molecule (52). These three
β-subunit residues are conserved in PGGT-I-β from eight species (Fig. 2),
suggesting a similar metal coordination environment in PFT and PGGT-I.

The crystal structure of rat PFT with bound CaaX substrate N-acetyl-
CVIM and FPP analog α-hydroxyfarnesylphosphonic acid has been solved
(67), and thus the residues that are in direct contact with both ligands are
known. Using the sequence alignment shown in Fig. 1, we have made a
simple homology model of rat PGGT-I by replacing side chains attached
to the backbone of PFT-β with those of PGGT-I-β (no attempt was made
to model loops and gaps of PGGT-I-β relative to PFT-β). Figure 3 shows
the residues of rat PFT that interact with the backbone of bound N-acetyl-
CVIM. The guanidino group of R202β of PFT donates a hydrogen bond
to the carbonyl oxygen of the tetrapeptide isoleucine. The side chains of
H149β and Y300β of PFT donate hydrogen bonds to water molecules,
which in turn are hydrogen bonded to the C-terminal carboxylate and N-
terminal cysteine SH of N-acetyl-CVIM, respectively. All three PFT-β
residues are conserved in PGGT-I-β. Two PFT-α residues interact with
the bound tetrapeptide: Q167α, which donates a side-chain amido NH_2 to
the C-terminal carboxylate oxygen, and H201α, which accepts a hydrogen
bond from a water molecule that donates a hydrogen bond to the carbonyl
oxygen of the valine residue of N-acetyl-CVIM. Presumably these interactions are the same for PGGT-I, which has the same α subunit. Thus, the

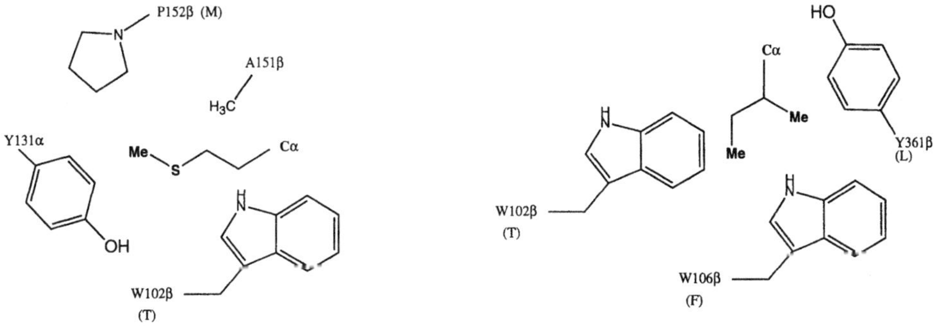

FIG. 4. Side chains of rat PFT that are in van der Waals contact with the methionine (*left*) and isoleucine (*right*) of bound *N*-acetyl-CVIM (*64*). Residues of rat PGGT-I modeled to be at the positions of the illustrated amino acid side chains and that are different from the PFT residues are indicated in parentheses (identical residues if not indicated). Mutation of yeast PFT residues at the positions corresponding to P152β and Y361β cause the discrimination of CIIS versus CIIL by the yeast enzyme to be lost (*68*).

interactions between enzyme and CaaX backbone may be identical for PFT and PGGT-I.

Residues of PFT that come in van der Waals contact with the side chain of methionine of *N*-acetyl-CVIM are Y131α, A151β, P152β, and W102β (Fig. 4). Significant differences are present in this region of PGGT-I; A151β is conserved, but P152β is replaced by methionine, and W102β is replaced by threonine (Fig. 4). These differences may be in part responsible for the preference of PGGT-I for L in the X position of CaaX (see above), but a detailed molecular understanding of the PFT versus PGGT-I CaaX specificity must await a high-resolution structure of the latter. It is interesting to note that yeast PFT has a serine at the position occupied by P152β of the rat enzyme, and the yeast mutant with asparagine replacing this serine has activity with CIIL substrate that is comparable to that with CIIS substrate (wild-type enzyme greatly prefers CIIS over CIIL) (*68*). Significant differences between PFT and PGGT-I also exist in the vicinity of the isoleucine of *N*-acetyl-CVIM. The isoleucine side chain is in van der Waals contact with W102β, W106β, and Y361β of PFT, and these positions are occupied by T, F, and L in PGGT-I-β (Fig. 4). Another yeast PFT mutant that allows for high activity on CIIS substrate is created by the replacement of tyrosine for leucine at the position occupied by rat PFT Y361β, which contacts the isoleucine side chain of bound CVIM (*68*). The side chain of the valine residue of N-acetyl-CVIM is in van der Waals contact with α-subunit residues of PFT, with no nearby β-subunit residues.

The X-ray structure of the rat PFT · FPP complex (*69*) provides a framework for understanding how PGGT-I accommodates the longer GGPP group. As shown in Fig. 5, FPP is in van der Waals contact with several hydrophobic residues from the β subunit (R202β, Y205β, G250β, Y251β, and C254β). The only exception is the apparent contact of the SH group of C254β with the FPP hydrocarbon chain, which seems unusual. This cysteine is also found in all species of PGGT-I listed in Fig. 2 except in the yeast enzyme, which has a threonine. Although the side chain of R202β contacts the FPP chain, it is only the hydrophobic methylene groups of the side chain that form the contact. Almost all of the residues of PFT-β near the FPP hydrocarbon chain are conserved in PGGT-I-β. An exception that may be important is that W102β of PFT is replaced by threonine in PGGT-I-β. It appears that W102β is too far from the FPP chain to contact it; the closest distance between heavy atoms is 5.9 Å. However, if the FPP chain were extended by one isoprenoid unit, it would likely clash with W102β. The much smaller side chain of threonine in this position of PGGT-I-β would likely make room for the extended chain of GGPP. Threonine is present in this position in all PGGT-I β-subunits except in the pea enzyme, which has a conservative valine-for-threonine change (Fig. 2). Y205β of PFT, which contacts the gem-dimethyl end of FPP (Fig. 5), is replaced by phenylalanine in three species of PGGT-I-β, and tyrosine occupies this position in the remaining species. Finally, Y251β in PFT is not conserved in PGGT-I, being replaced by S, T, A, and Q. This tyrosine does not make substantial contact with the FPP chain.

Residues that are likely to hydrogen bond to the pyrophosphate of FPP bound to PFT are H248β, R291β, K294β, and Y300β (69) (Fig. 1), and all of these are conserved in PGGT-I-β from all species listed in Fig. 2.

As discussed in [1] in this volume (28a), the mechanism of PFT-catalyzed farnesylation is thought to involve a Zn^{2+}-bound thiolate directly attacking carbon-1 of FPP. It is tempting to speculate that the same mechanism also applies for PGGT-I.

H. PROTEIN GERANYLGERANYLTRANSFERASE TYPE I INHIBITORS

Relatively few inhibitors of PGGT-I have been described compared with the large number of reported PFT inhibitors. Only those PGGT-I inhibitors that show some selectivity in favor of PGGT-I over PFT are described here. Figure 6 shows five different GGPP analogs (1–5) that have been reported to inhibit mammalian PGGT-I. No in vitro enzyme inhibition data are available for compounds 1–3, but all three at ~40 μM were found to block Rap 1 geranylgeranylation in cells when present at ~40 μM (70). Hydroxyphosphonate GGPP analog 4 binds to mammalian PGGT-I with a K_i of 1 μM (43), and 3-aza-GGPP, compound 5, binds with a K_i of 15

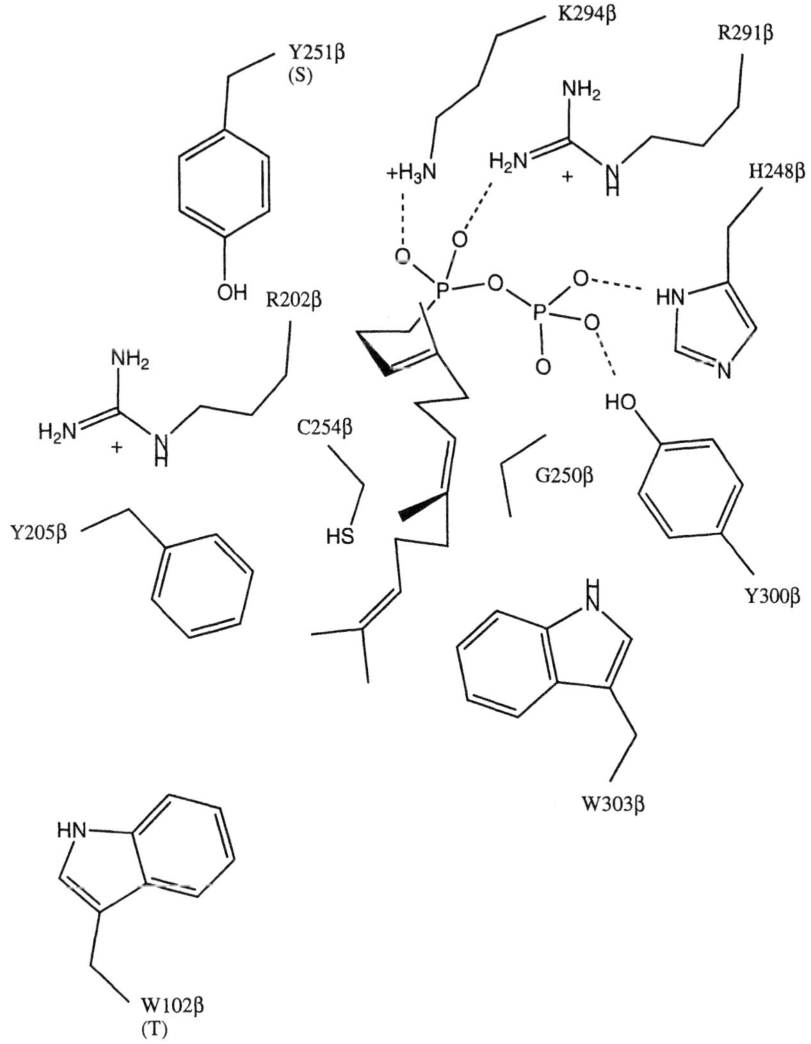

FIG. 5. Side chains of rat PFT that are in van der Waals contact with FPP (69). Residues of rat PGGT-I modeled to be at the positions of the illustrated amino acid side chains are identical except for W102B of PFT, which is replaced with threonine (shown in parentheses) in PGGT-I. Trp-102 appears to be too far from the FPP chain to be in van der Waals contact, and the smaller threonine residue at this position in PGGT-I may create space for the extra isoprene unit of GGPP.

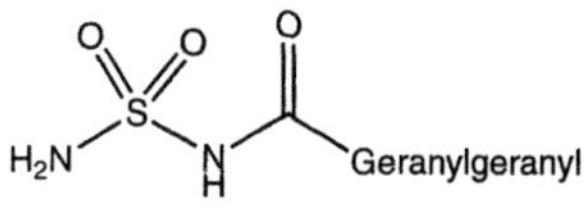

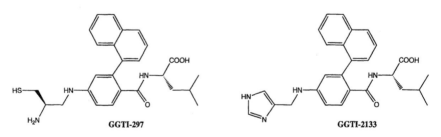

FIG. 6. Structures of GGPP analogs reported to inhibit mammalian PGGT-I.

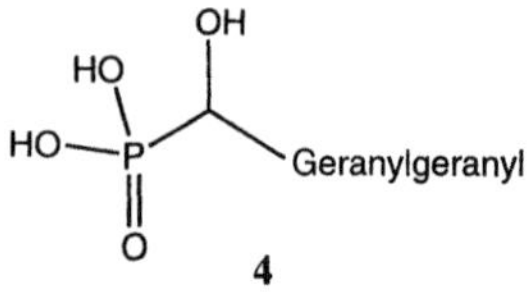

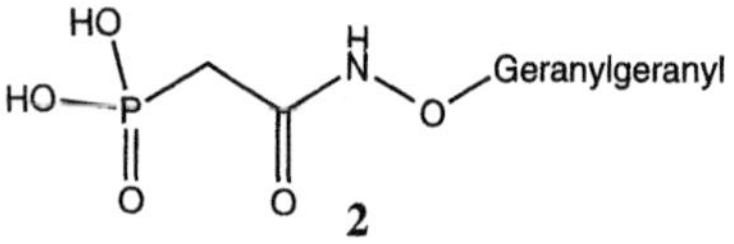

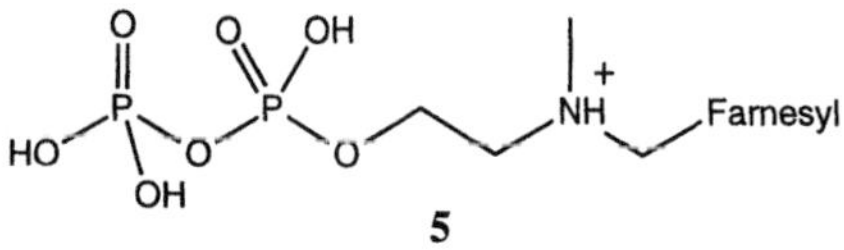

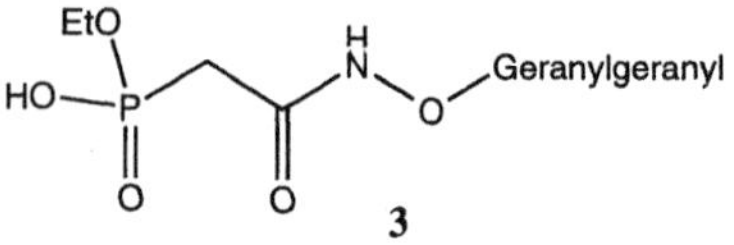

nM (*44*). One problem with these GGPP analogs is that they may inhibit other enzymes in cells that use GGPP as substrate.

A more promising approach is to design CaaX mimetics as PGGT-I inhibitors. The CaaX mimetic GGTI-297 (Fig. 7) is patterned after compounds that potently inhibit PFT, and contains a C-terminal leucyl residue for enhanced selectivity for PGGT-I versus PFT (*71*). GGTI-297 inhibits mammalian PGGT-I and PFT *in vitro* with IC_{50} values of 56 and 203 nM, respectively. The methyl ester derivative of GGTI-297, GGTI-298, is a cell-permeable prodrug that has been useful to block protein geranylgeranylation in cultured cells (see, e.g., Ref. *72*). PGGT-I inhibitors with improved selectivity have been reported. One example is GGTI-2133, which inhibits mammalian PGGT-I and PFT with IC_{50} values of 38 and 5400 nM, respectively (140-fold selective for PGGT-I) (*73*). A further advantage of GGTI-

FIG. 7. Structures of CaaX mimetics reported to selectively inhibit mammalian PGGT-I.

2133 is that it contains an imidazole ligand for Zn^{2+} rather than a thiol. The methyl ester prodrug of GGTI-2133 blocks Rap 1a geranylgeranylation in mammalian cells at 1 μM without effecting H-Ras farnesylation at 30 μM (73). Several other variants of GGTI-2133 also potently and selectively block PGGT-I action (73).

III. Future Prospects

Rapid progress toward understanding the properties of protein prenyl-transferases has been made since their discovery in the early 1990s. The availability of the X-ray structure of PGGT-I and PGGT-II will further add to our knowledge of these important enzymes. Key remaining concerns are to understand the CaaX motif substrate selection by PFT versus PGGT-I at the single-amino acid level and to understand how PGGT-II attaches two geranylgeranyl groups to cysteines that are in different motifs (CXC, CC, and CCXX). Given the clinical importance of PFT inhibitors for treating cancer and the fact that PGGT-I, in the presence of a PFT inhibitor, can geranylgeranylate certain proteins that are normally farnesylated, the clinical need for PGGT-I inhibitors remains to be considered. Also, PGGT-I inhibitors may be useful as drugs against eukaryotic pathogens such as *Candida albicans* and certain parasites that cause infections in the tropics.

ACKNOWLEDGMENTS

Our work on protein prenylation is supported by a grant from the National Institutes of Health (CA52874).

REFERENCES

1. Glomset, J. A., Gelb, M. H., and Farnsworth, C. C. (1990). *Trends Biochem. Sci.* **15,** 139.
2. Maltese, W. A. (1990). *FASEB J.* **4,** 3319.
3. Schafer, W. R., and Rine, J. (1992). *Annu. Rev. Genet.* **30,** 209.
4. Clarke, S. (1992). *Annu. Rev. Biochem.* **61,** 355.
5. Zhang, F. L., and Casey, P. J. (1996). *Annu. Rev. Biochem.* **65,** 241.
6. Gelb, M. H., Scholten, J., and Seboly-Leopold, J. S. (1998). *Curr. Opin. Chem. Biol.* (in press).
7. Sakagami, Y., Isogai, A., Suzuki, A., Tamura, S., Kitada, C., and Fujino, M. (1978). *Agric. Biol. Chem.* **42,** 1093.
8. Wolda, S. L., and Glomset, J. A. (1988). *J. Biol. Chem.* **263,** 5997.
9. Beck, L. A., Hosick, T. J., and Sinensky, M. (1988). *J. Cell. Biol.* **107,** 1307.

10. Farnsworth, C. C., Wolda, S. L., Gelb, M. H., and Glomset, J. A. (1989). *J. Biol. Chem.* **264,** 20422.
11. Schafer, W. R., Trueblood, C. E., Yang, C.-C., Mayer, M. P., Rosenberg, S., Poulter, C. D., Kim, S.-H., and Rine, J. (1990). *Science* **249,** 1133.
12. Hancock, J. F., Magee, A. I., Childs, J. E., and Marshall, C. J. (1989). *Cell* **57,** 1167.
13. Casey, P. J., Solski, P. A., Der, C. J., and Buss, J. E. (1989). *Proc. Nat. Acad. Sci. U.S.A.* **86,** 8323.
14. Farnsworth, C. C., Gelb, M. H., and Glomset, J. A. (1990). *Science* **247,** 320.
15. Yamane, H. K., Farnsworth, C. C., Xie, H., Howald, W., Fung, B. K.-K., Clarke, S., Gelb, M. H., and Glomset, J. A. (1990). *Proc. Natl. Acad. Sci. U.S.A.* **87,** 5868.
16. Mumby, S. M., Casey, P. J., Gilman, A. G., Gutowski, S., and Sternweis, P. C. (1990). *Proc Natl. Acad Sci U.S.A.* **87,** 5873.
17. Farnsworth, C. C., Kawata, M., Yoshida, Y., Takai, Y., Gelb, M. H., and Glomset, J. A. (1991). *Proc. Natl. Acad. Sci. U.S.A.* **88,** 6196.
18. Ma, Y.-T., and Rando, R. R. (1992). *Proc. Natl. Acad. Sci. U.S.A.* **89,** 6275.
19. Jang, G. F., Yokoyama, K., and Gelb, M. H. (1993). *Biochemistry* **32,** 9500.
20. Boyartchuck, V. L., Ashby, M. N., and Rine, J. (1997). *Science* **275,** 1796.
21. Schmidt, W. K., Tam, A., Fujimura-Kamada, K., and Michaelis, S. (1998). *Proc. Natl. Acad. Sci. U.S.A.* **95,** 11175.
22. Hrycyna, C., and Clarke, S. (1990). *Mol. Cell. Biol.* **10,** 5071.
23. Hrycyna, C. A., Sapperstein, S. K., Clarke, S., and Michaelis, S. (1991). *EMBO J.* **10,** 1699.
24. Romano, J. D., Schmidt, W. K., and Michaelis, S. (1998). *Mol. Biol. Cell* **9,** 2231.
25. Smeland, T., Seabra, M. C., Goldstein, J. L., and Brown, M. S. (1994). *Proc. Natl. Acad. Sci. U.S.A.* **91,** 10712.
26. Gelb, M. H., Reiss, Y., Ghomashchi, F., and Farnsworth, C. C. (1995). *Bioorg. Med. Chem. Lett.* **8,** 881.
27. Rando, R. R. (1996). *Biochem. Soc. Trans.* **24,** 682.
28. Reiss, Y., Goldstein, J. L., Seabra, M. C., Casey, P. J., and Brown, M. S. (1990). *Cell* **62,** 81.
28a. Spence, R. A., and Casey, P. J. (2000). *In* "The Enzymes," Vol. XXI, Chap. 1. Academic Press, San Diego, California. [This volume]
29. Yokoyama, K., Goodwin, G. W., Ghomashchi, F., Glomset, J. A., and Gelb, M. H. (1991). *Proc. Natl. Acad. Sci. U.S.A.* **88,** 5302.
30. Casey, P. J., Thissen, J. A., and Moomaw, J. F. (1991). *Proc. Natl. Acad. Sci. U.S.A.* **88,** 8631.
31. Joly, A., Popjak, G., and Edwards, P. A. (1991). *J. Biol. Chem.* **266,** 13495.
31a. Seabra, M. C. (2000). *In* "The Enzymes," Vol. XXI, Chap. 6. Academic Press, San Diego, California. [This volume]
32. Kinsella, B. T., Erdman, R. A., and Maltese, W. A. (1991). *J. Biol. Chem.* **266,** 9786.
33. Moores, S. L., Schaber, M. D., Mosser, S. D., Rands, E., O'Hara, M. B., Garsky, V. M., Marshall, M. S., Pompliano, D. L., and Gibbs, J. B. (1991). *J. Biol. Chem.* **266,** 14603.
34. Cox, A. D., Hisaka, M. M., Buss, J. E., and Der, C. J. (1992). *Mol. Cell. Biol.* **12,** 2606.
35. Pompliano, D. L., Rands, E., Schaber, M. D., Mosser, S. D., Anthony, N. J., and Gibbs, J. B. (1992). *Biochemistry* **31,** 3800.
36. Yokoyama, K., Zimmerman, K. Z., Scholten, J. D., and Gelb, M. H. (1997). *J. Biol. Chem.* **272,** 3944.
37. Stirtan, W. G., and Poulter, C. D. (1995). *Arch. Biochem. Biophys.* **321,** 182.
38. Pickett, W. C., Zhang, F. L., Silverstrim, C., Schow, S. R., Wick, M. M., and Kerwar, S. S. (1995). *Anal. Biochem.* **225,** 60.
39. Moomaw, J. F., and Casey, P. J. (1992). *J. Biol. Chem.* **267,** 17438.

40. Yokoyama, K., and Gelb, M. H. (1993). *J. Biol. Chem.* **268**, 4055.
41. Seabra, M. C., Reiss, Y., Casey, P. J., Brown, M. S., and Goldstein, J. L. (1991). *Cell* **65**, 429.
42. Finegold, A. A., Johnson, D. I., Farnsworth, C. C., Gelb, M. H., Judd, S. R., Glomset, J. A., and Tamanoi, F. (1991). *Proc. Natl. Acad. Sci. U.S.A.* **88**, 4448.
43. Yokoyama, K., McGeady, P., and Gelb, M. H. (1995). *Biochemistry* **34**, 1344.
44. Zhang, F. L., Moomaw, J. F., and Casey, P. J. (1994). *J. Biol. Chem.* **269**, 23465.
45. Bukhtiyarov, Y. E., Omer, C. A., and Allen, C. M. (1995). *J. Biol. Chem.* **270**, 19035.
46. Zhang, F. L., Diehl, R. E., Kohl, N. E., Gibbs, J. B., Giros, B., Casey, P. J., and Omer, C. A. (1994). *J. Biol. Chem.* **269**, 3175.
47. Dhawan, P., Yang, E., Kumar, A., and Mehta, K. D. (1998). *Gene* **210**, 9.
48. Segel, I. H. (1975). "Enzyme Kinetics." Wiley-Interscience, New York.
49. Stirtan, W. G., and Poulter, C. D. (1997). *Biochemistry* **36**, 4552.
50. Furfine, E. S., Leban, J. J., Landavazo, A., Moomaw, J. F., and Casey, P. J. (1995). *Biochemistry* **34**, 6857.
51. Armstrong, S. A., Hannah, V. C., Goldstein, J. L., and Brown, M. S. (1995). *J. Biol. Chem.* **270**, 7864.
52. Park, H.-W., Boduluri, S. R., Moomaw, J. F., Casey, P. J., and Beese, L. S. (1997). *Science* **275**, 1800.
53. Caplin, B. E., Hettich, L. A., and Marshall, M. S. (1994). *Biochim. Biophys. Acta* **1205**, 39.
54. Farnsworth, C. C., Seabra, M. C., Ericsson, L. H., Gelb, M. H., and Glomset, J. A. (1994). *Proc. Natl. Acad. Sci. U.S.A.* **91**, 11963.
55. Roskoski, R., Jr., and Ritchie, P. (1998). *Arch. Biochem. Biophys.* **356**, 167.
56. Adamson, P., Marshall, C. J., Hall, A., and Tilbrook, P. A. (1992). *J. Biol. Chem.* **267**, 20033.
57. Whitten, M. E., Yokoyama, K., Schieltz, D., Ghomashchi, F., Lam, D., Yates, J. R., Palczewski, K., and Gelb, M. (2000). *Methods Enzymol.* **316**, 436.
58. Carboni, J. M., Yan, N., Cox, A. D., Bustelo, X., Graham, S. M., Lynch, M. J., Weinmann, R., Seizinger, B. R., Der, C. J., Barbacid, M., and Manne, V. (1995). *Oncogene* **10**, 1905.
59. James, G. L., Goldstein, J. L., and Brown, M. S. (1995). *J. Biol. Chem.* **270**, 6221.
60. Rowell, C. A., Kowalczyk, J. J., Lewis, M. D., and Garcia, A. M. (1997). *J. Biol. Chem.* **272**, 14093.
61. Whyte, D. B., Kirschmeier, P., Hockenberry, T. N., Nunez, O. I., James, L., Catino, J. J., Bishop, W. R., and Pai, J. K. (1997). *J. Biol. Chem.* **272**, 14459.
62. Sun, J., Qian, Y., Hamilton, A. D., and Sebti, S. M. (1998). *Oncogene* **16**, 1467.
63. Lerner, E. C., Zhang, T. T., Knowles, D. B., Qian, Y., Hamilton, A. D., and Sebti, S. M. (1997). *Oncogene* **15**, 1283.
64. Strickland, C. L., Windsor, W. T., Syto, R., Wang, L., Bond, R., Wu, Z., Schwartz, J., Le, H. V., Beese, L. S., and Weber, P. C. (1998). *Biochemistry* **37**, 16601.
65. Kalman, V. K., Erdman, R. A., Maltese, W. A., and Robishaw, J. D. (1995). *J. Biol. Chem.* **270**, 14835.
66. Wilson, A. L., Erdman, R. A., Castellano, F., and Maltese, W. A. (1998). *Biochem. J.* **333**, 497.
67. Strickland, C. L., Windsor, W. T., Syto, R., Wang, L., Bond, R., Wu, Z., Schwartz, J., Le, H. V., Beese, L. S., and Weber, P. C. (1998). *Biochemistry* **37**, 16601.
68. Del Villar, K., Mitsuzawa, H., Yang, W., Sattler, I., and Tamanoi, F. (1997). *J. Biol. Chem.* **272**, 680.
69. Long, S. B., Casey, P. J., and Beese, L. S. (1998). *Biochemistry* **37**, 9612.
70. Macchia, M., Jannitti, N., Gervasi, G., and Danesi, R. (1996). *J. Med. Chem.* **39**, 1352.
71. Qian, Y., Vogt, A., Vasudevan, A., Sebti, S. M., and Hamilton, A. D. (1998). *Biorg. Med. Chem.* **6**, 293.

72. Sun, J., Qian, Y., Chen, Z., Marfurt, J., Hamilton, A. D., and Sebti, S. M. (1999). *J. Biol. Chem.* **274,** 6930.
73. Vasudevan, A., Qian, Y., Vogt, A., Blaskovich, M., Ohkanda, J., Sebti, S. M., and Hamilton, A. D. (1999). *J. Med. Chem.* **42,** 1333.

6

Biochemistry of Rab Geranylgeranyltransferase

MIGUEL C. SEABRA

Molecular Genetics, Division of Biomedical Sciences
Imperial College School of Medicine
London SW7 2AZ, United Kingdom

I. Protein Prenylation

Eukaryotic cells contain three distinct prenyltransferases that attach either a farnesyl group (15 carbons) or a geranylgeranyl group (20 carbons) in thioether linkage to C-terminal cysteines in a variety of cellular proteins.

131

These posttranslational modifications provide a mechanism for membrane attachment of proteins that lack a transmembrane domain.

Protein prenylation was discovered in animal cells in 1984 when Glomset and co-workers found that cultured cells incorporate derivatives of mevalonate via covalent linkage to various proteins (1). A few years earlier, it had been shown that peptide mating factors secreted by several fungi contain a farnesyl group attached in thioether linkage to C-terminal cysteine residues (2, 3), but this was felt to be a curiosity restricted to unusual fungi. In 1988 the nuclear envelope protein lamin B was identified as the first protein to be modified by a derivative of mevalonic acid, and a year later it was shown to contain a farnesyl group linked via a thioether bond to a C-terminal cysteine by mass spectroscopy (4, 5).

Meanwhile, it was reported that the mating a-factor from *Saccharomyces cerevisiae* and the mammalian Ras proteins are also farnesylated on a C-terminal cysteine (6–9). Importantly, the farnesyl group was shown to be absolutely critical for normal function of the modified protein by acting as the first and necessary step that leads to membrane binding and targeting of these proteins to their proper intracellular localization (7, 8). Particularly dramatic experiments by Marshall and co-workers showed that mutant oncogenic H-Ras transforms mammalian cells only when it is prenylated and thereby attached to the inner surface of the plasma membrane (8).

It is now established that protein prenylation is a frequent posttranslational modification that affects many proteins. Using quantitative methods, Epstein *et al.* estimated that approximately 0.5% of all proteins in mammalian cells are prenylated (10). Among those, the largest class of prenylated proteins is the Ras superfamily of proteins. Ras proteins are divided into five families: Ras, Rho/Rac, Rab, Arf, and Ran (11–14). Of those, all known members of the Ras, Rho, and Rab families comprising well over 70 distinct proteins in mammals have been shown or are predicted to be prenylated. Another large class of prenylated proteins is the $G\gamma$ subunits of heterotrimeric G proteins, which comprises at least 11 distinct proteins (15). Many other proteins including cGMP phosphodiesterase α and β subunits and nuclear lamins are also prenylated (15).

Only two types of isoprenoids are used for protein prenylation. These are either the C_{15} farnesyl (F) or the C_{20} geranylgeranyl (GG). Both isoprenoids are recognized by the enzymes as the pyrophosphate (PP) form, FPP or GGPP. The modification is usually a single prenyl modification but at least one class of substrates is modified by two GG groups. There have been reports of substrates such as RhoB and K-Ras where both farnesylation and geranylgeranylation have been documented (16, 17). However, there has been no demonstration yet that a single protein will contain both modifications simultaneously, and so these reports reflect alternative modification of the same cysteine residue by the two isoprenoids.

The enzymology of protein prenylation reviewed in this volume has been the subject of intense study. As discussed in the other chapters and previous reviews (see, e.g., Refs. *18–20*), there are three known protein prenyltransferases classified into two functional classes. Protein farnesyltransferase (PFT) and protein geranylgeranyltransferase type I (PGGT-I) form the CaaX-type prenyltransferases. They are so called because they recognize a prenylation motif called the CaaX box for a tetrapeptide sequence composed of a conserved cysteine, the site of prenyl transfer, followed by two aliphatic amino acids and a variable C-terminal residue. If the C-terminal residue is methionine, serine, glutamine, alanine, or cysteine, it signals recognition by PFT whereas if it is leucine or phenylalanine, it signals recognition by PGGT-I.

Rab geranylgeranyltransferase (RGGT), also known as protein geranylgeranyltransferase type II, forms a distinct class as this enzyme accepts only members of the Rab family of GTP-binding proteins as substrates. RGGT is the subject of this chapter.

II. Identification of Geranylgeranylated Rab Proteins

Earlier studies of protein prenylation focused on proteins that contained a CaaX box. However, it was soon recognized that members of the emerging Rab family did not contain classic CaaX boxes but nevertheless contained cysteines at or near the C terminus (*21*). This led to the suspicion that Rabs, too, were prenylated. In 1991, several simultaneous reports showed that Rab proteins are modified with GG groups both *in vitro* and *in vivo* (*22–24*). Also, some of these reports emphasized the functional importance of the prenyl modification. Mutants that lack the C-terminal cysteines were not able to associate with membranes and function properly (*23, 24*). Other Rabs with different double-cysteine motifs were later shown also to be geranylgeranylated (*25*), as were the yeast orthologs, *S. cerevisiae* and *Schizosaccharomyces pombe* Ypt/Sec4 proteins (*26, 27*).

Structural analysis of Rab3a purified from bovine brain membranes revealed, surprisingly, that both cysteines were modified by GG groups (*28*). Furthermore, the C-terminal residue showed modification with a methyl ester, as was found after proteolysis of the aaX peptide following prenylation on CaaX-containing substrates. Methylesterification is not, however, a general finding in Rabs. Rabs that contain an XXCysXCys motif are methylated whereas those that contain an XXXCysCys are not. If there is only one methyltransferase that recognizes primarily the prenylated cysteine (farnesylated or geranylgeranylated), it is possible that there is inhibition of methylation due to steric hindrance when two geranylgeranyl groups are modifying adjacent amino acids such as those in the XXXCysCys motif (29).

III. Prenyl Acceptor: Rab Family of Proteins

Rab proteins belong to the Ras superfamily of proteins, a large group of proteins involved in the regulation of cellular signaling (reviewed in Refs. 11–13). This superfamily of GTP-binding proteins is classified into five distinct families, according to structural and functional homologies (14, 30). These proteins are referred to as "small GTPases" because they possess GDP and GTP binding and enzymatic GTP-hydrolyzing activity, and are of relative low mass when compared with other GTP-binding proteins. This cycle of GTP binding and hydrolysis generates a molecular switch mechanism because the two guanine nucleotides induce different conformations on the GTPase on binding.

Many different Ras-like proteins have been cloned and sequenced. Their overall primary structure is conserved and shown to be composed of several distinct domains. Among them is a variable N-terminal region, four conserved GTP-binding domains, and a variable C-terminal region that may contain a prenylation motif (11).

The crystallization of many Ras-like proteins, including Rab3a, showed that their tertiary structure is also highly conserved (31–34). The elucidation of three-dimensional structures of the two alternative guanine nucleotide-bound conformations allowed the definition of two regions that change significantly with the binding of guanine nucleotide. These regions are designated switch I and switch II (35, 36). The former, comprising loop2 and $\beta 2$, is also known as the effector domain because it is the binding site of signaling proteins believed to be activated by binding the GTP-bound conformation of the GTPases. Switch II comprising loop4 and $\alpha 2$ has been less well characterized in terms of functional importance.

There are two types of C-terminal prenylation motifs in Ras-like proteins that form the basis for the functional classification of the protein prenyltransferases. Ras and Rho proteins contain CaaX sequences discussed above and Rab proteins contain a variable sequence that usually includes two cysteine residues in different combinations such as XXXCysCys, XXCysCysX, XCysCysXX, CysCysXXX, or XXCysXCys. In Rabs, both cysteines are modified by GG groups (28, 37). However, some Rabs present a classic CaaX box, where only a single cysteine residue is available for prenyl modification.

Rab proteins act as regulators of exocytic and endocytic pathways (38–

40). Almost 50 different Rabs have been identified in mammalian cells to date. In *S. cerevisiae*, these proteins are designated Ypt/Sec4 and represent a family of 10 distinct gene products (*41*). Each Rab exhibits a characteristic subcellular location (*42*). These observations led to the suggestion that subsets of Rabs regulate distinct membrane-trafficking steps.

Rab action involves a complex cycle of events (*38–40*). First, Rabs associate with a specific organelle or transport vesicle in the inactive GDP-bound form. Rabs associate with nascent vesicles and a GDP/GTP exchange factor (RabGEF) catalyzes the binding of GTP, which activates the Rab protein. When the transport vesicle finds the appropriate target organelle, Rabs promote their fusion with acceptor membranes. One of the mechanisms by which Rabs function may be to induce the interaction between the soluble *N*-ethylmaleimide-sensitive factor attachment protein receptors (SNARE). SNAREs specify the correct targeting of the transport vesicle with the appropriate acceptor compartment. After fusion of the vesicle with the acceptor compartment, Rab[GTP] is inactivated by the hydrolysis of GTP catalyzed by a Rab-specific GTPase-activating protein (RabGAP). To cycle back to the donor compartment, Rab[GDP] is removed from the membrane by Rab GDP dissociation inhibitor (RabGDI) (*43–45*). In response to subsequent signals yet unknown, RabGDI mediates the reassociation of Rab[GDP] with the donor membrane and the cycle resumes.

The participation of Rabs in this intermembrane cycle is absolutely dependent on their lipid modification with GG groups, as originally demonstrated for the yeast proteins Ypt1 and Sec4 (*46, 47*).

IV. Identification and Isolation of Rab Geranylgeranyltransferase

The functional importance of Rab geranylgeranylation led to efforts toward the identification of the enzyme involved. The first reports in 1991 described identification of a distinct RGGT activity by separation from the two previously recognized CaaX prenyltransferases and partial purification of the enzyme (*48, 49*). Those initial studies demonstrated that there were at least three protein prenyltransferases in animal cells. It was also shown that peptides corresponding to the C-terminal Rab sequence did not inhibit the reaction, unlike CaaX peptidomimetics for the CaaX prenyltransferases (*48*). Finally, it was reported that the enzyme was unstable and quickly lost activity after a few steps of conventional chromatography (*49*). It was clear that this enzyme was more divergent from CaaX prenyltransferases than originally expected and in practical terms these results meant that affinity chromatography used so efficiently for purification of PFT would not work.

These difficulties were overcome by biochemical complementation experiments. In 1992, the surprising observation that the RGGT activity required two chromatographically separable activities was reported (*50*). RGGT was shown to be composed of two components, termed components A and B (*50, 51*). On purification of both activities and their identification by molecular cloning, it became apparent that component B was the catalytic component whereas component A was a novel protein with homology to RabGDI (*50–53*). Component A was later renamed Rab escort protein (REP) when it was demonstrated that it served as an Rab-binding protein that assisted in the prenylation reaction and in the solubilization of digeranylgeranylated Rabs after the reaction (see Section V).

V.   Prenylation-Associated Factors: Rab Escort Protein Family

Two different REPs sharing 75% identity have been identified in mammals and designated REP1 and REP2 (*52, 54*). Both REPs are ubiquitously expressed even though the relative proportion of REP1 and REP2 varies from tissue to tissue (*55*). REPs appear to be functionally redundant with respect to the prenylation of most Rabs, with one notable exception (*54, 56*). Rab27 protein is more efficiently prenylated by REP1 than by REP2 (*56*). In yeast, REP is encoded by a single essential gene, designated *MRS6* or *MSI4* (*57, 58*). The yeast REP is essential to the geranylgeranylation of yeast Rabs (Ypts) and thus appears to serve the same function (*59*).

REP proteins are homologous to RabGDI, an Rab-binding protein whose function in Rab–membrane recycling is outlined above. There are at least three known isoforms of RabGDI in mammalian cells but more could exist (*43, 60, 61*). In yeast, GDI is encoded by a single essential gene designated GDI1 (*62*).

Sequence alignments of the known REPs and RabGDIs revealed significant structural homology, particularly in three regions termed sequence conserved regions (SCR1, -2, and -3) (*57*), suggesting that the members of this family may have similar three-dimensional structures. The crystal structure of RabGDI has been elucidated and found to be organized in two domains, a large complex multisheet domain I and a globular α-helical domain II (*63*). The SCRs are clustered on one face of the molecule and may represent the Rab-binding region. In particular, SCR1 and -3B are highly conserved and they form a compact region at the apex of domain I. Site-directed mutagenesis of residues in both SCRs dramatically reduced affinity for Rab3a (*63*). The three-dimensional structure of domain I revealed surprising similarity to FAD-containing enzymes, such as *p*-hydroxybenzoate hydroxylase, cholesterol oxidase, and glucose oxidase. This led to the speculation that FAD may be binding and modulating the activity of RabGDI. However, there is no indication that RabGDI binds FAD, and the region equivalent to the FAD binding pocket is a shallow groove. It is possible that the GG groups bind to this groove.

REP has not been crystallized yet despite intense efforts. It is likely that REP will fold into three domains rather than the two found in RabGDI. There seems to be enough sequence homology to predict that domains I and II will be present as in RabGDI. The third domain, tentatively called the insert region, may be formed by a stretch of 120 amino acids unique to REPs and positioned between SCR1 and SCR2. This insert region could fold between domains I and II to form a distinct third domain. Obviously in the absence of a three-dimensional structure, this is only a speculative view. It is interesting that the available sequences from invertebrate REPs do not contain an insert region (64, 65). The C terminus, on the other hand, is conserved and three small C-terminal regions have been termed REP conserved regions (RCR1, -2, and -3) and proposed to be characteristic of REPs only (66).

Functionally, REPs and RabGDIs also share many similarities. In semi-permeabilized cells, both REP and RabGDI can deliver Rab5 to endosomal membranes when added to the cells as an REP:Rab5 complex (67, 68). Both REP and RabGDI are able to extract Rab5 from membranes when added to cells in the free form (67). REP and GDI both also serve as GDP dissociation inhibitors that preferentially bind to GDP-bound Rabs (67, 69). The major difference between REP and RabGDI seems to be the ability of REP to assist in the prenylation reaction.

VI. Formation of Rab Escort Protein : Rab Complex

The first characterization studies of RGGT activity showed that the activity was not inhibited by peptides that mimicked the C-terminal sequence of Rabs (50). Furthermore, chimeric substrates containing as much as 75% of Rab sequence (50 residues from Ras at the N terminus and about 150 residues of Rab2 or Rab1b at the C terminus) were unable to incorporate GG groups or serve as inhibitors for wild-type Rab geranylgeranylation (70). Only mutant Rabs in which the C-terminal cysteine residues were substituted by nonprenyl acceptor serine residues (Rab-SS) were able to effectively inhibit RGGT activity (51). These studies suggested that recognition of Rab-specific conformational features was required for geranylgeranylation. When the sequence of REP revealed similarity to RabGDI, the possibility that this effect was due to a requirement for Rab to bind REP prior to binding to RGGT was considered. Binding studies showed that REP bound to Rabs prior to and after prenylation, and that the C-terminal cysteines were not important for the interaction (52, 71).

Formation of an REP:Rab complex seems to be a prerequisite for the geranylgeranylation of any Rab. In the absence of REP, there is no geranylgeranylation at all (71). However, it is not clear how this complex forms *in vivo*, because REP has a clear preference for Rabs in the GDP-bound conformation (72, 73). Newly synthesized Rabs likely exist in a GTP-bound state because the intracellular concentration of GTP is much higher than the concentration of GDP and the affinity for both nucleotides is similar. This suggests that GTP hydrolysis must occur prior to REP binding. The action of an RabGAP that stimulates GTP hydrolysis on newly synthesized Rabs seems essential because the rate of intrinsic GTP hydrolysis of Rabs is too slow to allow for the rapid and efficient prenylation observed in cells (56). This hypothesis may explain why a GTPase-deficient mutant of Rab1b, Rab1b^{Q67L}, is efficiently prenylated when expressed in mammalian cells but *in vitro* Rab1b^{Q67L} is efficiently prenylated only in the presence of GDP, not GTP (74). Alternatively, newly synthesized Rabs may be somehow prevented from assuming the GTP-bound conformation prior to prenylation. In this case, molecular chaperones may be required to help nascent Rabs to fold exclusively in the GDP-bound conformation so they could bind REP. There is no clear answer to this question yet but either hypothesis predicts the involvement of other proteins that promote the initial REP:Rab interaction.

A quantitative REP:Rab binding assay has been developed (71, 75). It relies on the quenching of a fluorescent analog of GDP, Mant-GDP, on REP binding. The equilibrium dissociation constants reported in two independent studies are surprisingly different. In one study where multiple REP:Rab complexes were studied, the K_d was found to be about 0.35 μM (71). In the other study, the reported K_d for REP:Rab7[GDP] was 20 nM (75). The reason for this discrepancy is not clear.

The REP:Rab interaction relies mostly on polar interactions because binding is disrupted by high salt concentrations but not detergents (71). However, the precise structural requirements for REP:Rab[GDP] interaction have not been elucidated. The preference for Rab in the GDP-bound conformation implies that the interaction involves a nucleotide-sensitive region of Rab. There is as yet no Rab[GDP] structure available but assuming that Rab and Ras undergo similar three-dimensional changes on GTP hydrolysis, two regions could change significantly on binding guanine nucleotide: the switch I (loop2/β2) region, and the switch II (loop4/α2) region (32). Either or both Rab regions could be involved in the interaction with REP. Point mutations in the α2 helix, which forms part of the switch II region, result in lack of association of the mutant Rabs with REP and consequently deficient prenylation (76). Mutations in the switch I/effector domain region have also been shown to reduce the geranylgeranylation efficiency (77), but this observation has been challenged (78).

Three other sequences in Rabs have been implicated in REP binding: the N terminus including a conserved lysine residue, the loop3/β3 region, and the C terminus. All these studies are indirect because they score alterations in the prenylation reaction rather than REP:Rab binding directly. Nevertheless, they probably reflect REP:Rab binding because mutations in Rab are unlikely to affect binding to RGGT (see Section X). Sanford et al. (79) showed that prenylation of Rab5 requires the presence of an N-terminal conserved sequence, YXXLFK, where the lysine residue (K) is absolutely required for prenylation and X is any amino acid. Beranger et al. (80) implicated a region in Rab6 (amino acids 60–67), corresponding to the loop3/β3 region. When the corresponding region of H-Ras was substituted into Rab6, it resulted in dramatic change in the K_m of the reaction. Other regions did not affect the kinetics of the reaction. Finally, the C-terminal region has been implicated despite the fact that the cysteine-rich prenylation motif itself is clearly not important for REP binding. A mutant Rab in which the two C-terminal cysteines were mutated to serines could no longer accept GG groups but retained wild-type affinity for REP (71). On the other hand, the C-terminal 10 amino acids of Rabs can affect the steady state kinetics of the prenylation reaction (54). Wild-type Rab3a has a lower V_{max} in the geranylgeranylation reaction as compared with wild-type Rab1a. When the 10 C-terminal amino acids were exchanged between these proteins, the kinetics were reversed. One interpretation of these experiments is that the C-terminal region of Rabs is affecting the rate of dissociation of prenylated Rab from REP, most likely the rate-limiting step in the reaction under the conditions used in that study.

It has been proposed that there may be significant conformational changes on REP on binding Rab (71). This possibility derives from inhibition studies described in more detail below (see Section X). Briefly, inhibition studies allowed a comparison between the affinity of free REP (3.5–7 μM) versus the affinity of REP:Rab (0.2–0.4 μM) for RGGT. The observed 10-fold difference may result from an important conformational change in REP on Rab binding. This possibility is further supported by analytical ultracentrifugation experiments (71, 81). Both REP and Rab have relatively large Stokes radii given the molecular weight of these proteins, indicating that both proteins deviate significantly from globular shape. It is known that the hypervariable C-terminal region of Rabs, comprising about 70 residues, is rather flexible and could be responsible for the observed Stokes radius (33, 34). Similarly, a large portion of about 160 residues in REP corresponding to the insert region may not contain regular secondary structure elements (71). In contrast, the complex between REP and Rab seems to be more compact compared with the individual proteins. Altogether, these results suggest that significant conformational changes accompany complex formation, which might include a partial ordering of the flexible regions in both proteins.

The two known REP proteins do not show different binding affinities for the limited number of Rab proteins tested so far, further suggesting that the interaction between REP and Rab is mediated by structural motifs that can be found in all Rab proteins (71). However, this general rule may not be applicable in all cases. We have proposed that choroideremia, the human retinal degenerative disease that results from mutations in REP1, may result from the incomplete geranylgeranylation of one Rab, Rab27 (56, 82). Because only REP2 is expressed in choroideremia cells, we suggested that this effect may be due to lower affinity of Rab27 for REP2, or that the complex may not be recognized properly by RGGT (56).

VII. Rab Geranylgeranyltransferase Genes and Primary Structure

The purification of RGGT from rat brain revealed that the enzyme was composed of two polypeptides, 60 and 38 kDa in apparent mass (50). Trypsin digestion patterns suggested that these polypeptides were distinct, which was confirmed on cDNA cloning of both gene products. The 60-kDa protein was designated α subunit and the 38-kDa protein was designated β subunit.

The α subunit has now been identified in mammals (human and rat), *Caenorhabditis elegans, Arabidopsis thaliana,* and in three species of yeast (*S. cerevisiae, S. pombe,* and *Candida albicans*) (53, 83, 84). The human and rat proteins are 91% identical. The N-terminal region contains seven copies of a repeat sequence of 34–35 amino acids centered about a trypto- phan (53). This repeat is characteristic of protein prenyltransferases since it was initially identified on PFT α subunit (85). This repetitive sequence may be related to the tetratricopeptide repeat motif (86). Proteins con- taining this motif are widely spread and serve diverse cellular functions, from cell cycle and signal transduction to protein secretion. The tetratrico- peptide repeat motif may be involved in protein–protein interactions. In prenyltransferases, it is mainly involved in dimer formation through interac- tions with the β subunit (87). The RGGT α subunit is significantly larger than PFT α subunit because of a C-terminal extension that is rich in leucine residues (see below, Section VIII). Interestingly, this extension is not pres- ent in the known yeast sequences but is present in the other eukaryotic se- quences.

The RGGT α-subunit gene (*RABGGTA*) is localized to chromosome 14q11.2. It is close (less than 2 kb away) to the transglutaminase-I gene, with the two genes organized in a head-to-tail orientation in relation to each other (*83*). This is a striking organization because the two genes are not functionally related (*88*).

The β subunit has been identified in mammals (human, rat, and mouse) as well as in *Drosophila melanogaster, C. elegans, A. thaliana,* and in three species of yeast (*S. cerevisiae, S. pombe,* and *C. albicans*) (*53, 83, 84, 89, 90*). The mammalian proteins are 93–97% identical and are about 50% identical to the yeast orthologs, revealing an exceptional degree of conservancy. The known protein prenyltransferase β subunits all present a repeated sequence that is less conserved than the α subunit repeat (*85*). The human β-subunit gene (*RABGGTB*) is localized to chromosome 1p31 and has not been associated with disease (*83, 91*).

VIII. Rab Geranylgeranyltransferase Three-Dimensional Structure

The crystal structure of RGGT at 2.0-Å resolution has been solved (Fig. 1) (*92*). The overall structure is similar to PFT with a few important differences (*87*). One of the striking differences is that the α subunit of RGGT is composed of three compact domains: a helical domain, an immunoglobulin-like (Ig-like) domain, and a leucine-rich repeat (LRR) domain whereas PFT-α is made up of a single helical domain.

The helical domain is structurally similar to the α subunit of PFT with a root mean square deviation (rmsd) of 2.1 Å between 216 superimposable C_α positions in the two proteins. The 15 helices in the helical domain form a crescent-shaped right-handed superhelix.

The Ig-like domain in RGGT-α is folded into an eight-stranded Greek-key β sandwich. All eight cysteine residues of the domain are in the reduced state. The domain is connected to the helical domain by two flexible loops where it inserts between helices α11 and α12. One face of this domain is packed against the LRR domain. The Ig-like domain is most similar to domain 4 of β-galactosidase (*93*) and they may be evolutionarily related.

The second additional domain in the α subunit is the LRR domain. The sequence of this domain contains five LRRs with lengths varying from 22 to 27 residues that fold into a right-handed superhelix of alternating β strands and 3_{10} helices. The conformation of each individual LRR unit is similar to that observed in porcine ribonuclease inhibitor (*94*). The hydrophobic face of helix α15 of RGGT-α is packed tightly against the hydrophobic interior of the LRR domain. This suggests that the position of the LRR domain is fixed relative to the helical domain of the same protein.

A   B

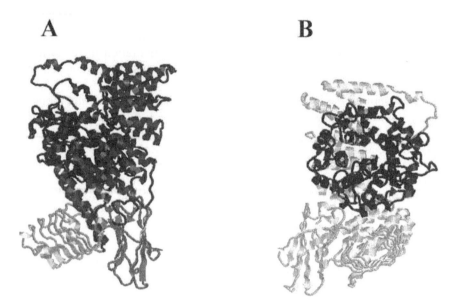

Fig. 1. Three-dimensional structure of Rab geranylgeranyltransferase. (A) Ribbon representation of the complete structure of Rab geranylgeranyltransferase with the α subunit
depicted in three colors, the helical domain in red, the Ig-like domain in orange, the leucine-
rich repeat in yellow, and the β subunit in blue. (B) Same as (A) except that the whole α
subunit is depicted in gray and the molecule is rotated along the z axis to illustrate the α–α
barrel present on the β subunit. (See color plate.)

The β subunit of RGGT contains an α–α barrel made up of 12 α helices
bearing close similarity to the PFT β subunit. There is a rmsd of 1.4 Å
between 280 superimposable C_α positions (87). The β subunit in RGGT is
smaller than the β subunit in PFT because it lacks the first α helix and the
C-terminal long loop. Like PFT β subunit, the center of the α–α barrel
forms a funnel-shaped pocket lined with mostly aromatic residues. The
bottom of this barrel is blocked by a turn, followed by a short α helix,
whereas the top of the barrel is open. The crescent-shaped helical domain
in RGGT α subunit embraces the β subunit around half its circumference
close to the open end of the barrel, while the Ig-like and LRR domains
do not make contacts with the β subunit.

A zinc ion is tightly bound to RGGT, primarily to the β subunit, where
it coordinates with Asp238β, Cys240β, and His290β. This arrangement is
similar to what was observed in PFT. Surprisingly, the fourth zinc ligand
is residue His-2 from the α subunit of the same heterodimeric molecule.
The N-terminal region of the α subunit is found in an extended conformation with His-2α coordinating the zinc ion and Lys-6α interacting with

Asp-272β, -283β, -284β, and -285β. Furthermore, the N terminus of the α subunit is formylated and the formyl group contacts Asp-287β. This intramolecular interaction may be functionally significant, possibly autoinhibitory to prevent binding of short substrate peptides to RGGT. This may be one reason why RGGT lost affinity for C-terminal tetrapeptides but other changes such as Leu-96β and Tyr-97α in the putative peptide-binding site may also be partially responsible for this effect.

The RGGT β subunit contains a central cavity lined with several hydrophobic residues including Trp-52β, Phe-147β, Tyr-195β, Trp-243β, Trp-244β, Phe-289β, and Phe-293β. A positively charged cluster formed by Arg-232β, Lys-235β, and Lys-105α is located near the opening of the cavity, close to the interface with the α subunit, and about 9 Å from the active site zinc ion. These features are also present in PFT, where FPP binds through its hydrophobic tail to the interior of the cavity and through the diphosphate head group to the positively charged cluster (95, 96). The conserved substitution of residues Trp-102β and Tyr-154β of PFT by residues Ser-48β and Leu-99β in RGGT-β leads to a significantly wider and deeper cavity in RGGT. Interestingly, in GGT-I-β, the tyrosine at the position corresponding to Tyr-154β in PFT is conserved, whereas the residue corresponding to Trp-102β in PFT is replaced by a smaller residue such as aspartate or serine. This indicates that the GGPP-binding pocket of the two GG transferases is not identical but the functional significance of this finding remains to be evaluated.

IX. Mechanism of Reaction: Binding of Geranylgeranyl Diphosphate

Biochemical characterization of PFT revealed that the enzyme possesses an FPP-carrier function (97). Stable PFT:FPP complexes were isolated by gel-filtration chromatography. The bound FPP was released unaltered on denaturation, suggesting that it is not covalently linked to the enzyme. Nevertheless, the FPP appears to bind in the active site because bound FPP can be specifically and rapidly transferred to a prenyl acceptor such as H-Ras. Comparable results were reported for GGPP binding to GGTase-I (98).

RGGT demonstrates similar properties, which is not surprising given the structural homology between protein prenyltransferases (99). The apparent K_d for the interaction is 0.6 nM, a value comparable with those reported for the other protein prenyltransferases (L. Desnoyers and M. C. Seabra, unpublished, 1998) (100). The enzyme has specific affinity for GGPP and little, if any, for FPP or GPP (50, 99). This result was expected because

RGGT is unable to transfer farnesyl to Rab substrates (M. C. Seabra, unpublished, 1992).

The stoichiometry of GGPP-binding to RGGT is 1:1, the same as for PFT and GGT-I (99). The three-dimensional structure is consistent with this result even though the structure of the RGGT–GGPP complex is not yet available. There is only one positively charged cluster that could bind a diphosphate moiety in the RGGT crystal structure, although the lipid-binding pocket could accommodate GG groups in alternative conformations (92). Residues Tyr-241β and Lys-105α are likely to play a crucial role in the correct positioning of the first and the second phosphate and thus of the GGPP C-1 atom near the protein substrate thiol for the subsequent catalytic attack. Mutations of the two corresponding residues, Tyr-310β of yeast PFT and Lys-164α of rat PFT, completely abolished enzyme activity (101, 102).

The finding that RGGT contains only one GGPP-binding site has important mechanistic implications because RGGT catalyzes double geranylgeranylation. When RGGT is preloaded with GGPP and then presented with REP:Rab under conditions where it is allowed to transfer only the bound GGPP, monoGG–Rab is produced (99). If the reaction is chased with more GGPP, then diGG–Rab is produced. The data thus suggest that double geranylgeranylation of Rabs proceeds via two independent reactions, in which the enzyme binds and transfers two consecutive GG groups to form monoGG–Rab as a reaction intermediate and then diGG–Rab as the final product.

X. Mechanism of Reaction: Binding of Rab Escort Protein: Rab Complex

As described above, REP was initially identified as component A of RGGT. The idea at the time was that the enzyme contained two subunits, a catalytic and a regulatory subunit, both required for activity (50, 51). Subsequent studies led to the finding that REP is absolutely required for recognition of Rab by RGGT (71). Therefore, the REP:Rab complex is the true substrate for RGGT. This is an important concept because REP should be regarded as a component of the substrate and not as a component of the enzyme.

So how is the REP:Rab complex recognized by RGGT? The observation that free REP inhibits the prenylation reaction led to the suggestion that free REP binds to RGGT and competes for binding with the REP:Rab complex possibly by forming nonproductive complexes with RGGT (71). Importantly, the reverse effect was not observed. Even high concentrations of free Rab did not result in enzyme inhibition (71). If free REP but not free Rab is able to bind RGGT, then the data imply that the REP:Rab complex is recognized primarily via an REP-binding site on the enzyme.

Nothing is known about the specific regions involved in REP(Rab): RGGT binding. It is tempting to speculate that the two motifs unique to RGGT-α, the Ig-like domain and the LRR domain, may serve as REP-anchoring regions. On the other hand, the insert region in REP is an attractive candidate region to mediate binding to RGGT. It must be pointed out that the respective yeast homologs do not contain these domains, and so an important question will be to determine whether the binding motifs responsible for the REP:Rab interaction are conserved in yeast and mammals.

XI. Mechanism of Reaction: Catalysis

To date, the most important insight into the mechanism of the reaction was the finding that double geranylgeranylation occurs in two distinct steps (71, 81, 99). The current evidence suggests the following order of events. RGGT loaded with GGPP binds REP:Rab, and catalyzes the transfer of the first GG group to Rab (71). After this transfer has occurred, the first GG group covalently attached to Rab may dissociate from the RGGT active site and interact with REP, Rab, the REP:Rab interface, or another RGGT-binding site. The displacement of the transferred GG group from the RGGT-binding site may be driven by the binding of another molecule of GGPP to RGGT. In PFT, product release is dependent on binding of new substrate, either FPP or Ras (103). The second GGPP seems to be used preferentially on monoGG–Rab rather than on a new unprenylated Rab, judging from the almost exclusive generation of digeranylgeranylated Rabs *in vitro* and *in vivo* (28, 37). A digeranylgeranylated product is also produced after *in vitro* reaction using recombinant yeast RGGT (Bet2p/Bet4p) and yeast REP (104).

In PFT, the first studies indicated that the reaction proceeds via a random sequential mechanism, based on substrate-binding studies and steady state kinetics (105, 106). Later, isotope partitioning studies indicated that the preferred catalytic pathway is through the enzyme–isoprenoid binary complex, while the pathway through enzyme–peptide binary complex is much slower (107). Steady state kinetic studies in RGGT have not been possible because of the complexity of the reaction. However, the finding that RGGT binds GGPP in a manner similar to the other prenyltransferases suggests that the kinetic mechanism may be similar and therefore GGPP binding to RGGT could be the first event in the reaction.

Once REP:Rab binds to the loaded RGGT, each of the two C-terminal cysteines is modified but the order in which they are attacked is not clear. Studies with Rab1 mutants, in which the C-terminal cysteines were replaced by serines, suggested that there is not an absolute order of addition of GG groups, because either mutant (Rab1-CS or Rab1-SC) can accept a prenyl group. However, the non-C-terminal cysteine seems to be the preferential target for the first addition, suggesting that there may be an order to the geranylgeranylation reaction (81).

The RGGT chemical mechanism of catalysis is unknown. In PFT, two hypotheses have been proposed, one electrophilic (108) and one nucleophilic (109). For more details, the reader is referred to Chapters 1 and 3 in this book. Given the similarities between protein prenyltransferases, it is likely that the catalytic mechanism is conserved.

It is not yet clear why RGGT processes preferentially monoGG–Rab substrates. Mechanistically, two possible scenarios could occur after the first geranylgeranyl transfer is accomplished. First, the monoprenylated product could remain bound to the enzyme, while the diphosphate head group dissociates from the active site. As soon as the PP-binding site becomes vacant, a second GGPP may be able to bind to the enzyme. The conversion of a thiolate-S ligand to a more weakly bound thioether-S ligand may drive ligand reorganization at the metal center, in which the other free cysteine at the Rab C terminus may become liganded to the zinc and ready for the second prenylation reaction. Another possibility is that after the first prenylation, the monoprenylated peptide dissociates from the active center temporarily, but because the catalytic ternary complex is still intact through interactions between RGGT and the Rab:REP complex, the dissociated peptide could bind to the active site again for the second prenylation. In both scenarios, rearrangement of the C-terminal peptide of Rab after the first prenylation must take place for prenylation of the second cysteine. This requires less specific and more flexible binding of the peptide in RGGT than in PFT and may explain the necessity for REP in the reaction.

The kinetic parameters measured to date suggest that the enzyme processes the different REP:Rab complex substrates with equal efficiency ($K_m = 0.35\ \mu M$) (71). The k_{cat} values obtained assuming double geranylgeranylation of the substrates are approximately twofold lower than that reported for PFT. Interestingly, if the rate of transfer of the first and second GG group by RGGT is the same, then the rate of each prenyl transfer would be almost identical between the two protein prenyltransferases. This may represent a striking conservation of reaction mechanism between the two enzymes.

As described above, a few Rabs contain a CaaX box containing a single prenylatable cysteine residue. This raised the question of which GG-transferase modifies these Rabs, because GGT-I could potentially act on these Rabs. This is particularly relevant for Rab8 and Rab18, in which the C-terminal amino acid is leucine, the sequence preferred by GGT-I. *In vitro,* both enzymes geranylgeranylate Rab8 (*110*). However, a specific inhibitor of GGT-I does not affect the rate of prenylation of Rab8 *in vivo.* When a point mutation that affects REP binding is introduced, Rab8^{Y78D}, *in vitro* prenylation of Rab8^{Y78D} by RGGT is abolished while prenylation by GGT-I is unaffected. Nevertheless, prenylation *in vivo* of Rab8^{Y78D} is severely reduced even when the protein is overexpressed, suggesting that RGGT is responsible for Rab8 prenylation (*110*). It is reasonable to predict that all Rabs are prenylated by RGGT under physiological circumstances.

One report suggested that RGGT may be regulated by phosphorylation (*111*). The RGGT α subunit was phosphorylated on insulin stimulation of cultured cells and this treatment led to increased Rab3 and Rab4 prenylation. While preliminary, these results raise the interesting possibility that RGGT activity is regulated by extracellular signals. Interestingly, insulin was also shown to induce phosphorylation of PFT α subunit (shared with GGT-I) and to activate farnesylation of Ras (*112*). If these results are confirmed, protein prenyltransferases may no longer be viewed simply as "housekeeping" enzymes but may be regulated by changing environmental conditions.

XII. Membrane Association of Geranylgeranylated Rabs

Why does the prenylation of Rab proteins require the presence of REP whereas the prenylation of other small G proteins does not require an accessory protein? The answer might lie in the fact that Rab proteins are doubly geranylgeranylated and consequently their hydrophobic character changes dramatically on prenylation. By binding to RGGT, REP may allow considerable flexibility of Rab around the RGGT active site and permit efficient digeranylgeranylation. After prenylation, REP may then be required to shield the hydrophobic GG groups to increase the solubility of prenylated Rab proteins, and might therefore be regarded as a specific prenylated protein-solubilizing factor.

REP was designated as an escort protein because it remains associated with diGG–Rab after completion of the double geranylgeranylation reaction and it may lead each Rab to its specific intracellular membrane compartment (*52, 81*). There are two pieces of evidence supporting this hypothesis. One is that REP seem to be competent for direct delivery of Rabs to their correct intracellular organelle (*67*). REP:Rab5–diGG complexes were assembled *in vitro* and incubated with semipermeabilized cells. The incubation led to the membrane association of Rab5 without requirement for RabGDI or other cytosolic factors. In another report, a mutant Rab1b containing a point mutation, Rab1b^{D44N}, that disrupted RabGDI binding was expressed in cultured cells (*113*). Rab1b^{D44N} was shown to be effectively targeted to intracellular membranes, demonstrating that REP can mediate this process. Interestingly, this mutant Rab1b was shown to be exclusively associated with membranes, as opposed to a mixed cytosol/membrane partition seen for the wild-type protein. Taken together, these results suggest that REP mediates the membrane insertion of newly synthesized and newly prenylated Rabs. RabGDI seems to act at a later step in Rab function, the recycling of Rabs from the acceptor compartment back to the donor compartment, so that Rabs can regulate multiple rounds of vesicular transport (*45*).

The mechanism of membrane association of geranylgeranylated Rabs is still unclear but may involve a distinct succession of steps such as targeting of Rab to the correct membrane, dissociation from an REP (or RabGDI) complex, partition into the membrane, and activation via nucleotide exchange. The fact that each Rab has a specific location restricted to a defined pair of donor and acceptor membranes suggests that membrane association of Rabs is mediated by a specific protein receptor that ensures proper targeting of Rabs. Targeting of Rabs to specific membranes is, at least in part, mediated by a sequence in the hypervariable domain just upstream of the Rab prenylation motif. Swapping these hypervariable regions between Rab5 and Rab7 reversed their localization (*114*). A possible membrane-bound receptor for Rab4 has been identified but has not been isolated yet (*115*). It is likely that the putative Rab targeting receptors constitute one large protein family, similar to the Rab family and that all receptors rely on the same mechanism to promote Rab membrane association. It is also likely that the putative receptors recognize multiple protein determinants, not only in the Rab proteins, but also in REP or RabGDI, because GG–Rabs are complexed with either REP or RabGDI when presented to membranes.

An activity named GDI displacement factor (GDF) has been identified (*116*). This GDF activity was partially purified from rat liver membranes and shown to meet many of the expected criteria for a promoter of Rab membrane association: GDF is tightly membrane associated, has high affinity for RabGDI:Rab9 complex, and stimulates dissociation from RabGDI and membrane association of prenylated Rab9. Also, GDF lacks specificity for a single Rab and works on a subset of Rabs present in endocytic membranes, suggesting that it does not work as a targeting receptor.

REP and RabGDI are the only proteins known to be able to extract Rabs from membranes and they possess low affinity for the GTP-bound form of Rabs. Therefore, nucleotide exchange catalyzed by a nucleotide exchange factor (RabGEF) may be the final step in the process of membrane association. Three RabGEFs, namely Rab3GEF acting on Rab3 (*117*), Sec2 acting on Sec4 (homologous to mammalian Rab8) (*118*), and Rabex-5 acting on Rab5 (*119*), have been identified to date. Surprisingly, all three are soluble proteins, raising the question of how they are targeted to the membrane themselves. Lack of RabGEF activity leads to mislocalization of the Rab substrate. Mutations in Sec2 affect targeting of Sec4 (*118*) and *aex-3* mutants of *C. elegans* show mislocalization of Rab3 (*120*). Aex-3 is postulated to be an Rab3GEF because it bears structural homology with mammalian Rab3GEF. It is not clear whether the mistargeting effect is a direct consequence of lack of GEF protein or an indirect consequence of lack of Rab activation leading to organelle membrane redistributions.

The association and dissociation of Rabs from REP or RabGDI in this cytosol/membrane cycle of Rabs may be tightly regulated. It was proposed that a cycle of RabGDI phosphorylation/dephosphorylation might regulate the association/dissociation of Rabs (*121*). Dephosphorylation leads to membrane association of Rabs, possibly by allowing binding to the putative Rab receptor at the membrane or, alternatively, by decreasing the affinity of the RabGDI : Rab complex.

Further evidence for posttranslational modifications affecting RabGDI came from genetic studies of *D. melanogaster*. A developmental mutation that affects chromosome separation in mitosis leads to a basic shift in the isoelectric point of three abundant proteins, one of which was identified as RabGDI (*122*). Although preliminary, these studies raise the possibility that posttranslational modifications regulate the function of RabGDI, and by analogy REP. Furthermore, Rab1a and Rab4 have been shown to be phosphorylated in a cell cycle-dependent manner (*123*). Rab4 phosphorylation on a serine residue close to the C terminus leads to its translocation to the cytosol during mitosis (*124*). Again, this study suggests regulation of the Rab cycle by posttranslational modifications that, as suggested above, may be influencing the affinity of the REP (or RabGDI) : Rab complex.

In conclusion, it is clear that the addition of GG groups to Rabs is essential to their function and that the geranylgeranylation reaction is intimately linked to Rab membrane association, an important event in vesicular transport that is not understood.

XIII. Conclusions and Future Work

This chapter summarizes the current knowledge on the geranylgeranylation of Rabs. This enzymatic reaction is rather complex because it involves a heterodimeric enzyme, a heterodimeric protein substrate, one lipid substrate, and two consecutive enzymatic transfers. While dramatic progress has been achieved since the first identification of this activity in 1991, much is still unknown.

The elucidation of the three-dimensional structures of RGGT, Rab, and RabGDI will determine a "second generation" of experiments, by allowing definition of questions toward a more deep understanding of the reaction. Among those are structure–function relationships including the details of REP:RGGT interaction, and the mechanism of the second geranylgeranylation, including understanding whether RGGT dissociates between the two GG additions.

On the other hand, the molecular dissection of membrane association of prenylated Rabs will focus on other critical issues concerning prenylated proteins: how are prenylated proteins interacting with membranes, what is the role of the prenyl groups, and how is membrane association regulated?

References

1. Schmidt, R. A., Schneider, C. J., and Glomset, J. A. (1984). *J. Biol. Chem.* **259,** 10175.
2. Kamiya, Y., Sakurai, A., Tamura, S., and Takahashi, N. (1978). *Biochem. Biophys. Res. Commun.* **83,** 1077.
3. Sakagami, Y., Yoshida, M., Isogai, A., and Suzuki, A. (1981). *Science* **212,** 1525.
4. Wolda, S. L., and Glomset, J. A. (1988). *J. Biol. Chem.* **263,** 5997.
5. Farnsworth, C. C., Wolda, S. L., Gelb, M. H., and Glomset, J. A. (1989). *J. Biol. Chem.* **264,** 20422.
6. Anderegg, R. J., Betz, R., Carr, S. A., Crabb, J. W., and Duntze, W. (1988). *J. Biol. Chem.* **263,** 18236.
7. Schafer, W. R., Kim, E., Sterne, R., Thorner, J., Kim, S. H., and Rine, J. (1989). *Science* **245,** 379.
8. Hancock, J. F., Magee, A. I., Childs, J. E., and Marshall, C. J. (1989). *Cell* **57,** 1167.
9. Casey, P. J., Solski, P. A., Der, C. J., and Buss, J. E. (1989). *Proc. Natl. Acad. Sci. U.S.A.* **86,** 8323.
10. Epstein, W. W., Lever, D., Leining, L. M., Bruenger, E., and Rilling, H. C. (1991). *Proc. Natl. Acad. Sci. U.S.A.* **88,** 9668.
11. Barbacid, M. (1987). *Annu. Rev. Biochem.* **56,** 779.
12. Bourne, H. R., Sanders, D. A., and McCormick, F. (1990). *Nature (London)* **348,** 125.
13. Bourne, H. R., Sanders, D. A., and McCormick, F. (1991). *Nature (London)* **349,** 117.
14. Kahn, R. A., Der, C. J., and Bokoch, G. M. (1992). *FASEB J.* **6,** 2512.
15. Cox, A. D., and Der, C. J. (1992). *Crit. Rev. Oncog.* **3,** 365.
16. James, G. L., Goldstein, J. L., and Brown, M. S. (1995). *J. Biol. Chem.* **270,** 6221.
17. Armstrong, S. A., Hannah, V. C., Goldstein, J. L., and Brown, M. S. (1995). *J. Biol. Chem.* **270,** 7862.
18. Casey, P. J., and Seabra, M. C. (1996). *J. Biol. Chem.* **271,** 5289.
19. Schafer, W. R., and Rine, J. (1992). *Annu. Rev. Genet.* **30,** 209.
20. Brown, M. S., and Goldstein, J. L. (1993). *Nature (London)* **366,** 14.

21. Zahraoui, A., Touchot, N., Chardin, P., and Tavitian, A. (1989). *J. Biol. Chem.* **264,** 12394.
22. Kinsella, B. T., and Maltese, W. A. (1991). *J. Biol. Chem.* **266,** 9540.
23. Khosravi-Far, R., Lutz, R. J., Cox, A. D., Conroy, L., Bourne, J. R., Sinensky, M., Balch, W. E., Buss, J. E., and Der, C. J. (1991). *Proc. Natl. Acad. Sci. U.S.A.* **88,** 6264.
24. Kinsella, B. T., and Maltese, W. A. (1992). *J. Biol. Chem.* **267,** 3940.
25. Peter, M., Chavrier, P., Nigg, E. A., and Zerial, M. (1992). *J. Cell Sci.* **102,** 857.
26. Rossi, G., Yu, J. A., Newman, A. P., and Ferro-Novick, S. (1991). *Nature (London)* **351,** 158.
27. Newman, C. M., Giannakouros, T., Hancock, J. F., Fawell, E. H., Armstrong, J., and Magee, A. I. (1992). *J. Biol. Chem.* **267,** 11329.
28. Farnsworth, C. C., Kawata, M., Yoshida, Y., Takai, Y., Gelb, M. H., and Glomset, J. A. (1991). *Proc. Natl. Acad. Sci. U.S.A.* **88,** 6196.
29. Smeland, T. E., Seabra, M. C., Goldstein, J. L., and Brown, M. S. (1994). *Proc. Natl. Acad. Sci. U.S.A.* **91,** 10712.
30. Valencia, A., Chardin, P., Wittinghofer, A., and Sander, C. (1991). *Biochemistry* **30,** 4637.
31. de Vos, A. M., Tong, L., Milburn, M. V., Matias, P. M., Jancarik, J., Noguchi, S., Nishimura, S., Miura, K., Ohtsuka, E., and Kim, S. H. (1988). *Science* **239,** 888.
32. Wittinghofer, A., and Pai, E. F. (1991). *Trends Biochem. Sci.* **16,** 382.
33. Dumas, J. J., Zhu, Z., Connolly, J. L., and Lambright, D. G. (1999). *Structure* **7,** 413.
34. Ostermeier, C., and Brunger, A. T. (1999). *Cell* **96,** 363.
35. Tong, L. A., de Vos, A. M., Milburn, M. V., Jancarik, J., Noguchi, S., Nishimura, S., Miura, K., Ohtsuka, E., and Kim, S. H. (1989). *Nature (London)* **337,** 90.
36. Krengel, U., Schlichting, L., Scherer, A., Schumann, R., Frech, M., John, J., Kabsch, W., Pai, E. F., and Wittinghofer, A. (1990). *Cell* **62,** 539.
37. Farnsworth, C. C., Seabra, M. C., Ericsson, L. H., Gelb, M. H., and Glomset, J. A. (1994). *Proc. Natl. Acad. Sci. U.S.A.* **91,** 11963.
38. Novick, P., and Zerial, M. (1997). *Curr. Opin. Cell Biol.* **9,** 396.
39. Schimmoller, F., Simon, I., and Pfeffer, S. R. (1998). *J. Biol. Chem.* **273,** 22161.
40. Olkkonen, V. M., and Stenmark, H. (1997). *Int. Rev. Cytol.* **176,** 1.
41. Lazar, T., Gotte, M., and Gallwitz, D. (1997). *Trends Biochem. Sci.* **22,** 468.
42. Chavrier, P., Parton, R. G., Hauri, H. P., Simons, K., and Zerial, M. (1990). *Cell* **62,** 317.
43. Sasaki, T., Kikuchi, A., Araki, S., Hata, Y., Isomura, M., Kuroda, S., and Takai, Y. (1990). *J. Biol. Chem.* **265,** 2333.
44. Araki, S., Kikuchi, A., Hata, Y., Isomura, M., and Takai, Y. (1990). *J. Biol. Chem.* **265,** 13007.
45. Pfeffer, S. R., Dirac-Svejstrup, A. B., and Soldati, T. (1995). *J. Biol. Chem.* **270,** 17057.
46. Molenaar, C. M., Prange, R., and Gallwitz, D. (1998). *EMBO J.* **7,** 971.
47. Walworth, N. C., Goud, B., Kabcenell, A. K., and Novick, P. J. (1989). *EMBO J.* **8,** 1685.
48. Moores, S. L., Schaber, M. D., Mosser, S. D., Rands, E., O'Hare, M. B., Garsky, V. M., Marshall, M. S., Pampliano, D. L., and Gibbs, J. B. (1991). *J. Biol. Chem.* **266,** 14603.
49. Horiuchi, H., Kawata, M., Katayama, M., Yoshida, Y., Musha, T., Ando, S., and Takai, Y. (1991). *J. Biol. Chem.* **266,** 16981.
50. Seabra, M. C., Goldstein, J. L., Südhoff, T. C., and Brown, M. S. (1992). *J. Biol. Chem.* **267,** 14497.
51. Seabra, M. C., Brown, M. S., Slaughter, C. A., Sudhof, T. C., and Goldstein, J. L. (1992). *Cell* **70,** 1049.
52. Andres, D. A., Seabra, M. C., Brown, M. S., Armstrong, S. A., Smeland, T. E., Cremers, F. P., and Goldstein, J. L. (1993). *Cell* **73,** 1091.
53. Armstrong, S. A., Seabra, M. C., Sudhof, T. C., Goldstein, J. L., and Brown, M. S. (1993). *J. Biol. Chem.* **268,** 12221.

54. Cremers, F. P. M., Armstrong, S. A., Seabra, M. C., Brown, M. S., and Goldstein, J. L. (1994). *J. Biol. Chem.* **269,** 2111.
55. Desnoyers, L., Anant, J. S., and Seabra, M. C. (1996). *Biochem. Soc. Trans.* **24,** 699.
56. Seabra, M. C., Ho, Y. K., and Anant, J. S. (1995). *J. Biol. Chem.* **270,** 24420.
57. Waldherr, M., Ragnini, A., Schweyer, R. J., and Boguski, M. S. (1993). *Nature Genet.* **3,** 193.
58. Fujimura, K., Tanaka, K., Nakano, A., and Toh-e, A. (1994). *J. Biol. Chem.* **269,** 9205.
59. Jiang, Y., and Ferro-Novick, S. (1994). *Proc. Natl. Acad. Sci. U.S.A.* **91,** 4377.
60. Shisheva, A., Sudhof, T. C., and Czech, M. P. (1994). *Mol. Cell Biol.* **14,** 3459.
61. Janoueix-Lerosey, I., Jollivet, F., Camonis, J., Marche, P. N., and Goud, B. (1995). *J. Biol. Chem.* **270,** 14801.
62. Garrett, M. D., Zahner, J. E., Cheney, C. M., and Novick, P. J. (1994). *EMBO J.* **13,** 1718.
63. Schalk, I., Zeng, K., Wu, S. K., Stura, E. A., Matteson, J., Huang, M., Tanodn, A., Wilson, I. A., and Balch, W. E. (1996). *Nature (London)* **381,** 42.
64. Ragnini, A., Teply, R., Waldherr, M., Voskova, A., and Schweyen, R. J. (1994). *Curr. Genet.* **26,** 308.
65. Dong, H., Jin, Y., Johansen, J., and Johansen, K. M. (1999). *Biochim. Biophys. Acta* **1449,** 194.
66. Bauer, B. E., Lrenzetti, S., Miaczynska, M., Bui, D. M., Schweyen, R. J., and Ragnini, A. (1996). *Mol. Biol. Cell* **7,** 1521.
67. Alexandrov, K., Horiuchi, H., Steele-Mortimer, O., Seabra, M. C., and Zerial, M. (1994). *EMBO J.* **13,** 5262.
68. Ullrich, O., Horiuchi, H., Bucci, C., and Zerial, M. (1994). *Nature (London)* **368,** 157.
69. Ullrich, O., Stenmark, H., Alexandrov, K., Huber, L. A., Kaibuchi, K., Sasaki, T., Takai, Y., and Zerial, M. (1993). *J. Biol. Chem.* **268,** 18143.
70. Khosravi-Far, R., Clark, G. J., Abe, K., Cox, A. D., McLain, T., Lutz, R. J., Sinensky, M., and Der, C. J. (1992). *J. Biol. Chem.* **267,** 24363.
71. Anant, J. S., Desnoyers, L., Machius, M., Demeler, B., Hansen, J. C., Westover, K. D., Deisenhofer, J., and Seabra, M. C. (1998). *Biochemistry* **37,** 12559.
72. Sandford, J. C., Pan, Y., and Wessling-Resnick, M. (1993). *J. Biol. Chem.* **268,** 23773.
73. Seabra, M. C. (1996). *J. Biol. Chem.* **271,** 14398.
74. Wilson, A. L., Sheridan, K. M., Erdman, R. A., and Maltese, W. A. (1996). *Biochem. J.* **318,** 1007.
75. Alexandrov, K., Simon, I., Iakovenko, A., Holz, B., Goody, R. S., and Scheidig, A. J. (1998). *FEBS Lett.* **425,** 460.
76. Overmeyer, J. H., Wilson, A. L., Erdman, R. A., and Maltese, W. A. (1998). *Mol. Biol. Cell* **9,** 223.
77. Wilson, A. L., and Maltese, W. A. (1993). *J. Biol. Chem.* **268,** 14561.
78. Beranger, F., Paterson, H., Powers, S., de Gunzberg, J., and Hancock, J. F. (1994). *Mol. Cell Biol.* **14,** 744.
79. Sanford, J. C., Pan, Y., and Wessling-Resnick, M. (1995). *Mol. Biol. Cell* **6,** 71.
80. Beranger, F., Cadwallader, K., Porfiri, E., Powers, S., Evans, T., de Gunzburg, J., and Hancock, J. F. (1994). *J. Biol. Chem.* **269,** 13637.
81. Shen, F., and Seabra, M. C. (1996). *J. Biol. Chem.* **271,** 3692.
82. Seabra, M. C. (1996). *Ophthal. Genet.* **17,** 43.
83. van Bokhoven, H., Rawson, R. B., Merkx, G. F., Cremers, F. P., and Seabra, M. C. (1996). *Genomics* **38,** 133.
84. Jiang, Y., Rossi, G., and Ferro-Novick, S. (1993). *Nature (London)* **366,** 84.
85. Boguski, M. S., Murray, A. W., and Powers, S. (1992). *New Biol.* **4,** 408.
86. Zhang, H., and Grishin, N. V. (1999). *Protein Sci.* **8,** 1658.

87. Park, H. W., Boduluri, S. R., Moomaw, J. F., Casey, P. J., and Beese, L. S. (1997). *Science* **275**, 1800.
88. Song, H.-J., Rossi, A., Ceci, R., Kim, I.-G., Anzano, M. A., Jang, S.-I., De Laurenzi, V., and Steinert, P. M. (1997). *Biochem. Biophys. Res. Commun.* **235**, 10.
89. Wei, L.-N., Lee, C.-H., Chinpaisal, C., Copeland, N. G., Gilbert, D. J., Jenkins, N. A., and Hsu, Y.-C. (1995). *Cell Growth Differ.* **6**, 607.
90. Godfrey, R., and Davey, J. (1996). *Yeast* **12**, 479.
91. Sanders, R., Islam, K. B., Betz, R., Larsson, C., and Edvard Smith, C. I. (1996). *Genomics* **35**, 633.
92. Zhang, H., Seabra, M. C., and Deisenhofer, J. (2000). *Structure* **8**, 241.
93. Jacobson, R. H., Zhang, X. J., DuBose, R. F., and Matthews, B. W. (1994). *Nature (London)* **369**, 761.
94. Kobe, B., and Deisenhofer, J. (1993). *Nature (London)* **366**, 751.
95. Dunten, P., Dammlott, U., Crowther, R., Weber, D., Palermo, R., and Birktoft, J. (1998). *Biochemistry* **37**, 7907.
96. Long, S. B., Casey, P. J., and Beese, L. S. (1998). *Biochemistry* **37**, 9612.
97. Reiss, Y., Seabra, M. C., Armstrong, S. A., Slaughter, C. A., Goldstein, J. L., and Brown, M. S. (1991). *J. Biol. Chem.* **266**, 10672.
98. Yokoyama, K., and Gelb, M. H. (1993). *J. Biol. Chem.* **268**, 4055.
99. Desnoyers, L., and Seabra, M. C. (1998). *Proc. Natl. Acad. Sci. U.S.A.* **95**, 12266.
100. Yokoyama, K., Zimmerman, K., Scholten, J., and Gelb, M. H. (1997). *J. Biol. Chem.* **272**, 3944.
101. Andres, D. A., Goldstein, J. L., Ho, Y. K., and Brown, M. S. (1993). *J. Biol. Chem.* **268**, 1383.
102. Dolence, J. M., Rozema, D. B., and Poulter, C. D. (1997). *Biochemistry* **36**, 9246.
103. Tschantz, W. R., Furfine, E. S., and Casey, P. J. (1997). *J. Biol. Chem.* **272**, 9989.
104. Witter, D. J., and Poulter, C. D. (1996). *Biochemistry* **35**, 10454.
105. Reiss, Y., Seabra, M. C., Armstrong, S. A., Slaughter, C. A., Goldstein, J. L., and Brown, M. S. (1991). *J. Biol. Chem.* **266**, 10672.
106. Pompliano, D. L., Rands, E., Schaber, M. D., Mosser, S. D., Anthony, N. J., and Gibbs, J. B. (1992). *Biochemistry* **31**, 3800.
107. Pompliano, D. L., Schaber, M. D., Mosser, S. D., Omer, C. A., Shafer, J. A., and Gibbs, J. B. (1993). *Biochemistry* **32**, 8341.
108. Dolence, J. M., and Poulter, C. D. (1995). *Proc. Natl. Acad. Sci. U.S.A.* **92**, 5008.
109. Huang, C. C., Casey, P. J., and Fierke, C. A. (1997). *J. Biol. Chem.* **272**, 20.
110. Wilson, A. L., Erdman, R. A., Castellano, F., and Maltese, W. A. (1998). *Biochem. J.* **333**, 497.
111. Goalstone, M. L., Leitner, J. W., Golovchenko, I., Stjernholm, M. R., Cormont, M., Le Marchand-Brustel, Y., and Draznin, B. (1999). *J. Biol. Chem.* **274**, 2880.
112. Goalstone, M. L., and Drazin, B. (1996). *J. Biol. Chem.* **271**, 27585.
113. Wilson, A. L., Erdman, R. A., and Maltese, W. A. (1996). *J. Biol. Chem.* **271**, 10932.
114. Chavrier, P., Gorvel, J. P., Stelzer, E., Simons, K., Gruenberg, J., and Zerial, M. (1991). *Nature (London)* **353**, 769.
115. Ayad, N., Hull, M., and Mellman, I. (1997). *EMBO J.* **16**, 4497.
116. Dirac-Svejstrup, A. B., Sumizawa, T., and Pfeffer, S. R. (1997). *EMBO J.* **16**, 465.
117. Wada, M., Nakanishi, H., Satoh, A., Hirano, H., Obaishi, H., Matsuura, Y., and Takai, Y. (1997). *J. Biol. Chem.* **272**, 3875.
118. Walch-Solimena, C., Collins, R. N., and Novick, P. J. (1997). *J. Cell Biol.* **137**, 1495.
119. Horiuchi, H., Lippe, R., McBride, H. M., Rubino, M., Woodman, P., Stenmark, H., Rybin, V., Wilm, M., Ashman, K., Mann, M., and Serial, M. (1997). *Cell* **90**, 1149.

120. Iwasaki, K., Staunton, J., Saifee, O., Nonet, M., and Thomas, J. H. (1997). *Neuron* **18,** 613.
121. Steele-Mortimer, O., Gruenberg, J., and Clague, M. J. (1993). *FEBS Lett.* **329,** 313.
122. Zahner, J. E., and Cheney, C. M. (1993). *Mol. Cell Biol.* **13,** 217.
123. Bailly, E., McCaffrey, M., Touchot, N., Zahraoui, A., Goud, B., and Bornens, M. (1991). *Nature (London)* **350,** 715.
124. van der Sluijs, P., Hull, M., Huber, L. A., Male, P., Goud, B., and Mellman, I. (1992). *EMBO J.* **11,** 4379.

7

Postisoprenylation Protein Processing: CXXX (CaaX) Endoproteases and Isoprenylcysteine Carboxyl Methyltransferase

STEPHEN G. YOUNG*,[†] • PATRICIA AMBROZIAK*,[†] •
EDWARD KIM*,[†] • STEVEN CLARKE[‡]
*Gladstone Institute of Cardiovascular Disease
San Francisco, California 94141
[†]Cardiovascular Research Institute and Department of Medicine
University of California, San Francisco
San Francisco, California 94143
[‡]Department of Chemistry and Biochemistry
Molecular Biology Institute
University of California, Los Angeles
Los Angeles, California 90095

THE ENZYMES, Vol. XXI

I. Introduction

Multiple eukaryotic proteins that terminate with a "CXXX" sequence undergo a series of posttranslational processing reactions. The CXXX proteins terminate with the amino acids –C–X–X–X, where the "C" is a cysteine, the next two "X" residues are frequently aliphatic amino acids, and the third "X" can be one of several amino acids. Because the middle two X residues are often aliphatic, these proteins have also been referred to as "CaaX" proteins. The first posttranslational modification is the attachment of a farnesyl or geranylgeranyl isoprenoid lipid to the thiol group of the cysteine residue (the "C" of the CXXX sequence) by specific cytosolic protein isoprenyltransferases. Second, the last three amino acids of the protein (i.e., the –XXX) are removed by an isoprenylprotein-specific endoprotease; this step is dependent on the isoprenylation step and is thought to take place on the cytoplasmic surface of the endoplasmic reticulum

(ER). Third, the carboxyl group of the newly exposed isoprenylcysteine is methylated by an ER-associated isoprenylcysteine carboxyl methyltransferase. These three protein-processing steps have been studied most intensively for the yeast mating pheromone **a**-factor and the Ras proteins, a group of small GTP-binding proteins that mediate signal transduction and affect cell growth. However, these modifications occur in a large number of CXXX proteins with diverse biological functions, including the nuclear lamins, the γ subunits of heterotrimeric guanine nucleotide-binding proteins (G proteins), some phosphodiesterases, and other small GTP-binding proteins such as the Rac and Rho proteins.

The Rab family of proteins also undergoes C-terminal isoprenylation. These proteins are important for vesicular trafficking within cells and can be divided into subgroups that contain a Cys–Cys (CC) or a Cys–Xaa–Cys (CXC) motif at the C terminus. Both the CC and CXC subgroups are geranylgeranylated at both cysteines, but neither undergoes endoproteolytic processing. The CXC proteins, but not the CC proteins, are methylated at the C-terminal isoprenylcysteine.

The enzymes responsible for the isoprenylation of the CXXX proteins, protein farnesyltransferase and protein geranylgeranyltransferase I, have been thoroughly characterized in yeast and in higher organisms. These enzymes have attracted attention because the isoprenylation of the Ras proteins is essential, both for their targeting to the plasma membrane and for the ability of mutationally activated Ras proteins to produce a transformed phenotype in cultured cells (*1, 2*). Interest in the protein isoprenyltransferases has been further fueled by reports that protein farnesyltransferase inhibitors retard the growth of cancers (*3–9*) and might hold promise in the treatment of parasitic diseases (*10*). Reviews (*11, 12*) and chapters in this volume have summarized the progress in understanding the protein isoprenyltransferases.

This chapter summarizes progress in understanding the "postisoprenylation" processing of CXXX proteins—the endoproteolytic processing step and the carboxyl methylation step. For much of the 1990s, getting a handle on these steps was slow, at least when compared with the rapid progress in understanding the protein isoprenyltransferases, and there was uncertainty about their physiologic importance. More recently, however, there have been exciting advances in understanding these steps, making it an attractive time to review this area. In a landmark study, Boyartchuk. Ashby, and Rine (*13*) identified two genes from *Saccharomyces cerevisiae, RCE1* and *AFC1,* that are involved in the proteolytic removal of the "–XXX" from two farnesylated CXXX proteins (Ras2p and the precursor to the yeast mating pheromone **a**-factor). This breakthrough made it possible to mine the expressed sequence tag (EST) databases and clone the mammalian

TABLE I

Properties of CXXX Endoprotease Activities

Source	Substrate and kinetic data	Conditions affecting activity	Comments	Ref.
Yeast membranes	N-Acetyl-KSKTK-(S-farnesyl-C)-VIM; specific activity: 158.3 pmol/min/mg protein	Sensitive to sulfhydryl reagents, N-ethylmaleimide. ZnCl₂; not sensitive to serine and aspartyl proteinase inhibitors or to o-phenanthroline	Sensitivity to sulfhydryl reagents and lack of sensitivity to o-phenanthroline are similar to the properties of yeast Rce1p (13, 33)	Hrycyna and Clarke (22)
Yeast membranes; rat liver microsomes	Dansyl-WDPA-(S-farnesyl-C)-V[³H]IA; specific activity: 974 ± 147 pmol/min/mg protein	Sensitive to zinc ions; not sensitive to PMSF, o-phenanthroline, unfarnesylated a-factor peptide, and a large panel of protease inhibitors	Lack of sensitivity to o-phenanthroline is similar to what Ashby and co-workers ultimately documented for yeast Rce1p (13, 33). Specificity for farnesylated peptides similar to human and yeast Rce1 (13, 15, 33)	Ashby et al. (24)
Dog pancreatic microsomes; bovine liver microsomes	N-[³H]Acetyl-(S-farnesyl-C)-VIS; K_m = 5.76 ± 0.71 μM; V_{max} = 251 ± 8.5 pmol/min/mg protein. N-[³H]Acetyl-(S-farnesyl-C)-VI and N-[³H]acetyl-(S-farnesyl-C)-V were also tested	Presence of D-farnesylcysteine in the peptide abolishes substrate activity, as does carboxyl methyl esterification of the peptide. Activity not inhibited by fivefold excess nonfarnesylated peptide	Specificity for farnesylated peptides similar to human and yeast Rce1 (13, 15, 33)	Ma and Rando (26)
Calf liver microsomes	³H-Labeled isoprenylated tripeptides in which the cysteine was modified with a 15-carbon farnesyl group, a 10-carbon geranyl group, or a 20-carbon geranylgeranyl group. Stereospecificity examined with isoprenylated peptides containing D-amino acids	Activity not inhibited by a wide variety of compounds known to inhibit serine proteases, cysteine proteases, metalloproteases, and aspartyl proteases. Substrates with 10-, 15-, and 20-carbon lipids were cleaved, but substrates with D-amino acids were cleaved poorly	Lack of sensitivity to o-phenanthroline is similar to yeast Rce1p (13, 33)	Ma et al. (27, 28)

Source	Substrate/kinetics	Inhibitor sensitivity	Comments	Reference
Rat liver microsomes. Activity located mainly in ER, as judged by cell fractionation and enzymatic markers	ECB-(S-farnesyl-C)-VI[³H]S (ECB, extended chain biotin); specific activity, 2.7 microunits/mg microsomes for VI[³H]S; $K_m = 1.1 \mu M$; $V_{max} = 7$ pmol/min/mg protein. ECB-NPFRQRRFC (GG)AI[³H]L: specific activity, 0.13 microunits/mg microsomes for AI[³H]L production. $K_m = 2.5 \mu M$. $V_{max} = 0.4$ pmol/min/mg protein	Sensitive to 4-(hydroxymercuri)-benzoate, leupeptin (partial), ZnCl₂; not sensitive to unfarnesylated peptide or to a wide panel of protease inhibitors including o-phenanthroline, PMSF, EDTA, aprotinin, pepstatin, DFP, and DTT	Sensitivity to sulfhydryl reagents and lack of sensitivity to o-phenanthroline are similar to sensitivity to o-phenanthroline are similar to yeast Rce1p (13, 33)	Jang et al. (30)
Bovine liver microsomal membranes. Activity solubilized with CHAPSO, eluted at high molecular weight end of gel-filtration column	N-[³H]Acetyl-(S-farnesyl-C)-VIM: $K_m = 0.65 \pm 0.08 \mu M$; $V_{max} = 1.96 \pm 0.07$ nmol/min/mg protein	Major peak of activity not inhibited by o-phenanthroline or wide spectrum of protease inhibitors. Partially purified enzyme is sensitive to thiol regents, RPI (reduced peptide inhibitor), PCMB, N_α-tosyl-L-phenylalanine (TPCK), N-tert-butyloxycarbonyl-(S-farnesyl-L-cysteine)-chloromethyl ketone (BFCCMK), N-benzyloxycarbonyl-glycylglycyl-(S-farnesyl-L-cysteine) chloromethyl ketone (ZGGFCCMK)	Lack of sensitivity to o-phenanthroline is similar to one of the key properties of yeast Rce1p (13, 33)	Chen et al. (32)
Bovine brain microsomal membranes; optimal pH 9.0; solubilized with sodium deoxycholate; appeared to be a 480-kDa protein or complex	Dansyl-KSKTK-(S-farnesyl-C)-VIM: $K_m = 1.0 \mu M$; $V_{max} = 14$ pmol/min/mg protein	Insensitive to many protease inhibitors, including leupeptin, chymostatin, E-64, DFP, EDTA; insensitive to o-phenanthroline, ZnCl₂	Insensitivity to o-phenanthroline is similar to one of key properties of yeast Rce1p (13, 33)	Nishii et al. (34)
Pig brain membranes; appeared to be a 70-kDa protein, loosely membrane associated	Propionyl-GSP-(S-farnesyl-C)-[³H]-VLM: $K_m = 32.5$ mM; $V_{max} = 60.8$ nmol/min/mg protein	Sensitive to o-phenanthroline, p-CMB, DTT, chymostatin, N-ethylmaleimide, ZnCl₂, CuCl₂, leupeptin (partial); insensitive to PMSF, leupeptin, E-64, pepstatin	Sensitivity to o-phenanthroline is similar to one of key properties of yeast Afc1p (13, 33)	Akopyan et al. (31)
Activity purified fivefold from rat liver microsomes after solubilization in CHAPS	Tested 64 tripeptides [N-acetyl-(S-farnesyl-C)-X₁X₂] as competitive inhibitors of hydrolysis of N-acetyl-(S-farnesyl-C)-VI[³H]S. Also tested radiolabeled peptides	Nonisoprenylated peptides do not inhibit enzymatic activity; pH optimum of ~6–6.5. Most potent inhibition is observed with farnesylated tripeptides that have most hydrophobic X₁X₂ dipeptide unit	Peptide hydrolysis results could reflect more than one enzymatic activity	Jang and Gelb (29)

orthologs for *RCE1* and *AFC1* (*14–17*). Similarly, the EST databases have been mined to identify the human isoprenylcysteine carboxyl methyltransferase (the ortholog of the *S. cerevisiae* gene *STE14*) (*18*). Strategies for defining the physiologic importance of the endoproteolysis and carboxyl methylation steps in higher organisms have also taken shape. Kim and co-workers (*14*) produced *Rce1* knockout mice and established that the *Rce1* gene is solely responsible for the endoproteolytic processing of the Ras proteins. In this chapter, we provide an overview of earlier work describing the endoproteolysis and carboxyl methylation enzymatic activities, as well as newer work on the identification of the gene products responsible for the endoproteolysis and carboxyl methylation steps in yeast and in mammals.

II. Characterization of Isoprenylprotein Endoprotease Activities

A. EARLY CHARACTERIZATIONS OF ISOPRENYLPROTEIN ENDOPROTEASE ACTIVITIES IN YEAST AND MAMMALS

During the 1980s, it became clear that certain yeast mating pheromones with a CXXX motif, including **a**-factor from *S. cerevisiae,* underwent modification with an isoprenoid lipid, endoproteolytic trimming of the C-terminal three amino acids, and carboxyl methylation. In 1988, Clarke and co-workers (*19*) reported that H-Ras, a mammalian CXXX protein, contained a C-terminal methyl ester and hypothesized that the processing of the mammalian Ras proteins and other CXXX proteins involved lipidation, endoproteolysis of the C-terminal three amino acids, and carboxyl methylation of the isoprenylcysteine. The existence of a specific proteolytic processing step received support in 1989, when Gutierrez *et al.* (*20*) provided evidence that the processing of mammalian Ras proteins involves the proteolytic release of the C-terminal three amino acids. At about the same time, Fujiyama *et al.* (*21*) provided direct evidence of the endoproteolytic release of the C-terminal three amino acids from yeast Ras2p.

Once it was clear that the processing of multiple CXXX proteins involved endoproteolytic processing, several groups sought to purify and characterize the enzymatic activity. Each group found that the relevant endoprotease activity was present in the membrane fractions. Table I summarizes the key findings from several of these studies.

Hrycyna and Clarke (*22*) identified three enzymatic activities from *S. cerevisiae* that were capable of removing the C-terminal three amino acids from the synthetic peptide *N*-acetyl-KSKTK-(*S*-farnesyl-C)-VIM. Two enzymatic activities were in the soluble fraction, and one was membrane associated. One soluble activity was due to carboxypeptidase Y, a vacuolar enzyme (22), and the other was due to a vacuolar/Golgi endoproteinase related to rat metalloendopeptidase 24.15, an enzyme with specificity for cleavage after hydrophobic residues (23). Those soluble enzymatic activities were judged not to be relevant to the processing of isoprenylated CXXX proteins because neither was localized to the cytosol, where the isoprenylated substrates would be expected to be present, and because they did not require the presence of the isoprenyl group for proteolytic activity. On the other hand, the membrane-associated activity appeared to be a good candidate for the CXXX endoprotease. That activity was markedly inhibited by 1 mM N-ethylmaleimide and by 0.5 mM p-hydroxymercuribenzoate, suggesting that the membrane associated enzyme might be a sulfhydryl-containing protease. The activity was unaffected by inhibitors of serine proteases such as phenylmethylsulfonyl fluoride (PMSF), dichloroisocoumarin, leupeptin, or the aspartyl protease inhibitor pepstatin. The enzyme activity was inhibited by 80% by 0.5 M zinc chloride, but was reduced by only 20% by 2.0 mM o-phenanthroline.

Ashby et al. (24, 25) also identified multiple enzymatic activities in yeast that were capable of removing the C-terminal three residues (–VIA) intact from the **a**-factor octapeptide dansyl-WDPA-(S-farnesyl-C)-VIA, demonstrating that the enzyme was an *endo*protease rather than a carboxypeptidase. The membrane-bound activity was unaffected by PMSF, o-phenanthroline, or nonfarnesylated **a**-factor peptide, but was inhibited by a high concentration of zinc ions. They also demonstrated that rat liver membranes contained an endoproteolytic activity that released the intact tripeptide–VIA from the farnesylated octapeptide.

Ma and Rando (26) reported an endoprotease activity from liver and pancreatic microsomes that cleaved the –XXX tripeptide from the synthetic tetrapetide substrate N-acetyl-(S-farnesyl-C)-VIS. The protease also cleaved the tripeptide N-acetyl-(S-farnesyl-C)-VI as well as the dipeptide N-acetyl-(S-farnesyl-C)-V (albeit with lower efficiency), but not the N-acetyl-(S-farnesyl-C)-amide or otherwise identical peptides containing cysteine in the D-configuration. These initial studies established that a dipeptide is required for enzymatic activity and that the cleavage at the scissile bond is stereoselective. Also, they demonstrated that endoproteolysis does not occur when the peptide contained a C-terminal methyl ester. The activity was not inhibited by a wide variety of compounds known to inhibit serine proteases, cysteine proteases, metalloproteases, and aspartyl proteases (27).

In a separate study, Ma et al. (28) tested the calf liver endoprotease activity against isoprenylated tripeptides (CVI) in which the cysteine was modified with a 15-carbon farnesyl group, a 10-carbon geranyl group, or a 20-carbon geranylgeranyl group. All three were cleaved, demonstrating broad substrate specificity with respect to isoprenoid side chains. In contrast, nonisoprenylated derivatives (the tripeptide CVI or the *tert*-butylthiol derivative of CVI) were not measurably processed. In a separate study, Jang and Gelb (*29*) found that replacement of the farnesyl group with a straight-chain hydrocarbon (*n*-pentadecyl) only modestly affected the endoprotease activity.

Further experiments by Ma *et al.* (*28*) were designed to examine the stereospecificity of the endoprotease for each of the amino acids within the CXXX sequence. The L-D-L and L-L-D analogs of the *N*-acetyl-(*S*-farnesyl-C)-VI tripeptide series were poor substrates for the endoprotease, as were tripeptides containing more than one D amino acid (D-D-D, D-D-L, and L-D-D). The stereospecificity of the X_3 position of CXXX sequence was also explored; only the L-L-L-L derivative was a good substrate for the protease, although the L-L-L-D derivative was processed minimally.

In 1993, Jang *et al.* (*30*) reported the presence of an isoprenyl-specific endoprotease from rat liver microsomes. The activity released a C-terminal tripeptide from the synthetic isoprenylated peptides ECB (extended chain biotin group)-NPFRQRRFF-(*S*-geranylgeranyl-C)-AI[^{3}H]L and ECB-(*S*-farsnesyl-C)-VI[^{3}H]S. A variety of experiments suggested that ^{3}H-labeled dipeptides were produced from tripeptides by secondary proteolysis. Non-isoprenylated peptides at concentrations 10- to 100-fold higher than those of the isoprenylated substrates did not affect the release of the tripeptide. Percoll density fractionation of rat liver membranes indicated that the endoprotease was localized mainly in the ER (in fractions containing the ER marker glucose-6-phosphatase). Enzymatic activity was inhibited by less than 20% by a large panel of inhibitors of serine, cysteine, and aspartic acid proteases and zinc metalloproteases.

B. Solubilization and Partial Purification of Enzymatic Activities

Several groups attempted to purify the isoprenylprotein endoprotease activity, but none was successful in purifying a protein to homogeneity. Akopyan *et al.* (*31*) purified an activity approximately 100-fold from the microsomal fraction of pig brain membranes by chromatography on DEAE Trysacryl M and Sephacryl S-200. The activity, which appeared to have a mass of about 70 kDa, cleaved [^{3}H]VLM from propionyl-GSP-(*S*-farnesyl-C)-[^{3}H]VLM. The activity was strongly inhibited by *p*-chloromercuribenzoate (PCMB), *N*-ethylmaleimide, chymostatin, zinc chloride, and *o*-phenanthroline and thus appeared to be distinct from the previously studied activities.

Chen *et al.* (*32*) partially purified a CXXX endoprotease activity about 10-fold from bovine liver microsomal membranes by solubilization in

CHAPSO   3[(3-cholamidopropyl)dimethylammonio]-2-hydroxylpropane sulfonic acid] and chromatography on Resource Q and Superose 12. They identified two activity peaks after fractionating the solubilized protein on a gel-filtration column. A major peak (with a molecular mass greater than 600 kDa) was not inhibited by o-phenanthroline, while a minor peak of activity (with a molecular mass of ~60 kDa) was inhibited by o-phenanthroline. The major peak shared most of the properties of the endoprotease in crude microsomal preparations but, unlike the activity described by Akopyan *et al.* (*31*), was not inhibited by chymostatin, dithiothreitol (DTT), or o-phenanthroline. The o-phenanthroline-sensitive minor peak was not extensively characterized. However, now that Boyartchuk and co-workers have demonstrated the existence of two different yeast CXXX endoproteases (Rce1p and Afc1p) with different sensitivities to o-phenanthroline (*13, 33*), it seems possible that the activities within the major and minor peaks produced by Chen *et al.* (*32*) represented complexes containing the *RCE1* and *AFC1* gene products, respectively.

Nishii *et al.* (*34*) purified an endoprotease activity 104-fold from bovine brain microsomes with Sepharose CL-6B gel filtration and DE-52 chromatography. Sodium deoxycholate and sodium cholate solubilized the activity, whereas other detergents (Lubrol PX, Triton X-100, Nikkol, n-heptyl-β-thioglucoside, and n-dodecyltrimethylammonium bromide) were less effective. The molecular mass of the protein was judged to be 480 kDa by gel filtration on a Sepharose CL-6B column. The partially purified activity cleaved the farnesylated peptide dansyl-KSKTK-(S-farnesyl-C)-VIM with a K_m of 1.0 μM and a V_{max} of 14 pmol/min/mg protein. The optimal pH of this activity was 9.0. Neither EDTA nor inhibitors of serine proteases, cysteine proteases, aspartate proteases, and aminopeptidases blocked enzymatic activity. However, the activity was inhibited by o-phenanthroline and zinc chloride. Interestingly, m- and p-phenanthroline (nonchelating isomers of o-phenanthroline) were also effective inhibitors, suggesting that the inhibition by o-phenanthroline might have nothing to do with its chelating action.

The various efforts to characterize and purify the isoprenylprotein endoprotease activity are confusing because there are abundant and significant discrepancies in the key properties of the enzymatic activity (e.g., sensitivities to inhibitors, molecular weight). Almost all these differences remain unexplained. Some of the differences could relate to differences in animal species or the tissue that was examined. Also, it is possible that the differences relate to the existence of more than one CXXX endoprotease (*13, 33*). In any case, now that two endoproteases have been cloned and can be overexpressed in cell culture systems, it should be possible to characterize their biochemical properties in more detail. Also, it is now possible to characterize CXXX endopeptidase activities from mammalian cells that lack either the *AFC1* or *RCE1* gene products (*14*).

C. Development of Endoprotease Inhibitor Compounds

The laboratory of R. Rando pioneered the development of specific CXXX endoprotease inhibitors (*27, 28, 32, 35*). Soon after it became clear that the endoprotease activity (or activities) trims CXXX proteins after the farnesylcysteine residue, Ma and co-workers (*28*) tested the ability of various farnesylcysteine derivatives to block the endoprotease activity. In the initial series of experiments, they demonstrated that an aldehyde analog, *N-tert*-butyloxycarbonyl-*S*-farnesylcysteine aldehyde, was an effective competitive inhibitor of the endoprotease. In an attempt to identify more potent inhibitors, they prepared farnesylcysteine derivatives that contained several amino acid residues. The potency of these compounds increased as the number of amino acids in the analog increased from two to four. *N-tert*-Butyloxycarbonyl-(*S*-farnesyl-C)-ψ(CH$_2$-NH)-VIM (also designated RPI, for reduced peptide inhibitor) inhibited the calf liver endoprotease activity with a K_i of 86 nM. As described further below, RPI inhibits the recombinant human RCE1 enzyme (*15*).

In a subsequent study, Chen *et al.* (*32*) identified structurally distinct compounds that irreversibly inhibited the endoprotease activity. In characterizing a bovine liver endoprotease, they found that N_α-tosyl-L-phenylalanine chloromethyl ketone (TPCK) was an irreversible inhibitor; this irreversible inhibition could be blocked with the reversible inhibitor RPI. To develop more potent irreversible inhibitors, they synthesized and tested farnesylcysteine analogs of the chloromethyl ketone inhibitors: BFCCMK [*N-tert*-butyloxycarbonyl-(*S*-farnesyl-C)-chloromethyl ketone] and ZGGFCCMK [*N*-benzyloxycarbonylglycylglycyl-(*S*-farnesyl-C)-chloromethyl ketone]. The former compound exhibited a 15-fold increase of the second-order rate constant ($K_{inh}/K_i = 1164\ M^{-1}\ min^{-1}$) compared with that of N_α-tosyl-L-phenylalanine chloromethyl ketone ($K_{inh}/K_i = 77\ M^{-1}\ min^{-1}$), and the K_i of its inhibitor–enzyme complex was 30 μM compared with 1.1 mM for N_α-tosyl-L-phenylalanine chloromethyl ketone.

More recently, Chen (*36*) has reported the development of new chloromethyl ketone inhibitors. One of these, UM96001 (*N-tert*-butyloxycarbonyl-2 amino-DL-hexadecanoyl-chloromethyl ketone), was based on the structure of BFCCMK, but the farnesyl group was replaced with a dodecyl group, and the thioether unit (which is chemically labile) was replaced with a methylene unit. The latter modification provided more chemical stability as well as increased cell permeability. As discussed in more detail below, UM96001 was reported to block the growth of Ras-transformed rodent and human cell lines (*36*). It is important to point out, however, that much more work is required to document the specificity of these compounds for the CXXX endoproteases.

D. INFLUENCE OF –XXX SEQUENCE ON ENDOPROTEOLYSIS

Jang and Gelb (*29*) tested the ability of a panel of 64 different tripeptides [*N*-acetyl-(*S*-farnesyl-C)-X$_1$X$_2$] to compete for the CXXX endoprotease-mediated hydrolysis of *N*-acetyl-(*S*-farnesyl-C)-VI[^{3}H]S. Their studies indicated that the endoprotease activity prefers large hydrophobic residues in the X$_1$ and X$_2$ positions. When X$_1$ was a large hydrophobic residue, potent inhibition was observed when X$_2$ was hydrophobic; less inhibition was seen when X$_2$ was more hydrophilic (e.g., glutamine or serine); and little if any inhibition was observed when X$_2$ was anionic (e.g., aspartate). Inhibition was also modest when both X$_1$ and X$_2$ were small hydrophobic residues, or when both were hydrophilic, uncharged residues. Tripeptides with aspartate in the X$_2$ position were weak inhibitors, regardless of what residue was in the X$_1$ position; tripeptides with lysine and arginine were tolerated in the X$_1$ position, as long as X$_2$ was a large hydrophobic residue. They also analyzed hydrolysis of tripeptides directly with radiolabeled tripeptides and tetrapeptides, and reached similar conclusions, except that the radiolabeled peptides containing arginine were poor substrates.

Kato and co-workers (*2*) analyzed posttranslational modifications with K-Ras4B constructs with a number of different amino acid substitutions in the CXXX sequence. Changing the wild-type –VIM sequence to –VYM principally affected the postisoprenylation processing steps. When the –VYM mutant was expressed in NIH 3T3 cells, a significant percentage of the protein appeared to be farnesylated but not further processed (i.e., no endoproteolysis or carboxyl methylation). It is difficult to understand why the –VYM mutant exhibited defective postisoprenylation processing, given that both X$_1$ and X$_2$ were hydrophobic amino acids.

Heilmeyer and co-workers (*37*) have demonstrated that the α and β subunits of phosphorylase kinase, which terminate in CAMQ and CLVS, respectively, are farnesylated but not further processed in muscle tissue. The absence of endoproteolytic processing in these proteins is mysterious, especially because they contain hydrophobic amino acids in the X$_1$ and X$_2$ positions. Jang and Gelb (*29*) analyzed the hydrolysis of a radiolabeled CLVS peptide in an *in vitro* system and found that it was a good substrate for the rat microsomal enzyme activity.

Ma and Rando (*38*) also explored the endoproteolysis of isoprenylated peptide *N*-acetyl-(*S*-farnesyl-C)-RPQ and *N*-acetyl-(*S*-geranylgeranyl-C)-RPQ modeled after an isoprenylated delta virus CXXX protein. This pep-

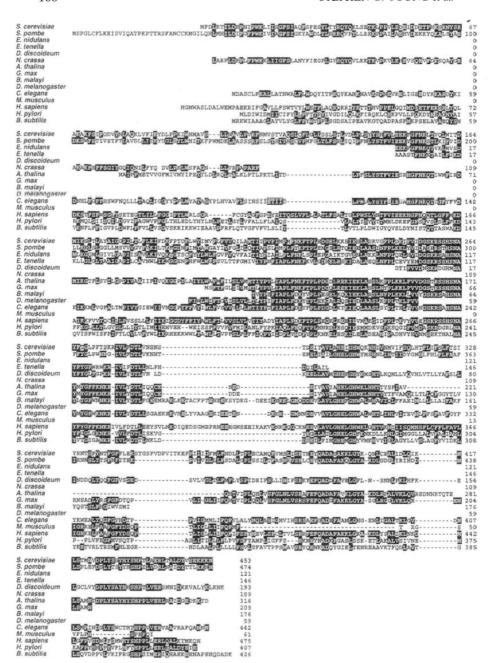

tide was referred to as a "non-CaaX" peptide because it has a nonaliphatic amino acid, arginine, in the X_1 position. They found that liver microsomes released an intact –RPQ peptide from the farnesylated substrate, and that none of the standard group-specific protease inhibitors (e.g., antipain, aprotinin, bestatin, chymostatin, EDTA, leupeptin, pepstatin, o-phenanthroline) affected the reaction. This result was similar to the properties of the endoprotease that processed peptides with conventional CXXX sequences (26, 28). However, the farnesylcysteine-based peptide inhibitors (e.g., RPI), which are potent inhibitors of the endoproteolysis of typical CXXX peptides, such as N-acetyl-(S-farnesyl-C)-VIM, did not block the hydrolysis of N-acetyl-(S-farnesyl-C)-RPQ or N-acetyl-(S-geranylgeranyl-C)-RPQ. This result raised the possibility of an additional endoprotease activity that participates in the hydrolysis of isoprenylated CXXX peptides with hydrophilic amino acids.

III. Identification of Two Yeast Genes, AFC1 (STE24) and RCE1, Involved in Endoproteolytic Processing of Isoprenylated CXXX Proteins

The fact that the "CXXX endoprotease" was refractory to biochemical purification prompted interest in a genetic screen for the responsible gene. Yeast defective in the farnesylation or methyl esterification of **a**-factor are sterile, and it seemed likely that a defect in the middle step of **a**-factor processing, the endoproteolysis step, would also produce sterility. Because methods for identifying sterile yeast mutants were well established, it might reasonably have been predicted that the identification of "**a**-factor endopro-

---

FIG. 1. Amino acid sequence alignment of yeast *AFC1* (*STE24*) and orthologs in other species. GenBank accession numbers for the full-length sequences: human, AAC68866; *S. cerevisiae*, P47154; *H. pylori*, AAD06444; *C. elegans*, CAB03839; *S. pombe*, Q10071; *A. thaliana* AAB61028. GenBank EST sequences: *D. melanogaster*, AA567990; *B. malayi*, AA585633; *E. tenella*, A1757461; *E. nidulans*, AA965341; mouse, AA498259; *G. max*, AI759796; *D. discoideum*, AUO37739; *N. crassa*, AI330202; *B. subtilis*, NC000964. The HEXXH sequence (residues 297–301 in the *S. cerevisiae* sequence) is conserved in all species. Protein sequences were aligned with Macvector 6.5, using a CLUSTALW alignment. Pairwise alignment was performed with a BLOSUM30 matrix, and multiple alignment was performed with a BLOSUM series matrix. The resulting alignment output was saved as a Word-98 document, and additional identities were shaded. The document was then imported into Illustrator.

tease" would have been a trivial task. Such optimism would, of course, have rested on the existence of a single CXXX endoprotease gene.

As it turned out, two yeast genes are involved in the removal of the C-terminal three amino acids from **a**-factor, and a genetic approach to the identification of responsible genes was vexatious. However, Boyartchuk and co-workers (*13*) overcame the obstacles and identified two genes, *RCE1* and *AFC1*, that are required for the endoproteolytic processing of **a**-factor as well as at least one other CXXX protein, Ras2p. Afc1p (for **a**-factor convertase) is a zinc protease that participates in the endoproteolytic processing of **a**-factor, while Rce1p (for Ras and **a**-factor-converting enzyme) participates in the processing of both Ras2p and **a**-factor. As outlined below (see Section IV), *AFC1* has also been identified as *STE24*, a gene involved in the N-terminal processing of **a**-factor (*39*). In this chapter, to avoid confusion, we have frequently designated the gene *AFC1* (*STE24*) [or *STE24* (*AFC1*)].

The key to discovering *RCE1* and *AFC1* rested on the identification of an **a**-factor substrate that was recognized by only one of the two endoproteases. Heilmeyer and co-workers (*37*) had shown that the α subunit of rabbit muscle glycogen phosphorylase kinase (a CXXX protein that terminates in –AMQ) is farnesylated but not further processed. Boyartchuk and colleagues (*13, 33*) produced farnesylated peptides that terminate in –AMQ and found that they were not endoproteolytically processed by *o*-phenanthroline-treated yeast microsomes. However, when the –AMQ sequence was incorporated into an **a**-factor (*MFA1*) construct and transformed into yeast, biologically active **a**-factor was produced and secreted! That observation suggested the possibility of more than one CXXX endoprotease in yeast and provided the basis for a sensitized genetic selection for CXXX endoprotease mutants. Yeast expressing –AMQ **a**-factor were mutagenized, and an autocrine arrest selection strategy was used to isolate sterile mutants (*13*). This approach resulted in the identification of a sterile yeast mutant (*afc1*) that was defective in processing the –AMQ form of **a**-factor. Of note, the *afc1* mutation had minimal effects on pheromone production in yeast that expressed wild-type **a**-factor, suggesting the existence of a second enzymatic activity capable of processing **a**-factor.

The *AFC1* gene was cloned by complementation (*13*). The gene specifies a 453-amino acid protein (Fig. 1) with multiple predicted transmembrane domains indicative of a polytopic integral membrane protein (Fig. 2A). At amino acids 297–301, Afc1p exhibits a perfect match with the HEXXH (H, His; E, Glu) motif of a group of zinc-dependent metalloproteases and shares other sequence similarities with neutral zinc metalloproteases (*13*). Mutating either of the conserved histidines in the HEXXH domain blocks

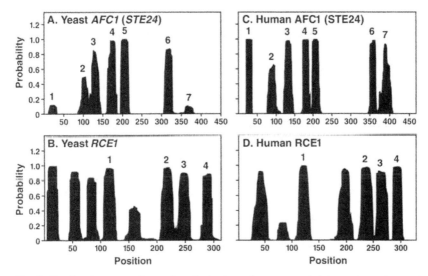

FIG. 2. Predicted transmembrane domains in the endoproteases. (A) Yeast Afc1p (Ste24p); (B) yeast Rce1p; (C) human *AFC1* (*STE24*); (D) human RCE1. Predicted transmembrane domains were determined with the TMHMM transmembrane domain analysis program (http://genome.cbs.dtu.dk/services/TMHMM-1.0/). To aid in the alignment, potential transmembrane domains with yeast/human sequence identities are indicated by numbers 1–7 for Afc1p and 1–4 for Rce1p.

the ability of *AFC1* to complement the mating defect of *AFC1*-deficient yeast (*afc1Δ*) expressing the –AMQ form of **a**-factor (*13*).

Yeast null mutants for *AFC1* produced reduced amounts of wild-type **a**-factor but were not sterile, indicating the existence of another gene capable of processing **a**-factor. Membranes from *afc1Δ* yeast exhibited a slight but reproducible decrease in their capacity to remove the last three amino acids from a farnesylated **a**-factor synthetic peptide. This slight decrease in endoprotease activity in the *afc1Δ* yeast was similar to the modest decrease observed by treating wild-type yeast membranes with *o*-phenanthroline. The residual CXXX endoprotease activity in the *afc1Δ* yeast membranes was insensitive to *o*-phenanthroline, suggesting that the remaining endoprotease activity was not zinc dependent. The properties of yeast *AFC1* (*STE24*) are summarized in Table II.

There were no *AFC1* (*STE24*) homologs in the yeast genome, indicating that the residual **a**-factor processing in *afc1Δ* yeast must have been due to a structurally distinct gene. To isolate the other CXXX endoprotease, *afc1Δ* yeast were mutagenized and screened for mutations that blocked the residual production of **a**-factor. In addition, yeast 2μ libraries were screened for plasmids that at high copy would partially restore **a**-factor production

TABLE II

PROPERTIES OF YEAST *AFC1* (*STE24*)-ENCODED PROTEINS

Cleaves C-terminal three amino acids (–VIA) from precursor to yeast mating pheromone **a**-factor. No other substrates yet reported

Cannot cleave –XXX from mutant mating factor terminating in sequence –TLM (C-terminal three amino acids from *STE18*-encoded γ subunit of heterotrimeric G protein), but can cleave C-terminal –XXX from mutant mating factor terminating in sequence –AMQ (C-terminal three amino acids from α subunit of rabbit muscle glycogen phosphorylase kinase)

Cleaves N-terminal seven amino acids from precursor to yeast mating pheromone **a**-factor

Enzymatic activity is zinc dependent and is blocked by *o*-phenanthroline

Located within ER by cell fractionation studies

Null mutant reduced mating efficiency in α cells. Overexpression of Ax11p suppresses mating defect

453-amino acid protein; 36% identity with human AFC1 amino acid sequence

Multiple predicted transmembrane domains

An HEXXH domain, which is shared with many other zinc proteases, is critical for enzymatic activity

Contains C-terminal dilysine motif, but that sequence is not required for ER localization

in *afc1*Δ yeast expressing –AMQ **a**-factor. Both approaches led to the identification of a second endoprotease gene, *RCE1*. *RCE1* encodes a 329-amino acid protein (Fig. 3) that is predicted to have multiple transmembrane domains indicative of an integral membrane protein (Fig. 2B). Unlike Afc1p, an analysis of the Rce1p protein sequence did not reveal sequences characteristic of any of the defined classes of proteases. However, there appeared to be remote similarities with the type IIb signal peptidase, which cleaves signal sequences from proteins containing nearby lipid modifications (*13*). Key properties of yeast *RCE1* are listed in Table III.

Yeast lacking the *RCE1* gene (*rce1*Δ) are viable, with a modest decrease in **a**-factor production, and there was little effect on mating efficiency. In these respects, the phenotypes of *rce1*Δ and *afc1*Δ were similar. The membranes from *rce1*Δ yeast manifested moderately decreased CXXX endoprotease activity in *in vitro* assays involving farnesylated **a**-factor peptides—more so than with the *afc1*Δ yeast. Of note, the residual CXXX endoprotease activity in *rce1*Δ yeast was sensitive to *o*-phenanthroline, consistent with Afc1p being a zinc protease. Yeast lacking both *RCE1* and *AFC1* (*afc1*Δ*rce1*Δ) are completely defective in the production of mature **a**-factor and are sterile, indicating that both *RCE1* and *AFC1* (*STE24*) process **a**-factor.

In contrast to the redundant functions of *RCE1* and *AFC1* in **a**-factor processing, genetic evidence indicates that *RCE1*, but not *AFC1*, is critical for the processing of Ras proteins, and that *RCE1* deficiency attenuates

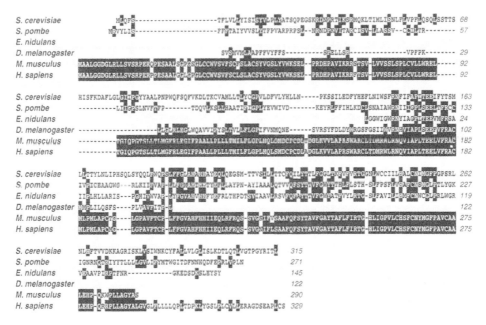

FIG. 3. Amino acid sequence alignment of yeast RCE1 and orthologs in other species. GenBank accession numbers: *E. nidulans,* Al210465; *D. melanogaster,* Al514743; human, NP005124; *S. cerevisiae,* Q03530; *S. pombe,* CAA22596; mouse, unpublished data (E. Kim and S. Young). Protein sequences were aligned as described in the legend to Fig. 1.

the activity of the Ras proteins (*13*). In yeast, Ras2p mutations that impair GTP hydrolysis (e.g., $RAS2^{Val19}$) induce sensitivity to heat shock and starvation (*13*). In *rce1*Δ yeast, the heat shock sensitivity elicited by a mutationally activated Ras2p is suppressed by a factor of 100, compared with wild-type yeast. The *afc1*Δ mutant had little if any effect on heat shock sensitivity, while the *afc1*Δ*rce1*Δ mutant was indistinguishable from that of the *rce1*Δ mutant. These experiments indicated that the endoproteolytic processing of Ras2p by Rce1p attenuated Ras function. However, the *rce1*Δ mutant almost certainly did not completely block all Ras functions, given that *rce1*Δ yeast are viable (*13*) and yeast lacking the Ras proteins are not (*40*).

To assess further the functional importance of *RCE1* in Ras function, Boyartchuk *et al.* (*13*) studied the effects of *RCE1* deficiency in yeast with a temperature-sensitive *RAS2* allele, *ras2-23*. Yeast harboring a *ras2-23* allele as their only Ras allele grow well at 30°C but exhibit a reduced growth rate at 34°C. The deletion of *RCE1* completely blocked the growth of the *ras2-23* mutant at 34°C (*13*). The deletion of *AFC1* had no effect on the growth of the *ras2-23* mutant.

TABLE III

Properties of Yeast RCE1

Cleaves C-terminal three amino acids (–VIA) from precursor to yeast mating pheromone
   a-factor. Required for endoproteolytic processing of Ras2p. No other substrates reported
Insensitive to inhibition by *o*-phenanthroline
Extremely low activity against nonfarnesylated proteins
Cannot cleave –XXX from a mutant mating factor terminating in sequence –AMQ (C-terminal three amino acids from α subunit of rabbit muscle glycogen phosphorylase ki-
   nase), but can cleave C-terminal –XXX from mutant mating factor terminating in se-
   quence –TLM (C-terminal three amino acids from *STE18*-encoded γ subunit of
   heterotrimeric G protein)
Unlike Afc1p (Ste24p), Rce1p does not cleave N terminus of **a**-factor
Appears to be located almost exclusively in ER
315-amino acid protein; 27.6% identical with the human RCE1 sequence
Multiple predicted transmembrane domains
Null mutant does not affect mating of α cells but causes slight decrease in **a**-factor production
Null mutant results in mislocalization of GFP–Ras2p fusion protein, with less reaching the
   plasma membrane. Null mutant largely blocks the phenotypes elicited by mutationally activated Ras2p

---

   The effect of endoproteolysis on Ras2p localization within yeast was
tested with a green fluorescent protein (GFP)–*RAS2* fusion construct in
which GFP was fused to the N terminus of Ras2p, leaving the CXXX motif
of Ras2p intact and available for processing. In wild-type yeast, fluorescence
microscopy revealed that the fusion construct was targeted to the periphery
of the cells (i.e., the plasma membrane). In keeping with that result, cell
fractionation revealed that the wild-type fusion protein sedimented with
the P100 membrane fraction. In contrast, the fusion protein in the *rce1*Δ cells
was widely dispersed within the cell, with the majority of the fluorescence
appearing inside the cell (i.e., either in the cytosol or associated with an
internal membrane compartment). Thus, the endoproteolytic processing of
the Ras proteins by Rce1p appears to contribute importantly to their proper
localization within the cell.
   RCE1 has few sequence similarities with *AFC1* or other proteases. Thus,
it is reasonable to question whether Rce1p is truly a protease or simply a
cofactor for another enzyme. Although this issue has not been settled
definitively, the evidence supports the idea that it is a protease. The levels
of endoprotease activity in yeast correlated with the amount of *RCE1*
expression (*13*). Moreover, as described below, overexpression of the human ortholog of *RCE1* in insect cells produced an enormous increase in
endoprotease activity (*15*).
   The studies by Boyartchuk and co-workers (*13, 41*) revealed that both

AFC1 and *RCE1* are active in processing **a**-factor, while only *RCE1* appears to be involved in Ras metabolism. Their *in vitro* endoproteolysis assay, which assessed the cleavage of a farnesylated **a**-factor peptide, indicated that ~35% of the total CXXX endoprotease activity in yeast membranes is due to *AFC1,* with *RCE1* accounting for the remainder. However, any conclusion regarding the extent to which Rce1p and Afc1p participate in "total endoproteolysis activity" should be viewed with caution, simply because the CXXX endoprotease activities of Afc1p and Rce1p are almost certainly dependent on the sequence of the CXXX peptide or protein used in the assay *(41)*.

There may be as many as 86 CXXX proteins in the yeast genome, as judged by search of the yeast protein database (http://genome-www2. stanford.edu/cgi-bin/SGD/PATMATCH/nph-patmatch?class = pept). A critical issue in the enzymology of Rce1p and Afc1p is to define their unique and overlapping protein substrates. It is unclear whether the majority of the yeast CXXX proteins are processed by a combination of both Rce1p and Afc1p, or whether many proteins are processed solely by Afc1p or solely by Rce1p. However, the tools are clearly in hand to address these issues.

IV. Role for *AFC1* (*STE24*) in Proteolytic Processing of N Terminus of **a**-Factor in *Saccharomyces cerevisiae*

The *S. cerevisiae* mating pheromone **a**-factor, a 12-amino acid farnesylated peptide, is the product of functionally redundant genes, *MFA1* and *MFA2*. These two genes encode precursor proteins of 36 and 38 amino acids, and the production of mature **a**-factor depends on a series of processing steps: isoprenylation, endoproteolytic removal of the C-terminal three amino acids (–VIA), carboxyl methylation of the isoprenylcysteine, and the proteolytic removal of the N-terminal 21 amino acids (following –KKDN). It has become clear that the removal of the N-terminal 21 residues involves two sequential steps: the initial removal of the first 7 amino acids (following –STAT), followed by the removal of 14 additional amino acids (following –KKDN) *(42–44)*. The first of these amino-terminal processing steps has been shown to be carried out by *STE24* *(39)*, while the second is carried out by *AXL1*- and *STE23*-encoded proteins. Remarkably, *STE24* was recognized to be identical to *AFC1*; thus, in addition to its role in cleaving the –XXX from the **a**-factor precursor, Ste24p (Afc1p) plays a distinct role in the N-terminal processing of **a**-factor. Figure 4 illustrates the five steps in the biogenesis of mature **a**-factor, showing the two distinct roles of Ste24p (Afc1p).

In 1995, Adames *et al.* *(45)* used a genetic screen for reduced mating

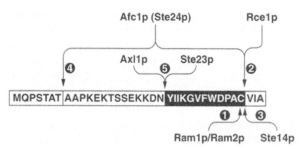

Fig. 4. A schematic illustrating the two distinct roles of Afc1p (Ste24p) in **a**-factor metabolism. First, the **a**-factor precursor is farnesylated by Ram1p/Ram2p. Second, the −VIA tripeptide is cleaved by both Afc1p (Ste24p) and Rce1p. Third, the isoprenylcysteine is carboxyl methylated. Fourth, the N-terminal seven amino acids are cleaved by Afc1p. Fifth, an additional 14 amino acids are cleaved by Axl1p and Ste23p. In the absence of Afc1p (Ste24p) and Rce1p, P0* accumulates. P0* is farnesylated but not further processed (e.g., no CXXX endoproteolysis and no carboxyl methylation of the isoprenylcysteine). P1, which is fully modified at the C terminus (i.e., having undergone isoprenylation, −XXX endoproteolysis, and carboxyl methylation), accumulates in the absence of Afc1p (Ste24p). P2, which lacks the N-terminal seven amino acids, accumulates in the absence of Axl1p and Ste23p. [Reproduced, with permission, from Ashby, M. N. (1998). *Curr. Opin. Lipidol.* **9**, 99–102.]

efficiency in *S. cerevisiae* to identify *AXL1* (or *STE22*). Axl1p has significant sequence identity to the insulin-degrading enzymes and belongs to a family of metalloproteases with a preference for small peptide substrates. In *axl1*Δ yeast, pulse–chase/metabolic labeling studies of **a**-factor formation revealed the early appearance of an **a**-factor intermediate, P1, that was fully modified at the C terminus (i.e., an **a**-factor intermediate that had undergone isoprenylation, −XXX endoproteolysis, and carboxyl methylation). P1 was converted to another **a**-factor intermediate, P2, which lacks the N-terminal seven amino acids. P2 accumulated in *axl1*Δ yeast, with only small amounts undergoing further processing to mature **a**-factor (*42, 44*). In wild-type yeast, P1 is converted to P2, but the P2 is then processed to mature **a**-factor (by the removal of 14 additional amino acids). These experiments, together with those from the laboratory of S. Michaelis (*42, 44*), revealed two important points about **a**-factor processing. First, the three sequential modifications of the C terminus of **a**-factor occur early and are completed before the proteolytic processing of the N terminus. Second, the Axl1p-mediated cleavage of the **a**-factor precursor represents the second of two N-terminal proteolytic processing steps. The removal of the initial seven amino acids occurs before the Axl1p proteolysis step.

The characterization and kinetic analysis of **a**-factor intermediates have been more completely described in a series of experiments by Chen *et al.* (*44*). A search of the *S. cerevisiae* sequence database revealed a gene homologous to *AXL1*, designated *STE23*, which also participates in the second N-terminal proteolysis step (*45*). Yeast mutants lacking *AXL1* (*axl1*Δ) retain a markedly reduced mating efficiency while mutants lacking both *AXL1* and *STE23* (*axl1*Δ*ste23*Δ) are sterile.

The identification of *AXL1* and *STE23* raised a key question: What protease cleaves the amino-terminal seven amino acids from the **a**-factor precursor? Multiple genetic screens for sterile yeast mutants in several laboratories failed to identify the protease. However, in 1997, 5 months before the report of the CXXX endoproteases by Boyartchuk *et al.* (*13*), Fujimura-Kamada *et al.* (*39*) solved the riddle. Using a sensitized genetic screen for sterile yeast, they identified a new gene, *STE24*, that was essential for the removal of the N-terminal seven amino acids from the **a**-factor precursor. *STE24* had multiple predicted membrane-spanning domains and shared important sequence similarities with a variety of different zinc-dependent proteases from diverse organisms ranging from bacteria to humans.

Yeast lacking *STE24* (*ste24*Δ) exhibited reduced mating efficiency but were not sterile. Pulse–chase experiments of **a**-factor maturation in *ste24*Δ yeast revealed a striking defect in the conversion of P1 (the intermediate that is fully modified at the C terminus) to P2 (the intermediate lacking the N-terminal seven amino acids). Why aren't the *ste24*Δ yeast sterile? The likely explanation is that some Axl1p/Ste23p-mediated cleavage of **a**-factor occurs, albeit at reduced efficiency, in the absence of Ste24p (Afc1p). Fujimura-Kamada *et al.* (*39*) demonstrated that overexpression of *AXL1* suppressed the mating defect in *ste24*Δ yeast, suggesting that the Ste24p-mediated cleavage of the first seven amino acids greatly facilitates, but is not absolutely required for, the subsequent proteolytic cleavage between amino acids 21 and 22.

With the publication of the CXXX endoprotease paper by Boyartchuk and co-workers (*13*), it was evident that the sequence of *AFC1* was identical to that of *STE24*. Thus, using a screen for reduced mating efficiency, the Michaelis group had identified *STE24* as a gene essential for the N-terminal processing of **a**-factor (*39*). Within months, Boyartchuk, Ashby, and Rine (*13*) used a screen for sterile mutants to identify *AFC1*, and concluded that it was one of two genes responsible for the C-terminal processing of **a**-factor. Thus, two groups applying similar genetic screens identified the same gene and reached different conclusions regarding its function. In a pair of follow-up papers (*14, 16*), this seeming discrepancy has been resolved. Both groups went on to show that Afc1p (Ste24p) is involved in two distinct steps of **a**-factor processing: the –XXX endoproteolytic cleavage, and the cleavage of the N-terminal seven amino acids.

Tam *et al.* (*16*) explored the dual roles of Ste24p in a-factor maturation by examining the processing of various *MFA1* truncations in yeast. Because conventional *MFA1* truncation mutants yield low levels of a-factor expression, they assembled ubiquitin–*MFA1* fusion constructs, which result in higher levels of expression. The basic strategy was to rely on the endogenous ubiquitin proteases within yeast to cleave the N-terminal ubiquitin sequences from the fusion proteins, releasing truncated a-factor proteins. When a ubiquitin fusion construct containing full-length *MFA1* protein (Ubi–P1) was transformed into *ste24Δmfa1Δ mfa2Δ* yeast, P1 accumulated, and only minute amounts of mature a-factor were produced. However, if the same yeast were transformed with ubiquitin fused to *MFA1* lacking the first seven amino acids (Ubi–P2), high levels of mature a-factor were produced and secreted. Thus, the removal of the first seven amino acids of a-factor obviated the requirement for Ste24p-mediated processing of the N terminus. As expected, *rce1Δste24Δmfa1Δmfa2Δ* yeast transformed with Ubi–P2 secreted no mature a-factor, because both CXXX endoproteases were absent. These experiments established that Ste24p has two distinct roles in a-factor processing and that the two roles could be uncoupled and characterized.

Tam and co-workers (*16*) went on to study a-factor biogenesis in wild-type, *ste24Δ*, *rce1Δ*, and *ste24Δrce1Δ* yeast with pulse–chase experiments. In wild-type yeast, P1 (the intermediate with only C-terminal modifications) appeared quickly and was rapidly converted to P2 and then to mature a-factor. As predicted, *ste24Δ* yeast accumulated P1, and there was minimal conversion of P1 to mature a-factor. In *ste24Δrce1Δ* yeast, a novel a-factor intermediate, designated P0*, accumulated. P0* is farnesylated but not further processed (e.g., no CXXX endoproteolysis and no carboxyl methylation of the isoprenylcysteine). Interestingly, P0* appeared transiently in *rce1Δ* yeast but was rapidly converted to P2 and mature a-factor, without the appearance of the P1 intermediate. Thus, in the absence of Rce1p, Ste24p appears to cleave the –XXX and the seven N-terminal amino acids in rapid succession, seemingly without releasing the intermediate (P1). The *rce1Δ* yeast secreted approximately half-normal amounts of mature a-factor.

Boyartchuk and Rine (*41*) likewise used ubiquitin–*MFA1* fusion constructs to dissect the dual roles of Afc1p (Ste24p) in a-factor maturation. In addition, they analyzed ubiquitin–*MFA1* fusion constructs in which the C-terminal –XXX sequence was mutated, so that the C-terminal cleavage reaction was carried out only by Afc1p or only by Rce1p, but not both proteins. Both Rce1p and Afc1p carry out the C-terminal processing of fusion proteins terminating in the wild-type *MFA1* sequence (–VIA). However, changing the C-terminal sequence to –AMQ (the C-terminal three amino acids from the α subunit of rabbit muscle glycogen phosphorylase kinase) eliminated C-terminal processing by Rce1p, while changing the sequence to –TLM (the C-terminal three amino acids from the *STE18*-encoded γ subunit of a heterotrimeric G protein) eliminated C-terminal processing by Afc1p.

By comparing mating efficiencies of *mfa1Δmfa2Δ*, *mfa1Δmfa2Δafc1*, and *mfa1Δmfa2Δafc1Δrce1Δ* yeast that had been transformed with various ubiquitin–*MFA1* fusion constructs, they demonstrated that Afc1p plays a role in the N-terminal processing of **a**-factor, in addition to its role in CXXX endoproteolysis. For example, a full-length –TLM **a**-factor construct produced less mature **a**-factor in *afc1Δ* yeast than in wild-type yeast. This result is likely explained by defective N-terminal processing in the *afc1Δ* cells. It cannot be explained by differences in C-terminal processing, since the –TLM construct is processed exclusively by Rce1p. Boyartchuk and Rine (*41*) observed less **a**-factor production with a ubiquitin fusion construct lacking the seven N-terminal amino acids than with a construct containing the entire **a**-factor protein, leading them to conclude that the N-terminal Afc1p-mediated cleavage step increases the efficiency of mature **a**-factor production.

The finding that the *AFC1* (*STE24*) gene product participates in a second proteolytic cleavage reaction in **a**-factor raises the possibility that this gene product could play dual processing roles for other yeast CXXX proteins, or maybe even that it has proteolytic processing roles for non-CXXX proteins. Thus far, however, no other substrates for Afc1p (Ste24p) have been reported.

V. Characterization of Mammalian Orthologs for *AFC1* (*STE24*) and *RCE1*

A. CHARACTERIZATION OF MAMMALIAN RCE1

When yeast *RCE1* was identified and reported, related sequences already existed in the human and mouse EST databases. Mammalian orthologs were cloned by Kim and Young, by Otto and Casey, and by Ashby. Ultimately, Kim, Young, and co-workers collaborated with Otto and Casey to define the roles of the mammalian orthologs in the endoproteolytic processing of mammalian CXXX proteins (*14, 15*). Kim, Young, and co-workers assessed the role of the mouse *Rce1* gene in the processing of Ras proteins by knocking out *Rce1* in mice (*14*), while Otto, Casey, and co-workers focused on the cloning, expression, and characterization of human

TABLE IV

PROPERTIES OF MAMMALIAN RCE1

RCE1 expressed in Sf9 cells processes farnesylated K-Ras, farnesylated H-Ras, farnesylated N-Ras, the farnesylated heterotrimeric $G\gamma_1$ subunit, geranylgeranylated K-Ras, and geranylgeranyl-Rap1B. There is no processing of these substrates by membranes from mouse fibroblasts lacking *Rce1* gene expression

250-fold greater specificity of the enzyme for isoprenylated peptides than for nonisoprenylated peptides

RPI, a reduced farnesyl-peptide analog, is an effective inhibitor of human RCE1, with an IC_{50} of approximately 5 nM

Located exclusively in membrane fractions of mammalian cells

RCE1 is expressed in all mouse and human tissues, including early-stage embryos, with highest levels in the heart and skeletal muscle

329-amino acid protein; 27.6% identity with *S. cerevisiae* Rce1p amino acid sequence; 42% identity and 68% similarity between residues 166 and 270

Protein has multiple predicted transmembrane domains

Achieving high levels of human Rce1 activity in Sf9 cells required deletion of N-terminal 22 amino acids of protein

Mice lacking *Rce1* expression (*Rce1$^{-/-}$*) die during embryonic development, beginning on day 15. No gross defects in organogenesis

Endoproteolytic processing of Ras proteins is absent in *Rce1$^{-/-}$* fibroblasts and embryos

Rce1 deficiency in fibroblasts produces a gross mislocalization of a GFP–K-Ras fusion protein within the cell

Membranes from *Rce1$^{-/-}$* fibroblasts cannot carry out C-terminal processing of farnesylated K-Ras, farnesylated H-Ras, farnesylated N-Ras, farnesylated heterotrimeric $G\gamma_1$ subunit, geranylgeranylated K-Ras, or geranylgeranyl-Rap1B

RCE1 in cultured cells (*15*). Key properties of human RCE1 and its mouse ortholog (Rce1) are listed in Table IV.

1. Expression and Characterization of Human RCE1

Using the yeast *RCE1* sequence, Otto and co-workers identified an apparent ortholog in a human EST database and then cloned the corresponding cDNA (*15*). Figure 3 shows the alignment of the human RCE1 protein, along with the RCE1 proteins from several other species. Like yeast Rce1p, the human protein is predicted to contain multiple transmembrane domains (Fig. 2D). Northern blots revealed that RCE1 is expressed ubiquitously, with the highest levels of expression in placenta, heart, and skeletal muscle. Although the predominant transcript length was 1.3 kb, larger transcripts were also observed, probably reflecting alternatively or aberrantly spliced transcripts. Heterogeneity in transcript length was even more obvious in mouse tissues (*15*). The significance of the larger transcripts, if any, is unknown.

High levels of yeast and human RCE1 proteins were expressed in Sf9

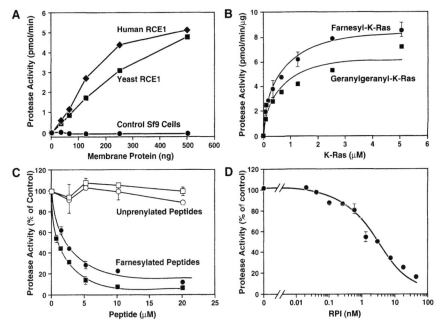

Fig. 5. Characterization of human RCE1 activity. (A) Yeast and human RCE1 gene products possess isoprenyl protein protease activity. Protease assays were conducted with increasing amounts of membranes prepared from noninfected Sf9 cells (●), or from Sf9 cells infected with recombinant baculoviruses containing the cDNA for either yeast (■) or human RCE1 (◆). Purified farnesyl-K-Ras (2 μM) was used as substrate. Isoprenyl protein protease activity was assessed with a coupled proteolysis/methylation assay (15). (B) Kinetics of proteolysis for farnesyl-K-Ras and geranylgeranyl-K-Ras. RCE1 membrane protein (15 ng) was incubated with increasing levels of farnesyl-K-Ras (●) and geranylgeranyl-K-Ras (■). (C) Competition of RCE1 activity by isoprenylated peptides. Isoprenylated peptides [farnesyl-CVIM (●) and farnesyl-GSPCVLM (■)] or nonisoprenylated peptides [CVIM (○) and GSPCVLM (□)] were included in the standard protease reaction mixture with 0.5 μM purified farnesyl-K-Ras. Activity is expressed as the percentage of that observed for an untreated control. (D) Inhibition of RCE1 activity by the farnesyl-peptide analog RPI. RPI was diluted in 10% DMSO, and added to the standard protease reaction mixture containing 0.5 μM purified farnesyl K-Ras. Activity is expressed as the percentage of that observed for an untreated control. [Reproduced, with permission, from Otto, J. C., Kim, E., Young, S. G., and Casey, P. J. (1999). *J. Biol. Chem.* **274,** 8379–8382.]

(*Spodoptera frugiperda* ovary) insect cells with recombinant baculoviruses (Fig. 5A). Of note, high-level expression of the human RCE1 protein could not be achieved without first deleting the first 22 amino acids of the protein. To assess the activity of yeast and human RCE1 proteins, extracts from the Sf9 cells were mixed with recombinant isoprenylated CXXX proteins. In this assay, the recombinant RCE1 proteins cleave the –XXX from the isoprenylated proteins, rendering the proteins susceptible to carboxyl methylation with recombinant Ste14p and S-adenosyl[*methyl*-^{3}H]methionine. Using this coupled endoproteolysis/methylation assay, Otto and co-workers demonstrated that human RCE1 processes farnesylated K-Ras, farnesylated H-Ras, farnesylated N-Ras, the farnesylated heterotrimeric G protein $G\gamma_1$ subunit, geranylgeranylated K-Ras, and geranylgeranyl-Rap1B. Both farnesylated and geranylgeranylated K-Ras exhibited a K_m value of approximately 0.5 μM and similar k_{cat} values (Fig. 5B). Isoprenylated CXXX peptides, but not nonisoprenylated peptides, were able to compete for the processing of isoprenylated K-Ras (Fig. 5C). In addition, a previously identified inhibitor of endoproteolytic processing, RPI (a reduced farnesylpeptide analog), was an extremely effective inhibitor of RCE1, with an IC_{50} of approximately 5 nM (Fig. 5D).

A perplexing aspect of the Sf9 expression system is that it required the truncation of the N-terminal 22 amino acids of the human protein. This finding has not been explained. One possibility is that the presence of the N-terminal 22 amino acids simply reduces the absolute level of RCE1 protein expression in Sf9 cells. Alternatively, it is possible that the N terminus of RCE1 might normally bind to a "protein partner." When that protein partner is absent (as might be the case in Sf9 cells), the amino-terminal domain might cause RCE1 to misfold, diminishing its activity. In any case, the cloning of human RCE1 and its expression in cultured cells represent important advances, as they lay the foundation for purifying the enzyme, performing structure–function studies, defining the catalytic domain of the protein, and identifying additional CXXX protein substrates.

Multiple RCE1 orthologs now exist in the databases. Comparison of the amino acid sequences encoded by *S. cerevisiae RCE1* and its human and mouse orthologs reveals the greatest sequence similarity in the C-terminal half of the protein (Fig. 3). When the predicted transmembrane helical regions of the human and yeast forms are compared (Fig. 2B and D) with the aligned sequences (Fig. 3), only four of the predicted transmembrane domains appear to match with aligned sequences—three near the C terminus and one in the central region of the molecule (designated 1, 2, 3, and 4 in Fig. 2, B and D).

2. *Assessing Physiologic Role of Rce1 Gene Product in Mice*

Kim, Young, and co-workers (*14*) sought to assess the physiologic role of *Rce1* in higher organisms by producing and characterizing *Rce1* knockout mice. The yeast *RCE1* sequences were used to identify potential orthologs within the mouse EST databases, and those were used to clone most of the mouse cDNA. The mouse and human proteins are highly conserved throughout their lengths, with 95% of the amino acid residues being identi-

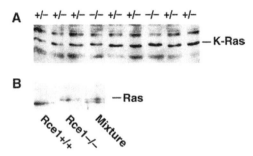

FIG. 6. Abnormal electrophoretic mobility of Ras proteins in lysates from Rce1 [?] embryos and $Rce1^{-/-}$ primary embryonic fibroblasts. (A) Western blot of lysates of $Rce1^{+/+}$, $Rce1^{+/-}$, and $Rce1^{-/-}$ embryos. (B) A Western blot of an equal mixture of $Rce1^{+/+}$ and $Rce1^{-/-}$ embryo lysates with antibody Ab-4, demonstrating the expected doublet band. [Reproduced, with permission, from Kim, E., Ambroziak, P., Otto, J. C., Taylor, B., Ashby, M., Shannon, K., Casey, P. J., and Young, S. G. (1999). *J. Biol. Chem.* **274**, 8383–8390.]

cal. Southern blots of mouse genomic DNA with one of the EST probes suggested the presence of a single copy of *Rce1* in the genome. Strain 129/Sv mouse genomic clones were identified and used to construct a sequence-replacement gene-targeting vector. *Rce1*-deficient mice were generated by standard gene-targeting techniques (*14*).

Heterozygous knockout mice ($Rce1^{+/-}$) were healthy and fertile. However, virtually all of the homozygous *Rce1* knockout embryos ($Rce1^{-/-}$) died, beginning on about embryonic day 15 (E15). $Rce1^{-/-}$ mice were born only rarely, and they invariably died within 1 week. The reason for the embryonic lethality remains mysterious. Inspection of the $Rce1^{-/-}$, $Rce1^{+/+}$, and $Rce1^{+/-}$ embryos after E15.5 revealed no consistent differences in morphology, organogenesis, stage of development, color, or size. The knockout mice did not appear to die from a defect in hematopoiesis. The $Rce1^{-/-}$ embryos were pink, and histologic analysis revealed abundant numbers of mature erythrocytes in their blood vessels. Moreover, lethally irradiated mice were successfully rescued with hematopoietic stem cells from the livers of $Rce1^{-/-}$ embryos (*14*).

Fibroblasts were cultured from the $Rce1^{-/-}$ embryos. Despite an absence of Ras endoproteolytic processing, the growth rates of $Rce1^{-/-}$ and $Rce1^{+/+}$ fibroblasts did not appear to be grossly different, at least when they were grown in medium containing a high concentration of fetal bovine serum.

To determine if mouse *Rce1* participates in the processing of Ras proteins, the electrophoretic mobilities of the Ras proteins from $Rce1^{-/-}$, $Rce1^{+/+}$, and $Rce1^{+/-}$ embryo lysates were analyzed on SDS–polyacrylamide gels (Fig. 6A). The electrophoretic mobility of the Ras proteins from $Rce1^{-/-}$

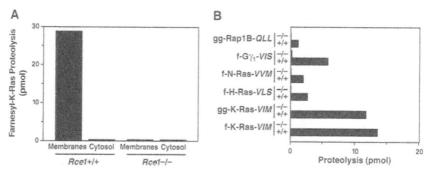

FIG. 7. Direct assays of the ability of $Rce1^{-/-}$ fibroblasts to endoproteolytically process isoprenylated CXXX proteins. (A) An assay of the ability of cytosolic and membrane fractions from $Rce1^{-/-}$ fibroblasts to endoproteolytically process farnesyl-K-Ras. Membrane and cytosolic fractions (80 μg) from $Rce1^{+/+}$ and $Rce1^{-/-}$ fibroblasts were incubated with farnesyl-K-Ras (2 μM). The processed farnesyl-K-Ras was then methylated with 5.0 μM [^{3}H]AdoMet (1.5 Ci/mmol) and 20 μg of membranes from Sf9 insect cells expressing high concentrations of the yeast isoprenyl protein carboxyl methyltransferase encoded by *STE14*. [^{3}H]Methyl-isoprenylated proteins were collected on filters, and methylation was quantified by scintillation counting. Proteolysis is described as the number of picomoles of [^{3}H]methyl groups transferred to the isoprenylated protein. (B) Endoprotease activity in membranes from $Rce1^{+/+}$ and $Rce1^{-/-}$ fibroblasts against a panel of isoprenylated proteins. The isoprenyl group attached to proteins is indicated by an "f-" for farnesyl or a "gg-" for geranylgeranyl. The –XXX sequence for each protein is indicated in italics. The concentrations of farnesyl-K-Ras and geranylgeranyl-K-Ras were 1.0 μM. The approximate concentration of geranylgeranyl-Rap1B was 0.2 μM, and the approximate concentrations of farnesyl-H-Ras, farnesyl-N-Ras, and farnesyl-Gγ_1 were 1 μM. [Reproduced, with permission, from Kim, E., Ambroziak, P., Otto, J. C., Taylor, B., Ashby, M., Shannon, K., Casey, P. J., and Young, S. G. (1999). *J. Biol. Chem.* **274**, 8383–8390.]

fibroblasts and embryos was distinctly abnormal. In fact, Ras proteins with normal electrophoretic mobilities were undetectable. When equal amounts of lysates from $Rce1^{-/-}$ and $Rce1^{+/+}$ embryos were mixed and analyzed on an SDS–polyacrylamide gel, a doublet Ras band was observed (Fig. 6B).

The abnormal electrophoretic mobility of the Ras proteins strongly suggests that the *Rce1* product is responsible for Ras endoproteolytic processing. Because endoproteolysis is required for subsequent carboxyl methylation of the isoprenylcysteine, it would be assumed that this would not occur with Ras proteins from $Rce1^{-/-}$ fibroblasts. To test this prediction, $Rce1^{+/+}$ and $Rce1^{-/-}$ fibroblasts were metabolically labeled with L-[*methyl-*^{3}H]methionine, and the Ras proteins were then immunoprecipitated from the cells. As expected, the Ras proteins from the $Rce1^{+/+}$ cells contained an ^{3}H-labeled cysteine methyl ester, whereas Ras proteins from the $Rce1^{-/-}$ cells did not (*14*). Membranes from $Rce1^{-/-}$ fibroblasts also lacked the capacity to process farnesylated K-Ras in a coupled proteolysis/carboxyl

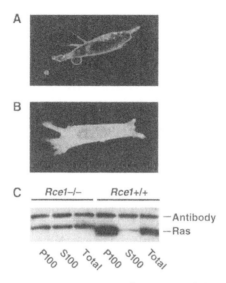

FIG. 8. Mislocalization of Ras proteins in $Rce1^{-/-}$ fibroblasts. (A) Fluorescence image of a GFP–K-Ras fusion protein in an $Rce1^{+/+}$ fibroblast, showing localization of the fusion protein at the plasma membrane. (B) Fluorescence image of a GFP–K-Ras fusion protein in an $Rce1^{-/-}$ fibroblast, showing localization of the fusion protein to the cytosol or internal membranes. (C) Mislocalization of Ras proteins as judged by cell fractionation. Cells were fractionated into cytosolic (S100) and membrane (P100) fractions by ultracentrifugation. The Ras proteins were immunoprecipitated from the S100 and P100 fractions, as well as from the total cellular lysates, and analyzed on a Western blot of an SDS–polyacrylamide gel. [Reproduced, with permission, from Kim, E., Ambroziak, P., Otto, J. C., Taylor, B., Ashby, M., Shannon, K., Casey, P. J., and Young, S. G. (1999). *J. Biol. Chem.* **274,** 8383–8390.]

methylation assay *(14, 15)* (Fig. 7A). The membranes from $Rce1^{-/-}$ fibroblasts lacked significant levels of processing activity for the farnesylated heterotrimeric $G\gamma_1$ subunit, farnesylated H-Ras, farnesylated N-Ras, geranylgeranylated K-Ras, and geranylgeranyl-Rap1B (Fig. 7B).

The Western blot studies indicated that Ras endoproteolytic processing was abolished in the *Rce1* knockout mice. To determine whether the loss of *Rce1* affected the intracellular localization of Ras, both immunofluorescence microscopy and cell fractionation studies were performed. An enhanced GFP–K-Ras fusion protein was transformed into $Rce1^{-/-}$ and $Rce1^{+/+}$ embryonic fibroblasts. In $Rce1^{+/+}$ cells, the fluorescence was localized to the plasma membrane (Fig. 8A). In contrast, most of the fluorescence in the $Rce1^{-/-}$ cells was cytosolic or associated with internal membranes (Fig. 8B). A substantial difference in the subcellular localization of Ras was also evident by immunoblot analysis of P100 and S100 fractions from $Rce1^{-/-}$ and $Rce1^{+/+}$ fibroblasts. Virtually all the Ras proteins in $Rce1^{+/+}$

cells were located in the P100 fraction (Fig. 8C), whereas a large fraction of the Ras proteins in $Rce1^{-/-}$ cells was in the S100 fraction (Fig. 8C).

The most reasonable interpretation of these results is that the mislocalization of the Ras proteins in the setting of $Rce1$ deficiency results from a failure to clip the C-terminal three amino acids from the proteins. The extent to which the mislocalization is due to the absence of isoprenylcysteine carboxyl methylation remains to be established. In the future, that issue could be clarified by examining the intracellular distribution of Ras proteins in mice lacking the isoprenylcysteine carboxyl methyltransferase. It is also possible that $Rce1$ deficiency impairs Ras palmitoylation. A significant part of the membrane association of the Ras proteins depends on palmitoylation (46–48), and it has been suggested that retardation in the endoproteolysis and methylation of Ras might affect the extent of Ras palmitoylation (49). It would not be particularly surprising if the absence of endoproteolysis and carboxyl methylation in $Rce1^{-/-}$ cells affected the extent of Ras palmitoylation.

3. Potential Relevance of Rce1 as Therapeutic Target for Treatment of Ras-Induced Cancers

The striking mislocalization of Ras2p in $rce1\Delta$ yeast, combined with the reduction in heat-shock sensitivity elicited by a mutationally activated Ras2p, led Boyartchuk et al. (13) to suggest that RCE1 might be an attractive target for the treatment of human cancers associated with activated forms of Ras. Could such a strategy be successful? On the one hand, some of the findings with the $Rce1$-deficient mice might lead one to be dubious about the potential of RCE1 inhibition to retard cell growth. For example, the $Rce1^{-/-}$ fibroblasts grew normally; $Rce1^{-/-}$ hematopoietic stem cells engrafted, and embryonic development appeared normal until late in gestation. On the other hand, other findings were more encouraging. The GFP–K-Ras fusion construct was profoundly mislocalized in mammalian cells—probably even more so than with the GFP–Ras2p fusion in $rce1\Delta$ yeast, making it plausible that Ras function would be abnormal in those cells. In yeast, the effects of the RCE1 knockout mutation were most impressive in the cells that expressed a mutationally activated Ras protein, so it is possible that the same will be true in humans. To date, no studies of Ras signaling in $Rce1^{-/-}$ fibroblasts have been reported, nor have there been any reports on the susceptibility of $Rce1^{-/-}$ fibroblasts to transformation with activated forms of Ras. Even when these experiments are completed, there is no guarantee that the results obtained in fibroblasts will be generalizable to the epithelial cell types most frequently involved in human cancers.

Potentially relevant to the suitability of RCE1 as a pharmacological target are the 1992 studies by Kato and co-workers (2), who examined the behavior of a Ras mutant that did not undergo efficient proteolysis/carboxyl methylation. They produced a panel of K-Ras4B constructs with a variety of amino acid substitutions in the CXXX sequence. Certain substitutions, such as changing the wild-type –VIM sequence to –VDM, abolished farnesylation, prevented membrane binding of the Ras, and eliminated Ras transforming activity in NIH 3T3 cells. A –VYM mutant principally affected the postisoprenylation processing steps. When that mutant was expressed in NIH 3T3 cells, some nonfarnesylated protein was produced, but a significant percentage of the protein was farnesylated but not further processed (i.e., no endoproteolysis or carboxyl methylation). No fully processed Ras was observed. Of the –VYM Ras protein that had undergone farnesylation, 50% was located in the membrane fractions (vs. >90% with wild-type Ras). Moreover, the –VYM construct displayed transforming capacity, although at a reduced efficiency compared with the wild-type construct. These results led the authors to conclude that isoprenylation, but not the subsequent processing events, were crucial for Ras transforming capacity. While these experiments appeared to be convincing, several caveats should be borne in mind. First, the expression vector that they used undoubtedly produced high levels of Ras expression, and it is possible that the transformation that they observed with the –VYM mutant represented a nonphysiologic consequence of flooding the system with high levels of Ras expression. The effects of blocking endoproteolysis might have been more significant at a physiologic level of Ras expression. Second, as noted earlier, it is not clear whether results with fibroblasts can be generalized to all cell types.

There has been one report suggesting that irreversible Ras endoprotease inhibitors block the growth of Ras-transformed rodent cells (36). At 5 μM concentrations, both BFCCMK [N-$tert$-butyloxycarbonyl-(S-farnesyl-C)-chloromethyl ketone] and UM96001 (N-$tert$-butyloxycarbonyl-2-amino-DL-hexadecanoylchloromethyl ketone) blocked the growth of three different Ras-transformed cells (two transformed with Kirsten murine sarcoma virus, and one human endometrial cancer cell line with an activating mutation in K-Ras) but did not affect the growth of nontransformed NIH 3T3 cells. Low concentrations of these compounds also blocked the anchorage-independent growth of transformed cells. While these results were encouraging, it is important to note that the report contained no evidence of the specificity of these compounds and no biochemical evidence that the drugs truly interfered with Ras endoproteolysis.

B. CHARACTERIZATION OF MAMMALIAN AFC1 (STE24)

Two groups have reported the cloning and initial characterization of the human ortholog of yeast *AFC1* (*STE24*) (16, 17). The human *AFC1* (*STE24*)

TABLE V

PROPERTIES OF HUMAN *AFC1* (*STE24*)

No protein substrates have been identified. Its role, if any, in processing isoprenylated CXXX proteins has not been established
The human enzyme is almost certainly capable of carrying out both N-terminal and C-terminal processing of precursor to yeast mating pheromone **a**-factor
Expressed in all mouse and human tissues examined, with the highest levels reported in kidney, prostate, testis, and ovary
Immunofluorescence microscopy with hemagglutinin-tagged AFC1 (STE24) indicated predominant localization in ER
An ~3-kb transcript encodes a 475-amino acid protein; 62% similarity and 36% identity with *S. cerevisiae AFC1* (*STE24*) amino acid sequence
Multiple predicted transmembrane domains
Contains an HEXXH domain, which is shared by the yeast enzyme and many other zinc proteases. Contains a degenerate dilysine ER retention sequence

cDNA encodes a 475-amino acid protein with 36% identity and 62% similarity to yeast Afc1p (Ste24p). Like its yeast counterpart, the putative human ortholog contains the zinc metalloprotease motif HEXXH, as well as multiple predicted transmembrane domains (Fig. 2C). Figure 1 shows the predicted sequence of the human AFC1 (STE24) protein in comparison with other related sequences in the databases. High levels of human AFC1 (STE24) transcripts were detected in all tissues, including heart, brain, kidney, spleen, intestine, testis, ovary, and prostate (*17*). Immunofluorescence studies in HEK-293 cells transiently transformed with a hemagglutinin-tagged AFC1 (STE24) construct revealed an ER-like staining pattern similar to that of calnexin (an ER protein), indicating that the human AFC1 (STE24) protein is localized to the ER and Golgi (*17*). These results are consistent with a prior report by Schmidt *et al.* (*50*) demonstrating an ER localization of Ste24p (Afc1p) in yeast. Key properties of mammalian AFC1 (STE24) are listed in Table V.

Tam *et al.* (*16*) initially cloned the human *STE24* cDNA ortholog as a part of their study to document the dual roles of Ste24p in the processing of **a**-factor. Interestingly, the human STE24 cDNA partially corrected the mating defect in *ste24Δrce1Δ* yeast and fully complemented the modest mating defect of single *ste24Δ* mutants. Those observations suggest strongly that the human protein can mediate both the C-terminal and N-terminal **a**-factor processing reactions.

While the human AFC1 (STE24) protein clearly has the capacity to cleave yeast **a**-factor, no **a**-factor orthologs have been identified in mammals. Thus, the natural substrate(s) for human AFC1 (STE24) is unknown. One possibility is that the natural substrate actually is an as-yet-unidentified **a**-

factor ortholog. Another possibility is that human AFC1 (STE24) participates in the C-terminal processing of a variety of CXXX proteins, both
alone and in combination with RCE1. A potential candidate for a substrate
is prelamin A, an isoprenylated CXXX protein that undergoes an N-terminal processing step after the removal of the –XXX sequence (51). Finally,
it is conceivable that human AFC1 (STE24) is not even involved in the
processing of mammalian CXXX proteins, but instead has a different role.
Along these lines, it is noteworthy that human AFC1 (STE24) shares important sequence similarities (e.g., the HEXXH domain and multiple trans-
membrane-binding domains) with a gene from *Helicobacter pylori*, a bacterial organism in which protein isoprenylation and CXXX protein
modifications are assumed to be nonexistent. It is conceivable that the role
of AFC1 in CXXX protein endoproteolysis is an evolutionary adaptation
unique to yeast, and that it plays a different and completely unrelated role
in mammalian cell biology. For example, for all we know, AFC1 could be
responsible for the γ-secretase activity that results in the production of β-
amyloid peptide; that endoproteolytic activity is thought to be located in
the ER. Knocking out the mouse *Afc1* gene ultimately might help to identify
the protein substrates and the physiologic importance of this protease in
higher organisms.

As illustrated in the sequence alignments in Fig. 1, *AFC1* (*STE24*) orthologs from a variety of organisms can be identified in the databases. Each
of the apparent orthologs contains the HEXXH sequence involved in the
binding of zinc, and each appears to be an integral membrane protein. As
judged by the TMHMM transmembrane domain analysis program, both
the yeast and the human enzymes appear to share seven transmembrane
helices, although a common high probability is found only for putative
helices 3, 4, 5, and 6 (Fig. 2A and C).

VI. Modification of C-Terminal Isoprenylcysteine Residues by Methyl Esterification Reactions in CXXX Proteins

The intellectual origin for the study of isoprenylated proteins, as
well as the realization that S-lipidated C-terminal cysteine residues in
polypeptides could be methyl esterified, can arguably be traced to the
structural determination of the peptidyl mating factors of two relatively
obscure jelly fungi. The A-10 mating factor of *Tremella mesenterica* was
found to contain an α-methyl ester of a C-terminal cysteine residue,
which was also modified by an oxidized C_{15} S-farnesyl group (52, 53).
A similarly methylated C-terminal residue was found on the *A*(Ia)
mating factor of *Tremella brasiliensis* (54). Although evidence of the

isoprenylation of mammalian proteins was available at the time (55), the connection between these two findings was clouded initially by the suggestion that mammalian proteins were modified by a distinct type of chemistry in which larger isoprenyl groups (C_{45} to C_{95}) were O-linked to hydroxyl-containing amino acids (56).

A milestone in the recognition of the similarity between the modifications of the fungal mating factors and mammalian proteins, and the eventual identification of methyl esterified C_{15} farnesylated and C_{20} gernanylgeranylated cysteines at the C termini of eukaryotic proteins, was the characterization of a mutant gene in *S. cerevisiae* (alternatively designated *STE16*, *DPR1*, or *RAM1*). That mutant gene, which was later shown to encode the β subunit of protein farnesyltransferase, was shown to be involved in the posttranslational modification of both the yeast mating pheromone **a**-factor and the Ras proteins (57). What common structural feature of the mating pheromone and the Ras proteins might allow them to be recognized by the same protein modification system? Powers *et al.* (57) suggested that it was within the C terminus of the proteins, where each contained a cysteine, followed by two aliphatic residues and then a C-terminal amino acid residue. This suggestion raised the possibility that a similar type of chemistry occurs with the *Tremella* mating factors and the yeast proteins. However, at that point, the lack of a precise structure for the *S. cerevisiae* **a**-factor and the C terminus of the Ras proteins precluded a direct comparison. In fact, it was initially speculated that the *Ram1p/Dpr1p*-catalyzed modification was a fatty acylation (57). In addition, evidence was presented that the mature yeast **a**-factor included a fatty acid-modified cysteine with no trimming of the terminal three amino acids (58) and that mammalian Ras proteins were similarly modified (59). What was clear was that the processing of the yeast Ras proteins was complex and that more than one reaction was likely involved (60).

Independently, the chemical nature of methyl esterified proteins in prokaryotic and eukaryotic cells was beginning to be established. It had been shown that bacterial chemoreceptors were modified on the side chain of L-glutamate residues by type I protein carboxyl methyltransferases, and that a variety of spontaneously damaged mammalian proteins was modified on the side chain of D-aspartyl and L-isoaspartyl residues by type II protein carboxyl methyltransferases [for a review, see Clarke (61)]. However, in two methyl esterified proteins, the site of methylation did not appear to fit with the products of either type of enzyme. These were the α subunit of the retinal cGMP phosphodiesterase (62) and nuclear lamin B (63). With the realization that the genes encoding both these proteins contained a C-terminal CXXX sequence, it was possible to hypothesize that the yeast mating pheromone **a**-factor, the Ras proteins, cGMP phosphodiesterase, and nuclear lamin B might all share a similar posttranslational modification pathway, and all might have a C-terminal structure similar to that in the *Tremella* mating factors (*19*). If this were the case, a key prediction would be that the **a**-mating factor and the Ras proteins would be methyl esterified. This hypothesis was first tested with the mammalian Ha-Ras protein, and the prediction was upheld. Figure 9, reproduced from an article by Clarke and co-workers (*19*), demonstrates the incorporation of methyl groups into the Ha-Ras protein. These results made it possible to formulate a proposal that the previously recognized CXXX motif (*57*) would direct separate lipidation, proteolytic, and methylation reactions (*19*).

The demonstration that the C-terminal residue of yeast **a**-factor was in fact a farnesylcysteine α-methyl ester (*64*) and the determination that the conserved cysteines in mammalian CXXX proteins were isoprenylated rather than fatty acylated (*65–67*) provided support for the proposed three-part posttranslational modification pathway for CXXX proteins, with the initial lipidation reaction being the addition of a short isoprenyl chain to the conserved cysteine residue. It was difficult, however, to demonstrate directly the presence of the methyl ester in many of the CXXX proteins (*68*). Although evidence of a methyl ester consistent with a C-terminal cysteine residue was presented for both mammalian Ha-Ras (*19*) and yeast Ras2p (*69*), the first direct demonstration of the methyl ester was accomplished for a group of 23- to 29-kDa proteins in retinal rods. With these proteins, a combination of proteolysis and oxidation was used to obtain free cysteic acid methyl ester (*70*). This approach was then used to show that the rod cGMP phosphodiesterase, the γ subunit of large G proteins, and the small G protein G25K also contained α-methyl esters of C-terminal cysteines (*71–73*). Subsequently, high-performance liquid chromatography (HPLC) methods were developed for the direct identification of farnesyl-cysteine methyl ester and geranylgeranyl methyl ester after complete proteolysis of proteins (*74, 75*).

The discovery of the C-terminal methyl esterified cysteines set off a search for the enzymes that would catalyze these reactions. A synthetic peptide related to the C terminus of the *Drosophila* Ras protein (LARYKC) was used to show that membrane fractions of rat liver, kidney, brain, and spleen could catalyze the methyl esterification of the S-geranylated, S-farnesylated, and S-geranylgeranylated forms of the peptide, but not the unmodified form (*76*). This result showed that the isoprenylation step occurred before the methylation step. Subsequent studies showed that the crucial feature for the recognition of the protein substrate by the methyltransferase was the isoprenylated cysteine residue itself and that a variety of peptides, modified cysteine residues, and nonamino acid analogs could serve as substrates (*77–80*). The availability of these synthetic substrates

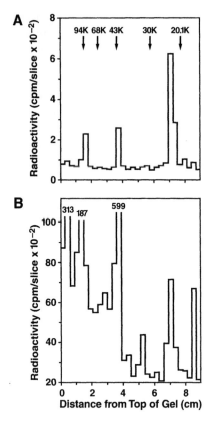

FIG. 9. Immunoprecipitation of Ha-Ras proteins from transformed fibroblasts labeled with [*methyl*-³H]methionine. Immunoprecipitates were denatured in SDS and the polypeptides were size fractionated on a 10% polyacrylamide–SDS gel. Dried gel slices were assayed for base-labile volatile radioactivity (A) and for total radioactivity (B). The amount of methyl groups incorporated into the Ras polypeptide was calculated from the radioactivity migrating from 6.9 to 7.5 cm from the top of the gel. Assuming a background of 40 cpm and 83% efficiency in methanol transfer from the gel slice to the scintillation fluid, the Ras polypeptide band contains 1000 cpm as methyl esters. The total radioactivity (which includes both methyl esters and methionine residues in the protein) in these same fractions is 6987 cpm if a background of 2000 cpm is assumed. These results suggest that approximately one radiolabeled methyl group is present for every six labeled methionine residues. The positions of the molecular weight standards (in $M_r \times 10^{-3}$) are indicated by arrows. [Reproduced, with permission, from Clarke, S., Vogel, J. P., Deschenes, R. J., and Stock, J. (1988). *Proc. Natl. Acad. Sci. U.S.A.* **85**, 4643–4647.]

then allowed for the characterization of both the yeast and mammalian methyltransferases, which are discussed in the following sections.

VII. Isoprenylation and Carboxyl Methylation in Proteins Containing Cys–Xaa–Cys and Cys–Cys Motifs at Their C Termini

Not all proteins that are isoprenylated at the C terminus have the canonical precursor CXXX sequence. Some members of the Rab small G protein family involved in vesicular transport have a C-terminal Cys–Xaa–Cys sequence that is geranylgeranylated at both cysteine residues and methyl esterified. Examples of these proteins include the YPT5, Rab3a, and Rab4 proteins (81–84). Some differences have been reported in the affinity of inhibitors for the CXXX and CXC methyl-accepting substrates, and it has been suggested that the methyltransferase responsible for methylating the CXC proteins is distinct from the *STE14* isoprenylcysteine methyltransferase described below, which is responsible for the methylation of CXXX proteins (85). On the other hand, Beranger *et al.* (86) have presented evidence suggesting that only a single isoprenylcysteine methyltransferase exists, at least in yeast. They expressed mammalian Rab6, a CXC protein, into wild-type yeast and *ste14Δ* mutant yeast, and then grew the yeast in the presence of [^{35}S]methionine and S-adenosyl-L-[*methyl*-^{3}H]methionine. As expected, both the wild-type and *ste14Δ* yeast incorporated [^{35}S]methionine into Rab6. The wild-type yeast incorporated ^{3}H-labeled methyl esters into Rab6, as judged by the release of ^{3}H-labeled methanol from the Rab6 gel slices. In contrast, no ^{3}H-labeled methanol was released from Rab6 gel slices from *ste14Δ* mutant cells. Their results strongly suggest that the isoprenylcysteine methyltransferase for the CXC proteins is identical to the one for the CXXX proteins, at least in yeast. Further work is needed in this area. It will be important to confirm these results and also determine whether the yeast CXC proteins such as the *YPT5* gene product are methyl esterified in *ste14Δ* yeast.

Other Rab proteins, including YPT1, YPT3, and Rab2, have C-terminal Cys–Cys sequences that are also modified by geranylgeranylation at both of the cysteine residues. Interestingly, these proteins do not appear to be methylated by Ste14p or any other methyltransferase, and they appear to have a free α-carboxyl group on the C-terminal cysteine (82, 83, 87). Smeland and co-workers (88) demonstrated that changing the CXC sequence on Rab3a to a CC sequence abolished carboxyl methylation, while replacing the CC sequence of Rab1a with a CXC motif resulted in a carboxyl methylated protein. When the CC terminus of Rab1a was changed to Ser–Cys, the protein was methylated. Thus, it appears that two adjacent geranylgeranylcysteines in the CC proteins can prevent the methyltransferase from catalyzing methyl ester formation at the C-terminal residue (*88*).

VIII. Characterization of Yeast Protein Modification Catalyzed by STE14-Encoded Methyltransferase

A. IDENTIFICATION OF *Saccharomyces cerevisiae STE14* GENE AS STRUCTURAL GENE FOR METHYLTRANSFERASE

Methyltransferase activity was initially detected in a crude membrane fraction of wild-type *S. cerevisiae* at levels similar to those observed previously in rat microsomes, using an assay containing *S*-adenosyl-L-methionine and the methyl-accepting peptide *S*-farnesyl-LARYKC (*89*). As with the mammalian enzyme (*76*), little or no activity was detected in cytosolic fractions, and the nonfarnesylated peptide was not found to be a methyl acceptor. With the knowledge that the methyl ester on **a**-factor was important for its activity (*64*), it was possible to ask whether any of the previously described yeast **a**-sterile mutants might lack the methyltansferase. Normal methyltransferase activity was present in *STE6*- and *STE16*-deficient yeast, but no activity was found in the *STE14*-deficient mutants (*89*). Because the sequence of the peptide substrate in these experiments was derived from a Ras protein, this experiment suggested that one methyltransferase catalyzed both **a**-factor peptide and Ras methylation reactions. Furthermore, transformation of wild-type and mutant cells with a plasmid containing the *STE14* gene resulted in the overproduction of active methyltransferase (*89*).

Subsequent studies confirmed that the *S. cerevisiae STE14* gene was the structural gene for the isoprenylcysteine methyltransferase [protein-*S*-isoprenylcysteine *O*-methyltransferase (EC 2.1.1.100)] (*90*). Normally, bacteria such as *Escherichia coli* lack the isoprenylcysteine methyltransferase. However, when the yeast *STE14* gene was expressed as a fusion protein in *E. coli,* the membrane fraction of the bacteria catalyzed the methyl esterification of a synthetic peptide (*90*).

Key properties of yeast STE14 are listed in Table VI.

B. METHYLATION OF **a**-MATING FACTOR, *RAS1* AND *RAS2* GENE PRODUCTS, AND OTHER YEAST POLYPEPTIDES BY *STE14*-ENCODED METHYLTRANSFERASE

By overexpressing **a**-factor in yeast, it was possible to show that neither the intracellular nor the extracellular forms of **a**-mating factor are methylated in *ste14Δ* yeast (*90*). Similarly, when the *RAS1* and *RAS2* genes were

TABLE VI

PROPERTIES OF YEAST *STE14*

Major, and likely only, methyltransferase that methylates the carboxyl group of isoprenyl-cysteine of CXXX proteins
Likely responsible for methylating the C-terminal isoprenylcysteine of proteins that terminate in Cys–Xaa–Cys
Methylates *N*-acetyl farnesylcysteine and short farnesylated peptides efficiently
Methylates both farnesylated and geranylgeranylated substrates
Null mutants do not methylate **a**-factor and are sterile
Null mutants lack methylation of Ras1p and Ras2p. Nevertheless, the heat shock sensitivity and glycogen accumulation phenotypes are identical in *ras2^{val19}* mutants and *ras2^{val19}/ste14Δ*) double mutants
Localized in the ER membranes by immunofluorescence and subcellular fractionation
239-amino acid protein, apparent molecular mass of 24 kDa by SDS–polyacrylamide gels
Protein has multiple predicted transmembrane domains, but no *S*-adenosylmethionine-binding motif

overexpressed in *ste14Δ* yeast, no methyl esters could be detected in either protein (*90*). To examine the possible methylation of other polypeptides, Hrycyna *et al.* (*91*) incubated wild-type and *ste14Δ* yeast with *S*-adenosyl-L-[*methyl*-³H]methionine. Yeast incubated with *S*-adenosyl-L-[*methyl*-³H]methionine take up the label and use it for endogenous transmethylation reactions. Hrycyna *et al.* (*91*) size fractionated the resulting ³H-methylated polypeptides (representing proteins that have been methylated by Ste14p as well as by other protein methyltransferases) on SDS–polyacrylamide gels, and then analyzed the [³H]methyl ester content of the various proteins. In wild-type yeast, methyl esterified polypeptides were detected at 49, 38, 35, 33, 31, and 26 kDa (*91*). In *ste14Δ* yeast, methylated proteins were not observed in the 38-, 33-, 31-, and 26-kDa regions, suggesting that polypeptides migrating at these positions are substrates for the *STE14*-encoded isoprenylcysteine methyltransferase. Under the conditions of this experiment, **a**-mating factor would not be detected and the *RAS1* and *RAS2* gene products would be expected to make up at least a portion of the radioactivity at 38 kDa (*91*). Thus, it is clear that the *STE14* methyltransferase can modify a variety of yeast proteins in addition to the Ras proteins. As described above, there are perhaps more than 80 yeast CXXX proteins, including the products of the *RHO1, RHO2, RHO3, RHO4, BUD1/RSR1, STE18,* and *CDC42* genes (*90.*) All the *STE14*-dependent methylated polypeptides were localized in the membrane fraction, and their methylation was inhibited by the protein synthesis inhibitor cyclohexamide (*91*).

Pulse–chase experiments showed no turnover of methyl esters in the 38-kDa polypeptide and only slow turnover in the 33/31-kDa species. These

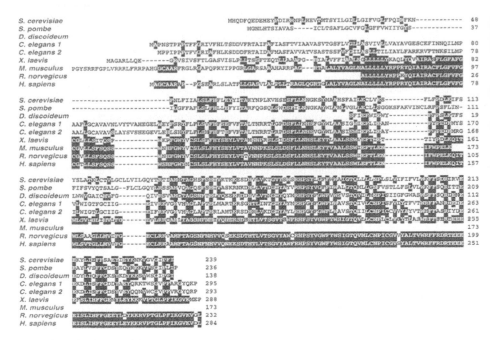

Fig. 10. Amino acid sequence alignment of yeast Ste14p and orthologs in other eukaryotes. GenBank accession numbers: *C. elegans* 1, AAB42280; *C. elegans* 2, AAB37832; *S. pombe,* BAA18999; *X. laevis,* BAA19000; *S. cerevisiae,* AAA16840; human, AF064084. GenBank EST accession numbers: mouse, AA022288; rat, AAD42926; *D. discoideum,* C89921. Protein sequences were aligned as described in the legend to Fig. 1.

results suggest that the isoprenylcysteine methylation reaction largely occurs immediately after protein translation and that the methylation is not readily reversible. These studies also indicate that a single isoprenylcysteine methyltransferase is present in *S. cerevisiae* and can methylate **a**-factor and other protein products.

C.   Characterization of *STE14* Gene Product

The DNA sequence of the yeast gene encoding the isoprenylcysteine methyltransferase was initially presented at a meeting (*92*) and was then published in full form (*43, 93*). Yeast Ste14p contains 239 residues and is rich in hydrophobic amino acids. Its amino acid sequence is shown in Fig. 10, along with orthologs in other species. The yeast enzyme, like its human counterpart, is predicted to have multiple transmembrane domains (Fig.

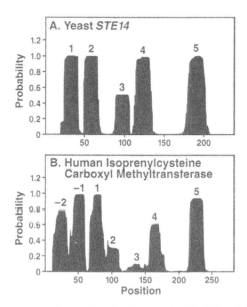

Fig. 11. Predicted transmembrane domains for yeast Ste14p (A) and its human ortholog, human isoprenylcysteine carboxyl methyltransferase (B). Predicted transmembrane domains were determined with the TMHMM transmembrane domain analysis program (http://genome.cbs.dtu.dk/services/TMHMM-1.0/). To aid in the alignment, potential transmembrane domains with yeast/human sequence identities are indicated by numbers 1–5. The regions labeled −1 and −2 in the human sequence are not found in the yeast sequence.

11). By immunofluorescence microscopy and cellular fractionation studies, the enzyme is localized to the ER (94). A similar localization was found for the mammalian enzyme (see below) (18, 100).

D. PHYSIOLOGICAL ROLE OF ISOPRENYLCYSTEINE METHYLTRANSFERASE IN YEAST

Several general possibilities have been suggested for the function of the methylation reaction. First, the conversion of a carboxylate anion to a methyl ester residue would be expected to make the C terminus more hydrophobic, rendering it more likely to partition into the hydrophobic membrane (for reviews, see Refs. 23 and 95). Second, once the three C-terminal amino acids have been removed, methylation of the isoprenylcysteine might protect the precursor polypeptide from proteolytic digestion (22). Finally, methyl esterification could serve as a recognition signal for specific receptor proteins in the plasma membrane or other intracellular membranes (68). Most of the isoprenylated proteins are involved in signal transduction, and it seems possible that methylation could modulate specific protein–protein interactions and thereby affect upstream or downstream signaling partners.

The availability of the *ste14Δ* mutants in *S. cerevisiae* provided an opportunity to address the function of methylation in this organism. Methylation of **a**-factor appears to be important for several reasons. First, the wild-type **a**-factor is at least 200-fold more active than the nonmethylated **a**-factor in *ste14Δ* yeast (*90*). That result was not particularly surprising, given that *ste14Δ* mutants were originally isolated because of a sterile phenotype (*96*). In addition, the lack of **a**-factor methylation results in enhanced proteolytic degradation inside cells, and an essentially complete block in the export of the protein from cells (*43*).

Aside from the dramatic loss of mating ability, *ste14Δ* yeast are viable and seem to be phenotypically normal (*43, 90*). Considering the importance of Ras1p and Ras2p in cell growth and survival, this result is rather surprising. To assess whether *ste14Δ* yeast exhibit a partial loss of Ras activity, *ste14Δ* yeast were constructed in a $ras2^{Val19}$ strain. The latter mutation results in the expression of a constitutively active Ras protein, which causes heat shock sensitivity and the loss of ability to accumulate glycogen. Mutations in the genes for protein farnesyltransferase have been found to block the expression of the phenotypes elicited by the $ras2^{Val19}$ mutant because farnesylation of the Ras proteins is essential for activity (*57, 97*). However, this was not the case with the *ste14Δ*/$ras2^{Val19}$ double mutants, indicating that the loss of methylation did not significantly affect the activity of the mutationally activated Ras proteins (*90*). These results suggest that the yeast Ras proteins retain functional activity in the absence of methyl esterification.

Even though the *ste14Δ* mutation had no major effect on the activity of the activated yeast Ras protein, a significant effect was found on the processing, stability, and membrane attachment, of Ras2p (Fig. 12) (*90*). In wild-type cells, the processing involves the conversion of a nonfarnesylated 41-kDa cytosolic precursor (p41) to a 40-kDa farnesylated species (p40) that can become palmitoylated and associate with the membrane (*60, 98*). In *ste14Δ* cells, the conversion of p41 to p40 appeared to be inhibited. Only a small amount of the p40 polypeptide is made in the cytosol, and even less becomes associated with the membrane (Fig. 12) (*90*). In addition, the total amount of protein was decreased and the fraction of protein in the membrane fraction was greatly reduced.

The inhibition of the p41-to-p40 conversion was especially puzzling because this conversion is thought to involve the farnesylation of an unmodified precursor protein, a reaction that should be unaffected in *ste14Δ* cells. There are several possible solutions to this paradox. One possibility is that

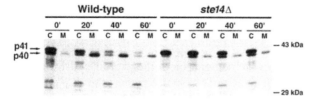

Fig. 12. Kinetics of Ras2p membrane localization in wild-type and *ste14Δ* yeast. Wild-type and *ste14Δ* yeast containing the plasmid pADH-*RAS2* were labeled with [^{35}S]methionine for 2 min at 30°C, and chased for the indicated times. Total cell extracts from the pulse–chase experiment were prepared and immunoprecipitated with the anti-Ras monoclonal antibody Y13-259 and immunoprecipitates were analyzed by SDS–PAGE on a 10% polyacrylamide gel. [Reproduced, with permission, from Hrycyna, C. A., Sapperstein, S. K., Clarke, S., and Michaelis, S. (1991). *EMBO J.* **10**, 1699–1709.]

the isoprenyltransferases or the *RCE1* protease or both are in a membrane-bound complex with the *STE14* methyltransferase, so that the loss of the methyltransferase might adversely affect the activity of the farnesyltransferase or the endoprotease. Thus, the accumulation of p41 protein might represent the unmodified precursor or the farnesylated but nonproteolyzed protein. In mammalian cells, the loss of RCE1 activity results in the production of Ras proteins with distinctly reduced electrophoretic mobility (*14*). A second possibility is that isoprenylation and protease reactions occur normally in the *ste14Δ* mutant but that the absence of methylation of the farnesylcysteine renders the protein more susceptible to proteolytic cleavage by some element of the intracellular protein degradation pathway (*22, 23*). Thus, the p41 in the *ste14Δ* mutant could represent a C-terminal truncated protein in which the farnesylcysteine (and perhaps additional residues) have been exoproteolytically cleaved. This form may migrate in the same position as the unmodified precursor p41 in the wild-type cells. This hypothesis would account for the apparent persistence of the p41 precursor species in the *ste14Δ* mutant yeast (Fig. 12) (*22*). With state-of-the-art mass spectroscopy, it should be possible to isolate the p41 protein from *ste14Δ* cells and determine its precise structure, distinguishing between these possibilities. Other possibilities could also be proposed to explain these data (*90*).

Finally, the absence of overt phenotypes in *ste14Δ* cells, such as retarded cell growth, indicates that the other protein substrates for the isoprenylcysteine methyltransferase can function in the absence of the C-terminal methyl ester or are not essential for viability. Nevertheless, it would be of interest to determine whether the absence of carboxyl methylation affects the intracellular stability of other CXXX proteins or their ability to become associated with membranes.

IX. Methyl Esterification of Isoprenylated Mammalian Proteins

A. SPECIFICITY OF ISOPRENYLCYSTEINE METHYLTRANSFERASE

A key element in the recognition of the protein substrate is the isoprenyl group. For the rat liver microsomal enzyme, the K_m for the unmodified LARYKC peptide was greater than 45 mM. In contrast, the K_m values for the S-geranyl derivative (C_{10}), the S-farnesyl derivative (C_{15}), and the S-geranylgeranyl (C_{20}) derivative were 389, 2.2, and 10.9 μM, respectively. When the peptide was modified at the sulfhydryl group with C_8, C_{10}, C_{13}, and C_{15} *n*-alkyl derivatives, some activity was observed but the estimated K_m values (based on a V_{max} comparable to that of the farnesyl peptide) were 480–1760 μM, reflecting relatively weak substrate recognition compared with the isoprenylated derivatives (76).

N-Acetylfarnesylcysteine is itself a good substrate for the methyltransferase (77, 78, 80, 99). The minimal recognition element is simply an isoprenylated thiopropionate moiety, and no amino acid or peptide sequences are needed at all (80, 99). K_m values for *N*-acetylfarnesylcysteine have been measured at 20 μM (78) and 22 μM (99), while values of 14 μM (80) and 20 μM (77) have been reported for *S*-farnesylthiopropionic acid. Thus, the affinity of the methyltransferase for these substrates is only about an order of magnitude less than that of a farnesylated peptide substrate (76) and suggests that most of the recognition by the methyltransferase is due to the carboxyl group being in the correct spatial relationship to the isoprene group. The K_m value for *N*-acetylgeranylgeranylcysteine has been reported as 7 μM (77), a value similar to that of the farnesyl analog. Kinetic evidence has been presented suggesting that the mammalian methyltransferase can recognize both farnesyl and geranylgeranyl substrates (79).

B. INTRACELLULAR LOCALIZATION OF ENZYME

From the initial report of the isoprenylcysteine methyltransferase activity, it was clear that the enzyme was membrane associated; the highest specific activities were in the microsomal fraction, containing largely ER membranes but also contains lesser amounts of plasma membrane and other internal membranes (76). Subsequent fractionation of rat liver microsomal membranes by Percoll density gradient centrifugation revealed that the methyltransferase largely comigrated with ER markers and not with markers for the plasma membrane or the Golgi apparatus (*100*). Studies have confirmed its localization to the ER (*18*). One report suggested that human neutrophils possess a plasma membrane-localized enzyme that is activated by anionic phospholipids (*101*), but this activity has not been characterized further. There appears to be only a single *STE14* gene in mammals, as judged by analyses of the EST databases. If a distinct plasma membrane-associated enzyme truly exists, it may be encoded by a structurally distinct gene. Further work is required to establish the existence of a second type of isoprenylcysteine methyltransferase activity in mammalian cells.

Attempts to solubilize the *STE14*-related enzyme in active form have not been successful (*100*). This enzyme appears to form an integral part of the membrane and its activity is lost when the membranes are treated with a variety of mild detergents that have been used successfully to solubilize other membrane proteins in active form (*100*).

C. Effect of GTP on Methylation Reactions

In a number of *in vitro* systems, methylation reactions attributable to the *STE14*-related methyltransferase have been enhanced by the presence of a nonhydrolyzable GTP analog, GTPγS (*102–105*). GTPγS does not appear to stimulate the methyltransferase directly (*100*) but serves to activate a class of small G proteins, such as G25K, for membrane attachment, which then allows for more efficient methylation by the membrane-bound Ste14p (*106*).

D. Methyltransferase Inhibitors as Functional Probes *in Vivo*

There has been an intensive effort to utilize specific inhibitors of the isoprenylcysteine methyltransferase to understand the functional roles of this enzyme. For example, 5'-methylthioadenosine has been reported to inhibit the carboxyl methylation and assembly of nuclear lamin B (a CXXX protein) (*107*), as well as a class of small G proteins (*108*). The origin of this effect may be complex, however, because the isoprenylcysteine methyltransferase does not appear to be inhibited by this compound in *in vitro* assays (*100*).

Other studies have utilized compounds that result in the accumulation of *S*-adenosylhomocysteine and other derivatives that inhibit both the iso-prenylcysteine methyltransferase (*76*) and other types of *S*-adenosyl-L-methionine-dependent methyltransferases. From the time when it was established that receptor function in bacterial chemotaxis requires the methylation of a group of L-glutamate residues [for a review, see (*61*)], attempts have been made to show that mammalian cell chemotaxis also requires methylation. Early work showed that chemoattractants could transiently increase the level of methyl esterification of rabbit neutrophils (*109, 110*). Treatment of a mouse macrophage cell line with 3-deaza-adenosine, a precursor of the methyltransferase inhibitor 3-deaza-adenosylhomocys-

teine, inhibited their chemotaxis (*111*), although this effect was later attrib-
uted largely to an effect on protein synthesis (*112*). The discovery that
methyl-esterified G proteins mediate mammalian chemotaxis sparked a
renewed interest in this topic. It was found, for example, that the treatment
of human HL-60 cells for 24 or 48 hr with peroxidate-oxidized adenosine,
a compound that results in the intracellular accumulation of *S*-adenosyl-
homocysteine, inhibits the fMet-Leu-Phe-dependent release of superoxide
ion (*113*). However, while suggestive of possible functional roles of the
isoprenylcysteine methyltransferase reaction, these studies must be interpreted with caution because of the large number of other methylation
reactions that could also be inhibited.

Another approach to more specifically inhibit the mammalian isoprenylcysteine methyltransferase in intact cells involves the use of methyl-accepting small molecules such as *N*-acetylfarnesylcysteine (*77–80*) and *N*-
acetylgeranylgeranylcysteine (*79, 114*) as cell-permeable competitive inhibitors. These compounds would be expected to inhibit the isoprenylcysteine
methyltransferase without inhibiting other types of methyltransferases.
However, these compounds may affect other cellular processes. For example, the fact that they contain isoprenyl groups could lead them to bind to
other cellular proteins that normally interact with isoprenylated proteins.

The potential benefits and risks of this approach are illustrated in studies
on the effect of *N*-acetylfarnesylcysteine on the chemotactic response of
mammalian cells. It was initially reported that macrophage chemotaxis was
inhibited about 80% in the presence of *N*-acetylfarnesylcysteine (*78*). This
study suggested that the methylation of an isoprenylated residue may be
a crucial component of the signaling system. Similar results were found in
human HL-60 cells (*115*) and with the chemoattractant-mediated activation
of the G_i protein in neutrophils (*116*). In another study, however, it was
shown that *N*-acetylfarnesylcysteine and its analogs could either inhibit
the chemotactic peptide-stimulated release of superoxide ion in human
neutrophils or stimulate peptide-independent release (*117*). Because similar
results were obtained with analogs that inhibit the isoprenylcysteine methyltransferase (e.g., *N*-butyrylfarnesylcysteine) and those that do not (e.g., *N*-
pivaloylfarnesylcysteine), it did not appear that these effects are mediated
entirely or in large part by their effect on the methyltransferase (*117*).
These results were supported by the finding that general methyltransferase
inhibitors did not affect superoxide ion release (*117*), although this result
does conflict with that reported in a similar experiment by another group
(*113*). Finally, Scheer and Gierschik (*118*) found that stimulation of GTPγS
binding and GTP hydrolysis by fMet-Leu-Phe in well-washed membranes
of HL-60 granulocytes (which were presumably free of the methyl donor
S-adenosylmethionine) were still affected by *N*-acetylfarnesylcysteine, sug-

gesting again that N-acetylfarnesylcysteine affected a cellular mechanism unrelated to the isoprenylcysteine methyltransferase. Subsequent work by these authors suggested that N-acetylfarnesylcysteine exerted its activity by interfering with the interaction of activated receptors with G proteins, probably with their $\beta\gamma$ subunits (119).

The N-acetylfarnesylcysteine inhibitor approach has also been used to study the role of isoprenylcysteine methylation in platelets. It was initially shown that N-acetylfarnesylcysteine inhibits platelet aggregation in response to collagen, ADP, and arachidonic acid (120). However, it was then demonstrated that N-benzoylfarnesylcysteine and N-pivaloylfarnesylcysteine, neither of which is a substrate or an inhibitor of the enzyme, could inhibit platelet aggregation as effectively as, or more effectively than, N-acetylfarnesylcysteine (121). This latter result again points to the possibility that these methylation inhibitors may also affect other pathways, and suggests that the isoprenylcysteine methyltransferase may not be involved in a direct signal transduction pathway (121).

Inhibitor studies with N-acetylfarnesylcysteine and related compounds are potentially attractive because of their specificity for the isoprenylcysteine methyltransferase. Unfortunately, these compounds appear to affect "nonmethyltransferase" pathways in a nonspecific fashion. The fact that the inhibitor compounds contain isoprenyl groups may lead them to bind to cellular proteins that interact with isoprenylated proteins. To the extent that this is the case, it might be difficult to distinguish the effect of methyltransferase inhibition from other nonspecific effects. Overall, the take-home lesson from these studies is that it may not be possible to conclude much about the *in vivo* action of the methyltransferase by treating cells with this class of compounds, except perhaps when the result is negative (i.e., when there is no effect on a function).

E. *In Vitro* STUDIES OF EFFECT OF METHYL ESTERIFICATION ON PROTEIN FUNCTION

A number of experimental procedures have been used to compare directly the activity of proteins with and without the C-terminal methyl ester and quantitate the effect of this modification on protein function. However, with the notable exception of studies on the activity of **a**-factor in *S. cerevisiae* (43, 64, 90, 122), these investigations have largely shown that methylation either does not affect, or affects minimally, the activities of isoprenylated proteins. These studies are described in detail below.

One approach has been to use isolated proteins and membrane fractions from cells preincubated with general methyltransferase inhibitors. With that approach, it has been possible to show that methylation of the γ subunit of the $\beta\gamma$ complex can enhance the ability of pertussis toxin to ADP-ribosylate the α subunit by about twofold while not enhancing the membrane attachment of the $\beta\gamma$ subunit itself (*113, 123*).

Another approach has taken advantage of the ability of a pig liver esterase preparation to catalyze the hydrolysis of C-terminal isoprenylcysteine methyl esters in the intact protein (*85*). With this approach, it was possible to produce the nonmethylated form of the transducin $\beta\gamma$ subunits and then to compare the capacities of the nonmethylated and fully methyl esterified proteins to stimulate GTPγS exchange in the presence of the transducin α subunit and photoactivated rhodopsin. In the presence of a detergent, no differences in GTPγS exchange were observed with the methylated and nonmethylated proteins (*85*). However, when membrane preparations were used, the methylated transducin $\beta\gamma$ was about twofold more efficient than the nonmethylated form (*85*). Of note, the methylated transducin $\beta\gamma$ was more than 10-fold more potent in activating phosphatidylinositol-specific phospholipase C and phosphoinositide 3-kinase. Similar experiments showed that methylation of the γ subunit of transducin $\beta\gamma$ can enhance its affinity for phosducin by about twofold (*124*). It has also been possible to separate methylated and nonmethylated forms of transducin $\beta\gamma$ by chromatography and then examine their *in vitro* activites (*125*). Here, it was found that methylation could faciliate functional association of transducin $\beta\gamma$ with the light-activated receptor (*125*). Interestingly, transducin $\beta\gamma$ is generally fully methylated in rod cells, raising the possibility that its activity may not be modulated *in vivo* by interconverting the methylated and nonmethylated forms (*125*).

The *in vitro* activity of the $\beta\gamma$ subunit of the nontransducin large G proteins has been assayed by using the pig liver esterase to produce demethylated geranylgeranylcysteine-containing γ subunits (*95*). Methylation had no effect on the ability of the $\beta_1\gamma_2$ subunit to interact with the α subunit, as shown by a pertussis toxin-catalyzed ADP-ribosylation assay. A small effect was seen on the ability of $\beta\gamma$ to activity phosphoinositol-specific phospholipase C or phosphoinositide 3-kinase (*95*). Once again, no striking effects of methyl esterification on protein function were observed.

Functional studies of the mammalian smg-25A/rab3A protein (a CXC isoprenylated protein) have also been reported (*126*). Although this protein may be modified by a methyltransferase distinct from the *STE14*-encoded enzyme (see earlier discussion of this topic), the C terminus still contains an α-methyl ester of an isoprenylated cysteine. It is possible to produce the unmodified precursor of this polypeptide in bacterial cells and then to isoprenylate it *in vitro* so that it contains the geranylgeranyl group but not the methyl ester. When the properties of the nonmethylated protein were compared with those of the fully methylated cellular-derived species, no differences were observed in its membrane-binding properties or in its interactions with its GDP/GTP exchange protein (126).

What conclusions can we draw from the general lack of a large-scale effect of methylation on the functions of proteins? Two avenues can be explored. First, the effect of methylation, although subtle in in vitro studies, might be crucial in fine-tuning the responses of these proteins in vivo, especially in signal transduction cascades. Second, the effect of the methylation reaction might simply ensure the stability of the protein and its isoprenylated C terminus. As hypothesized above for the S. cerevisiae Ras and as demonstrated for a-factor (43), the methylation reaction can provide a C-terminal cap that may prevent proteolytic "trimming" of the C terminus. Of note, RhoA, a small G protein, was found to have a decreased half-life when methylation was inhibited (134). The availability of a mouse knockout of the isoprenylcysteine methyltransferase will probably make it possible to address these hypotheses more directly.

F. REVERSIBILITY OF PROTEIN ISOPRENYLCYSTEINE METHYLATION

There is little evidence that C-terminal methyl esters on isoprenylated proteins can be physiologically hydrolyzed to the free C-terminal carboxyl group. Pulse–chase experiments have indicated slow if any turnover (20, 91) and where measured, it appears that the bulk of the cellular protein is fully modified (125). In vitro, methyl esters are readily cleaved by pig liver esterase (127), and while this observation has been nicely exploited as an experimental technique (see discussion above), it is not clear that the enzyme catalyzes this type of reaction in vivo. Yeast carboxypeptidase Y can also cleave the methyl ester linkage (22), but this vacuolar enzyme probably does not come in contact with isoprenylated proteins in vivo. No nonvacuolar activities were found in yeast that could catalyze the hydrolysis of N-acetylfarnesylcysteine methyl ester (22). The hydrolysis of N-acetylfarnesylcysteine methyl ester and methylated transducin can be catalyzed by unidentified proteins in the membrane fractions of bovine retinal rods (128, 129), and at least two activities have been found in rabbit brain that catalyze the hydrolysis of methyl esters from the small G protein G25K (130). Interestingly, one of the activities found in the latter report was identified as the lysosomal protease cathepsin B (130). As with the yeast vacuolar carboxypeptidase Y, it would not appear that this hydrolytic activity would be physiologically important. However, the second soluble enzyme activity (and a similar activity in the membrane fraction) may be more physiologically relevant (130). There are a large number of potential methylated substrates in cells, and a large number of potential methylesterases. Matching potential physiological substrates and enzymes to determine whether some isoprenylated proteins undergo reversible methylation will be a challenge. If reversible methylation occurs, modulation by the relative activity of methyltransferases and methylesterases could be used to effect distinct biological responses. Further work in this area is warranted.

G. Comparison of Amino Acid Sequences of Mammalian and Yeast Methyltransferases

After Ste14p from *S. cerevisiae* was identified and characterized, a 426-base pair (bp) mouse cDNA with homology to *STE14* appeared in the EST databases. This sequence was used to clone the human isoprenylcysteine carboxyl methyltransferase cDNA (*18*). The 3595-bp cDNA encoded a 284-amino acid protein that is 26% identical to yeast Ste14p (Fig. 10). Transfection of the human isoprenylcysteine carboxyl methyltransferase clone into *ste14Δ* yeast partially restored the mating defect, indicating that that the human enzyme can carry out the carboxyl methylation of a-factor. Like Ste14p, the human enzyme is predicted to have multiple transmembrane domains (Fig. 11).

Transient transfection of the human isopreylcysteine carboxyl methyltransferase expression vector into Cos-1 cells strikingly increased the methylation of *N*-acetylfarnesylcysteine and *N*-acetyl-*S*-geranylgeranylcysteine. Transfection of the carboxyl methyltransferase enhanced the methylation of the Rho GTPase in Cos-1 cells, and membranes from transfected Cos-7 cells increased the methylation of partially purified neutrophil Rac2 and RhoA. Key properties of the human isoprenylcysteine carboxyl methyltransferase are listed in Table VII.

In Fig. 10, we provide an updated alignment for the encoded amino acid sequences of the *S. cerevisiae STE14* gene, its human ortholog, as well as encoded sequences for the *Schizosaccharomyces pombe* (*94, 131*), *Xenopus laevis* (*131*), mouse, and rat isoprenylcysteine methyltransferase, and the sequences for putative *Dictyostelium discoidium* and *Caenorhabditis elegans* enzymes [compare with Fig. 7 of Romano *et al.* (*94*)]. It is interesting to note that two closely related genes are present in the nematode *C. elegans*. What is most impressive, however, is the conservation of amino acid sequences in all but a relatively small N-terminal region of each protein. The encoded proteins are rich in hydrophobic amino acids, consistent with the tight membrane association of the yeast and mammalian methyltransferases. Analysis of putative transmembrane helical regions by the TMHMM posterior program revealed seven potential regions for the human enzyme and five for the *S. cerevisiae* enzyme (Fig. 11). Two potential N-terminal helices in the human sequence were not found in the yeast sequence. The other five do align, although common high probabilities are found only for

TABLE VII

PROPERTIES OF HUMAN ISOPRENYLCYSTEINE CARBOXYL METHYLTRANSFERASE

Methylates C-terminal isoprenylcysteine in CXXX proteins
Might be responsible for methylating C-terminal cysteine of Rab proteins that terminate in cys–Xaa–Cys
When expressed in Cos-1 cells, methylates neutrophil Rho GTPases and isoprenylcysteine analogs N-acetyl-S-farnesylcysteine and N-acetyl-S-geranylgeranylcysteine
The human enzyme complements biochemical defect in $ste14\Delta$ yeast, indicating that human enzyme can carboxyl methylate **a**-factor
Located in ER
Human enzyme (284 amino acids) is 26% identical to the yeast protein Ste14p. Molecular mass is 33 kDa, as judged by SDS–polyacrylamide gels
Predicted to contain multiple transmembrane domains; the first 65 amino acids are 36% identical to amino acids 750–821 of the human band 3 anion transporter
Sequences homologous to S-adenosylmethionine-binding regions that are found in aspartyl and glutamyl protein carboxyl methyltransferases are not apparent in sequence of enzyme

the first and fifth of these sequences. This program does not clearly predict the location (cytosolic or intraluminal) of the domains between the transmembrane domains. Given a localization of the enzyme in the ER (*18, 94, 100*), it would be expected that the residues important for binding the precursor isoprenylated protein and the cosubstrate S-adenosyl-L-methionine would be located on the cytoplasmic rather than the luminal side of the membrane. This is because the precursor proteins are modified by cytosolic protein isoprenyltransferases and because no evidence has been presented that there is an intraluminal pool of S-adenosyl-L-methionine in addition to the cytoplasmic pool. The sequence of the isoprenylcysteine methyltransferase does not contain any of the four sequence motifs (regions I, post-I, II, and III) that have been described for a number of soluble methyltransferases (*132*).

We have been intrigued by the high conservation in the encoded amino acid sequences of the genes of the various organisms between putative transmembrane helices 4 and 5 and in the C terminus of the human isoprenylcysteine methyltransferase. The putative transmembrane helix 5 may actually represent a hairpin structure so that both the region between helices 4 and 5 and the C terminus are cytosolic and may form the active site of the enzyme. We are especially intrigued by the sequence corresponding to amino acids 205–221 in the *S. cerevisiae* protein, where eight residues are conserved in all species (FFXXRXXXEEXXLXXFF). The four aromatic phenylalanine residues can represent sites of $\pi-\pi$ interactions with the aromatic purine base of S-adenosylmethionine; interactions of the positively charged sulfonium atom of S-adenosylmethionine with aromatic amino acid residues (cation–π interactions) are also possible. Further evidence of the role of the C-terminal region in the methyltransferase activity is its disruption when a hemagglutinin-epitope tag is inserted at residue 226 (94). Closer to the N terminus, the conserved sequences are largely found in the putative transmembrane regions and may not be associated with the active site of the enzyme.

Because successful three-dimensional structure determination of intrinsic membrane proteins such as the isoprenylcysteine methyltransferase is problematic (at least at the present time), a combination of sequence comparisons, in vitro mutagenesis studies, and chemical modification studies may be useful for determining the amino acid residues important in the recognition of isoprenylated substrate proteins and the catalysis of methyl ester formation. For example, chemical modification studies have been used to try to identify active site residues. The mammalian enzyme has been shown to be inactivated by N-ethylmaleimide in the absence but not in the presence of S-adenosylmethionine or S-adenosylhomocysteine (100). N-Ethylmaleimide preferentially modifies cysteine residues although it can also modify lysine and histidine residues, suggesting that one or more of these residues is present in the binding site for S-adenosylmethionine. A similar protection from inactivation by S-adenosylhomocysteine by reagents with specificity for arginine and tryptophan residues has been reported (133). It would be useful to ask if the yeast enzyme may also be similarly inactivated so that specific residues may be targeted for analysis by site-directed mutagenesis techniques.

We have also been interested in the presence of genes in two bacterial species that appear to be related to the genes for the isoprenylcysteine methyltransferases. As shown in Fig. 13, STE14-like sequences are found in the prokaryotes Pseudomonas denitrificans and Mycobacterium tuberculosis. An analysis of these sequences with the TMHMM transmembrane domain analysis program (http://genome.cbs.dtu.dk/services/TMHMM-1.0/) revealed that P. denitrificans contains four transmembrane domains, while the M. tuberculosis sequence contains two. Biochemical studies will be needed to determine if these enzymes are indeed isoprenylcysteine methyltransferases or whether they may represent distinct methyltransferases or simply play other enzymatic functions. It is clear that E. coli does not contain an isoprenylcysteine methyltransferase (90), and it has been assumed that this type of methylation occurs only in eukaryotic cells. If some prokaryotes do have an isoprenylcysteine methyltransferase activity, it would indicate that protein isoprenylation reactions are more widespread in nature than previously thought.

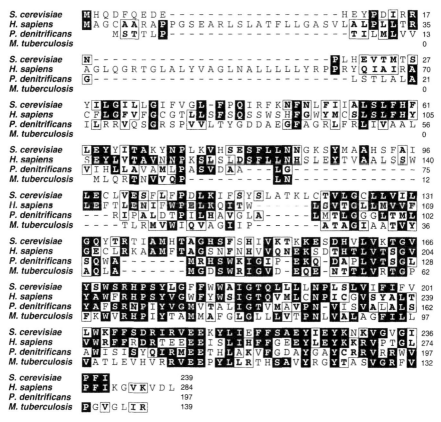

Fig. 13. Amino acid sequence alignment of yeast Ste14p and similar sequences in two prokaryotes. Protein sequences were aligned with Macvector 6.5, using a CLUSTALW alignment. Pairwise alignment was performed with a BLOSUM30 matrix and multiple alignment was performed with a BLOSUM series matrix. The document was then imported into Illustrator. Identical residues are shaded; similar residues are boxed. An analysis of the *P. denitrificans* and *M. tuberculosis* sequences with the TMHMM transmembrane domain analysis program revealed that the *P. denitrificans* contained four predicted transmembrane domains (amino acids 4–22, 43–65, 84–106, and 143–165), while the *M. tuberculosis* sequence contained two (amino acids 18–40 and 76–98). GenBank accession numbers: *P. denitrificans*, P29940; *M. tuberculosis*, CAB03782; *S. cerevisiae*, AAA16840; human, AF064084.

X. Conclusions

There has been substantial progress in understanding the postisoprenylation processing of CXXX proteins. The yeast genes responsible for the endoproteolytic processing of **a**-factor and the Ras proteins have been reported, and orthologs in other species have been identified. The importance of *RCE1* for the endoproteolysis of mammalian Ras proteins has been established with the *Rce1* knockout experiment and also with biochemical studies. However, many questions remain. For example, almost nothing has been reported regarding structure–function relationships for RCE1 in any organism. It is not clear why *Rce1* knockout mice die during embryonic development. The list of protein substrates for Rce1 is undoubtedly incomplete. Moreover, the protein substrates for which the endoproteolytic processing step carries significant functional consequences has not been established. Nor is it clear whether AFC1 (STE24) plays a role in the endoproteolytic processing of isoprenylated proteins in higher organisms. If it does, the unique and overlapping roles of RCE1 and AFC1 (STE24) in cleaving CXXX proteins need to be established. In addition, the role of AFC1 (STE24), if any, in the proteolytic processing of non-CXXX proteins needs to be investigated.

Advances have also been made in our understanding of isoprenylcysteine methyltransferase. The identity of the mammalian isoprenylcysteine methyltransferase has been established, and orthologs have been identified in multiple organisms. Many of its protein substrates have been established. Once again, however, there are many unanswered questions. There is little information on the structure of this enzyme (e.g., the location of the binding sites for the isoprenylcysteine or *S*-adenosylmethionine). In addition, its importance in mammals needs to be investigated with a gene-knockout experiment.

Much of the interest in the endoproteolysis and carboxyl methylation steps has stemmed from the fact that they are involved in the processing of the Ras proteins, which have been strongly implicated in the development of human cancers. Inhibitors of these steps might be effective in retarding the growth of cancers. It is quite possible that inhibition of these steps could inhibit Ras function in some cell types, but not others. We also believe that effective endoproteolysis/methylation inhibitors might limit cell growth by interfering with the function of "non-Ras" CXXX proteins. In the future, these possibilities could be addressed decisively by using animal models in which these processing steps have been selectively inhibited in certain tissues.

ACKNOWLEDGMENTS

This work was supported by NIH Grants HL47660 and AG15451-01 (to S.G.Y.) and GM26020 (to S.C.) and by a grant from the University of California Tobacco-Related Disease Research Program (to S.G.Y.). We thank M. Ashby, S. Michaelis, and J. Otto for comments on the manuscript.

References

1. Jackson, J. H., Cochrane, C. G., Bourne, J. R., Solski, P. A., Buss, J. E., and Der, C. J. (1990). *Proc. Natl. Acad. Sci. U.S.A.* **87**, 3042.
2. Kato, K., Cox, A. D., Hisaka, M. M., Graham, S. M., Buss, J. E., and Der, C. J. (1992). *Proc. Natl. Acad. Sci. U.S.A.* **89**, 6403.
3. Gibbs, J. B., Graham, S. L., Hartman, G. D., Koblan, K. S., Kohl, N. E., Omer, C. A., and Oliff, A. (1997). *Curr. Opin. Chem. Biol.* **1**, 197.
4. Gibbs, J. B., Oliff, A., and Kohl, N. E. (1994). *Cell* **77**, 175.
5. James, G. L., Goldstein, J. L., Brown, M. S., Rawson, T. E., Somers, T. C., McDowell, R. S., Crowley, C. W., Lucas, B. K., Levinson, A. D., and Marsters, J. C., Jr. (1993). *Science* **260**, 1937.
6. Kohl, N. E., Wilson, F. R., Mosser, S. D., Giuliani, E., DeSolms, S. J., Conner, M. W., Anthony, N. J., Holtz, W. J., Gomez, R. P., Lee, T.-J., Smith, R. L., Graham, S. L., Hartman, G. D., Gibbs, J. B., and Oliff, A. (1994). *Proc. Natl. Acad. Sci. U.S.A.* **91**, 9141.
7. Kohl, N. E., Omer, C. A., Conner, M. W., Anthony, N. J., Davide, J. P., deSolms, S. J., Giuliani, E. A., Gomez, R. P., Graham, S. L., Hamilton, K., Handt, L. K., Hartman, G. D., Koblan, K. S., Kral, A. M., Miller, P. J., Mosser, S. D., O'Neill, T. J., Rands, E., Schaber, M. D., Gibbs, J. B., and Oliff, A. (1995). *Nature Med.* **1**, 792.
8. Tamanoi, F. (1993). *Trends Biochem. Sci.* **18**, 349.
9. Qian, Y., Sebti, S. M., and Hamilton, A. D. (1997). *Biopolymers* **43**, 25.
10. Yokoyama, K., Trobridge, P., Buckner, F. S., Van Voorhis, W. C., Stuart, K. D., and Gelb, M. H. (1998). *J. Biol. Chem.* **273**, 26497.
11. Casey, P. J., and Seabra, M. C. (1996). *J. Biol. Chem.* **271**, 5289.
12. Zhang, F. L., and Casey, P. J. (1996). *Annu. Rev. Biochem.* **65**, 241.
13. Boyartchuk, V. L., Ashby, M. N., and Rine, J. (1997). *Science* **275**, 1796.
14. Kim, E., Ambroziak, P., Otto, J. C., Taylor, B., Ashby, M., Shannon, K., Casey, P. J., and Young, S. G. (1999). *J. Biol. Chem.* **274**, 8383.
15. Otto, J. C., Kim, E., Young, S. G., and Casey, P. J. (1999). *J. Biol. Chem.* **274**, 8379.
16. Tam, A., Nouvet, F. J., Fujimura-Kamada, K., Slunt, H., Sisodia, S. S., and Michaelis, S. (1998). *J. Cell Biol.* **142**, 635.
17. Kumagai, H., Kawamura, Y., Yanagisawa, K., and Komano, H. (1999). *Biochim. Biophys. Acta* **1426**, 468.
18. Dai, Q., Choy, E., Chiu, V., Romano, J., Slivka, S. R., Steitz, S. A., Michaelis, S., and Philips, M. R. (1998). *J. Biol. Chem.* **273**, 15030.
19. Clarke, S., Vogel, J. P., Deschenes, R. J., and Stock, J. (1988). *Proc. Natl. Acad. Sci. U.S.A.* **85**, 4643.
20. Gutierrez, L., Magee, A. I., Marshall, C. J., and Hancock, J. F. (1989). *EMBO J.* **8**, 1093.
21. Fujiyama, A., Tsunasawa, S., Tamanoi, F., and Sakiyama, F. (1991). *J. Biol. Chem.* **266**, 17926.
22. Hrycyna, C. A., and Clarke, S. (1992). *J. Biol. Chem.* **267**, 10457.
23. Hrycyna, C. A., and Clarke, S. (1993). *Pharmacol. Ther.* **59**, 281.
24. Ashby, M. N., King, D. S., and Rine, J. (1992). *Proc. Natl. Acad. Sci. U.S.A.* **89**, 4613.
25. Ashby, M. N., and Rine, J. (1995). *Methods Enzymol.* **250**, 235.
26. Ma, Y.-T., and Rando, R. R. (1992). *Proc. Natl. Acad. Sci. U.S.A.* **89**, 6275.
27. Ma, Y.-T., Gilbert, B. A., and Rando, R. R. (1993). *Biochemistry* **32**, 2386.
28. Ma, Y.-T., Chaudhuri, A., and Rando, R. R. (1992). *Biochemistry* **31**, 11772.
29. Jang, G.-F., and Gelb, M. H. (1998). *Biochemistry* **37**, 4473.

30. Jang, G.-F., Yokoyama, K., and Gelb, M. H. (1993). *Biochemistry* **32,** 9500.
31. Akopyan, T. N., Couedel, Y., Orlowski, M., Fournie-Zaluski, M.-C., and Roques, B. P. (1994). *Biochem. Biophys. Res. Commun.* **198,** 787.
32. Chen, Y., Ma, Y.-T., and Rando, R. R. (1996). *Biochemistry* **35,** 3227.
33. Ashby, M. N. (1998). *Curr. Opin. Lipidol.* **9,** 99.
34. Nishii, W., Muramatsu, T., Kuchino, Y., Yokoyama, S., and Takahashi, K. (1997). *J. Biochem.* **122,** 402.
35. Gilbert, B. A., Ma, Y.-T., and Rando, R. R. (1995). *Methods Enzymol.* **250,** 206.
36. Chen, Y. (1998). *Cancer Lett.* **131,** 191.
37. Heilmeyer, L. M. G., Jr., Serwe, M., Weber, C., Metzger, J., Hoffman-Posorske, E., and Meyer, H. E. (1992). *Proc. Natl. Acad. Sci. U.S.A.* **89,** 9554.
38. Ma, Y.-T., and Rando, R. R. (1993). *FEBS Lett.* **332,** 105.
39. Fujimura-Kamada, K., Nouvet, F. J., and Michaelis, S. (1997). *J. Cell Biol.* **136,** 271.
40. Kataoka, T., Powers, S., McGill, C., Fasano, O., Strathern, J., Broach, J., and Wigler, M. (1984). *Cell* **37,** 437.
41. Boyartchuk, V. L., and Rine, J. (1998). *Genetics* **150,** 95.
42. Chen, P. (1997). "Biogenesis of the Yeast Mating Pheromone a-Factor." Ph.D. Thesis. Johns Hopkins University School of Medicine, Baltimore, Maryland.
43. Sapperstein, S., Berkower, C., and Michaelis, S. (1994). *Mol. Cell. Biol.* **14,** 1438.
44. Chen, P., Sapperstein, S. K., Choi, J. D., and Michaelis, S. (1997). *J. Cell Biol.* **136,** 251.
45. Adames, N., Blundell, K., Ashby, M. N., and Boone, C. (1995). *Science* **270,** 464.
46. Booden, M. A., Baker, T. L., Solski, P. A., Der, C. J., Punke, S. G., and Buss, J. E. (1999). *J. Biol. Chem.* **274,** 1423.
47. Dudler, T., and Gelb, M. H. (1996). *J. Biol. Chem.* **271,** 11541.
48. Willumsen, B. M., Cox, A. D., Solski, P. A., Der, C. J., and Buss, J. E. (1996). *Oncogene* **13,** 1901.
49. Dudler, T., and Gelb, M. H. (1997). *Biochemistry* **36,** 12434.
50. Schmidt, W. K., Tam, A., Fujimura-Kamada, K., and Michaelis, S. (1998). *Proc. Natl. Acad. Sci. U.S.A.* **95,** 11175.
51. Kilic, F., Dalton, M. B., Burrell, S. K., Mayer, J. P., Patterson, S. D., and Sinensky, M. (1997). *J. Biol. Chem.* **272,** 5298.
52. Sakagami, Y., Yoshida, M., Isogai, A., and Suzuki, A. (1981). *Science* **212,** 1525.
53. Sakagami, Y., Isogai, A., Suzuki, A., Tamura, S., Kitada, C., and Fujino, M. (1979). *Agric. Biol. Chem.* **43,** 2643.
54. Ishibashi, Y., Sakagami, Y., Isogai, A., and Suzuki, A. (1984). *Biochemistry* **23,** 1399.
55. Schmidt, R. A., Schneider, C. J., and Glomset, J. A. (1984). *J. Biol. Chem.* **259,** 10175.
56. Bruenger, E., and Rilling, H. C. (1986). *Biochem. Biophys. Res. Commun.* **139,** 209.
57. Powers, S., Michaelis, S., Broek, D., Santa Anna, S., Field, J., Herskowitz, I., and Wigler, M. (1986). *Cell* **47,** 413.
58. Becker, J. M., Marcus, S., Kundu, B., Shenbagamurthi, P., and Naider, F. (1987). *Mol. Cell. Biol.* **7,** 4122.
59. Chen, Z. Q., Ulsh, L. S., DuBois, G., and Shih, T. Y. (1985). *J. Virol.* **56,** 607.
60. Fujiyama, A., Matsumoto, K., and Tamanoi, F. (1987). *EMBO J.* **6,** 223.
61. Clarke, S. (1985). *Annu. Rev. Biochem.* **54,** 479.
62. Swanson, R. J., and Applebury, M. L. (1983). *J. Biol. Chem.* **258,** 10599.
63. Chelsky, D., Olson, J. F., and Koshland, D. E., Jr. (1987). *J. Biol. Chem.* **262,** 4303.
64. Anderegg, R. J., Betz, R., Carr, S. A., Crabb, J. W., and Duntze, W. (1988). *J. Biol. Chem.* **263,** 18236.

65. Hancock, J. F., Magee, A. I., Childs, J. E., and Marshall, C. J. (1989). *Cell* **57,** 1167.
66. Mumby, S. M., Casey, P. J., Gilman, A. G., Gutowski, S., and Sternweis, P. C. (1990). *Proc. Natl. Acad. Sci. U.S.A.* **87,** 5873.
67. Casey, P. J., Solski, P. A., Der, C. J., and Buss, J. E. (1989). *Proc. Natl. Acad. Sci. U.S.A.* **86,** 8323.
68. Clarke, S. (1992). *Annu. Rev. Biochem.* **61,** 355.
69. Deschenes, R. J., Stimmel, J. B., Clarke, S., Stock, J., and Broach, J. R. (1989). *J. Biol. Chem.* **264,** 11865.
70. Ota, I. M., and Clarke, S. (1989). *J. Biol. Chem.* **264,** 12879.
71. Ong, O. C., Ota, I. M., Clarke, S., and Fung, B. K.-K. (1989). *Proc. Natl. Acad. Sci. U.S.A.* **86,** 9238.
72. Fung, B. K.-K., Yamane, H. K., Ota, I. M., and Clarke, S. (1990). *FEBS Lett.* **260,** 313.
73. Yamane, H. K., Farnsworth, C. C., Xie, H., Howald, W., Fung, B. K.-K., Clarke, S., Gelb, M. H., and Glomset, J. A. (1990). *Proc. Natl. Acad. Sci. U.S.A.* **87,** 5868.
74. Stimmel, J. B., Deschenes, R. J., Volker, C., Stock, J., and Clarke, S. (1990). *Biochemistry* **29,** 9651.
75. Xie, H., Yamane, H. K., Stephenson, R. C., Ong, O. C., Fung, B. K.-K., and Clarke, S. (1990). *Methods* **1,** 276.
76. Stephenson, R. C., and Clarke, S. (1990). *J. Biol. Chem.* **265,** 16248.
77. Volker, C., Lane, P., Kwee, C., Johnson, M., and Stock, J. (1991). *FEBS Lett.* **295,** 189.
78. Volker, C., Miller, R. A., McCleary, W. R., Rao, A., Poenie, M., Backer, J. M., and Stock, J. B. (1991). *J. Biol. Chem.* **266,** 21515.
79. Pérez-Sala, D., Gilbert, B. A., Tan, E. W., and Rando, R. R. (1992). *Biochem. J.* **284,** 835.
80. Tan, E. W., Pérez-Sala, D., Cañada, F. J., and Rando, R. R. (1991). *J. Biol. Chem.* **266,** 10719.
81. Farnsworth, C. C., Kawata, M., Yoshida, Y., Takai, Y., Gelb, M. H., and Glomset, J. A. (1991). *Proc. Natl. Acad. Sci. U.S.A.* **88,** 6196.
82. Farnsworth, C. C., Seabra, M. C., Ericsson, L. H., Gelb, M. H., and Glomset, J. A. (1994). *Proc. Natl. Acad. Sci. U.S.A.* **91,** 11963.
83. Newman, C. M. H., Giannakouros, T., Hancock, J. F., Fawell, E. H., Armstrong, J., and Magee, A. I. (1992). *J. Biol. Chem.* **267,** 11329.
84. Giannakouros, T., Newman, C. M. H., Craighead, M. W., Armstrong, J., and Magee, A. I. (1993). *J. Biol. Chem.* **268,** 24467.
85. Parish, C. A., and Rando, R. R. (1994). *Biochemistry* **33,** 9986.
86. Beranger, F., Cadwallader, K., Porfiri, E., Powers, S., Evans, T., de Gunzburg, J., and Hancock, J. F. (1994). *J. Biol. Chem.* **269,** 13637.
87. Wei, C., Lutz, R., Sinensky, M., and Macara, I. G. (1992). *Oncogene* **7,** 467.
88. Smeland, T. E., Seabra, M. C., Goldstein, J. L., and Brown, M. S. (1994). *Proc. Natl. Acad. Sci. U.S.A.* **91,** 10712.
89. Hrycyna, C. A., and Clarke, S. (1990). *Mol. Cell. Biol.* **10,** 5071.
90. Hrycyna, C. A., Sapperstein, S. K., Clarke, S., and Michaelis, S. (1991). *EMBO J.* **10,** 1699.
91. Hrycyna, C. A., Yang, M. C., and Clarke, S. (1994). *Biochemistry* **33,** 9806.
92. Sapperstein, S., Berkower, C., and Michaelis, S. (1989). *In* "DNA Sequence Analysis of *STE14,* a Gene Required for Biogenesis of the *Saccharomyces cerevisiae* **a**-Factor Mating Pheromone," p. 139. Cold Spring Harbor Press, Cold Spring Harbor, New York.
93. Ashby, M. N., Errada, P. R., Boyartchuk, V. L., and Rine, J. (1993). *Yeast* **9,** 907.

94. Romano, J. D., Schmidt, W. K., and Michaelis, S. (1998). *Mol. Biol. Cell* **9,** 2231.
95. Parish, C. A., Smrcka, A. V., and Rando, R. R. (1996). *Biochemistry* **35,** 7499.
96. Herskowitz, I. (1987). *Nature (London)* **329,** 219.
97. Goodman, L. E., Judd, S. R., Farnsworth, C. C., Powers, S., Gelb, M. H., Glomset, J. A., and Tamanoi, F. (1990). *Proc. Natl. Acad. Sci. U.S.A.* **87,** 9665.
98. Fujiyama, A., and Tamanoi, F. (1990). *J. Biol. Chem.* **265,** 3362.
99. Gilbert, B. A., Tan, E. W., Pérez-Sala, D., and Rando, R. R. (1992). *J. Am. Chem. Soc.* **114,** 3966.
100. Stephenson, R. C., and Clarke, S. (1992). *J. Biol. Chem.* **267,** 13314.
101. Pillinger, M. H., Volker, C., Stock, J. B., Weissman, G., and Philips, M. R. (1994). *J. Biol. Chem.* **269,** 1486.
102. Backlund, P. S., Jr., and Aksamit, R. R. (1988). *J. Biol. Chem.* **263,** 15864.
103. Huzoor-Akbar, Winegar, D. A., and Lapetina, E. G. (1991). *J. Biol. Chem.* **266,** 4387.
104. Gingras, D., Boivin, D., and Béliveau, R. (1993). *Am. J. Physiol.* **265,** F316.
105. Desrosiers, R. R., and Béliveau, R. (1998). *Arch. Biochem. Biophys.* **351,** 149.
106. Backlund, P. S., Jr. (1992). *J. Biol. Chem.* **267,** 18432.
107. Chelsky, D., Sobotka, C., and O'Neill, C. L. (1989). *J. Biol. Chem.* **264,** 7637.
108. Haklai, R., and Kloog, Y. (1991). *Cell. Mol. Neurobiol.* **11,** 415.
109. O'Dea, R. F., Viveros, O. H., Axelrod, J., Aswanikumar, S., Schiffmann, E., and Corcoran, B. A. (1978). *Nature (London)* **272,** 462.
110. Venkatasubramanian, K., Hirata, F., Gagnon, C., Corcoran, B. A., O'Dea, R. F., Axelrod, J., and Schiffman, E. (1980). *Mol. Immunol.* **17,** 201.
111. Aksamit, R. R., Falk, W., and Cantoni, G. L. (1982). *J. Biol. Chem.* **257,** 621.
112. Aksamit, R. R., Backlund, P. S., Jr., and Cantoni, G. L. (1983). *J. Biol. Chem.* **258,** 20.
113. Lederer, E. D., Jacobs, A. A., Hoffman, J. L., Harding, G. B., Robishaw, J. D., and McLeish, K. R. (1994). *Biochem. Biophys. Res. Commun.* **200,** 1604.
114. Philips, M. R., Pillinger, M. H., Staud, R., Volker, C., Rosenfeld, M. G., Weissmann, G., and Stock, J. B. (1993). *Science* **259,** 977.
115. McLeish, K. R., Gierschik, P., and Jakobs, K. H. (1989). *Mol. Pharmacol.* **36,** 384.
116. Philips, M. R., Staud, R., Pillinger, M., Feoktistov, A., Volker, C., Stock, J. B., and Weissmann, G. (1995). *Proc. Natl. Acad. Sci. U.S.A.* **92,** 2283.
117. Ding, J., Lu, D. J., Pérez-Sala, D., Ma, Y. T., Maddox, J. F., Gilbert, B. A., Badwey, J. A., and Rando, R. R. (1994). *J. Biol. Chem.* **269,** 16837.
118. Scheer, A., and Gierschik, P. (1993). *FEBS Lett.* **319,** 110.
119. Scheer, A., and Gierschik, P. (1995). *Biochemistry* **34,** 4952.
120. Huzoor-Akbar, Wang, W., Kornhauser, R., Volker, C., and Stock, J. B. (1993). *Proc. Natl. Acad. Sci. U.S.A.* **90,** 868.
121. Ma, Y.-T., Shi, Y.-Q., Lim, Y. H., McGrail, S. H., Ware, J. A., and Rando, R. R. (1994). *Biochemistry* **33,** 5414.
122. Marcus, S., Caldwell, G. A., Miller, D., Xue, C.-B., Naider, F., and Becker, J. M. (1991). *Mol. Cell. Biol.* **11,** 3603.
123. Rosenberg, S. J., Rane, M. J., Dean, W. L., Corpier, C. L., Hoffman, J. L., and McLeish, K. R. (1998). *Cell. Signal.* **10,** 131.
124. Chen, F., and Lee, R. H. (1997). *Biochem. Biophys. Res. Commun.* **233,** 370.
125. Fukada, Y., Matsuda, T., Kokame, K., Takao, T., Shimonishi, Y., Akino, T., and Yoshizawa, T. (1994). *J. Biol. Chem.* **269,** 5163.
126. Musha, T., Kawata, M., and Takai, Y. (1992). *J. Biol. Chem.* **267,** 9821.
127. Parish, C. A., and Rando, R. R. (1996). *Biochemistry* **35,** 8473.

128. Pérez-Sala, D., Tan, E. W., Cañada, F. J., and Rando, R. R. (1991). *Proc. Natl. Acad. Sci. U.S.A.* **88,** 3043.
129. Tan, E. W., and Rando, R. R. (1992). *Biochemistry* **31,** 5572.
130. Dunten, R. L., Wait, S. J., and Backlund, P. S., Jr. (1995). *Biochem. Biophys. Res. Commun.* **208,** 174.
131. Imai, Y., Davey, J., Kawagishi-Kobayashi, M., and Yamamoto, M. (1997). *Mol. Cell. Biol.* **17,** 1543.
132. Kagan, R. M., and Clarke, S. (1994). *Arch. Biochem. Biophys.* **310,** 417.
133. Boivin, D., Lin, W., and Béliveau, R. (1997). *Biochem. Cell Biol.* **75,** 63.
134. Backlund, P. S., Jr. (1997). *J. Biol. Chem.* **272,** 33175.

8

Reversible Modification of Proteins with Thioester-Linked Fatty Acids

MAURINE E. LINDER

Department of Cell Biology and Physiology
Washington University School of Medicine
St. Louis, Missouri 63110

I. Introduction

Protein palmitoylation was the first covalent lipid modification described in eukaryotic cells. Reports that brain myelin proteolipid protein contained ester-linked palmitic, oleic, and stearic acids appeared in the early 1970s (*1–3*), but were not widely recognized as harboring the discovery of a new protein modification. Schmidt and Schlesinger in 1979 rediscovered protein palmitoylation as a modification of viral glycoproteins (*4*) and, importantly, established that it is a ubiquitous event in eukaryotic cells (*5*). Today dozens

THE ENZYMES, Vol. XXI

TABLE I

SELECTED THIOACYLATED PROTEINS[a]

Protein	Modified sequence	Ref.
Class I		
CD36	^{1}MGC$\underline{\text{D}}$RN$\underline{\text{C}}^{7}$---------^{464}CA$\underline{\text{C}}$RSKTIK^{472}	9
β_2-Adrenergic receptor	----^{337}QELL$\underline{\text{C}}$LRR^{344}---	10
α_{2A}-Adrenergic receptor	----^{338}KKIL$\underline{\text{C}}$RGDR-COOH	11
Rhodopsin	----^{319}TTL$\underline{\text{CC}}$GKN-COOH	12
Class II		
CD 4	---^{390}GIFFCV/R$\underline{\text{C}}$RHRRR^{402}---	13
Cation-dependent mannose 6-phosphate receptor	----^{25}RPKSR$\underline{\text{C}}$VFD$\underline{\text{C}}^{34}$---	14
p63	----^{95}SSSASCSRRLGR^{106}----	15, 16
Transferrin receptor	----^{59}K$\underline{\text{R}}$$\underline{\text{C}}$SGSI$\underline{\text{C}}$/YGTIAV^{73}	17, 18
Vesicular stomatitis virus G	----^{481}VL/RVGIHL$\underline{\text{C}}$IKLK^{493}---	4, 19
Class II		
$G_z\alpha$, $G_i\alpha_{1,2,3}$, $G_o\alpha$	^{1}MG$\underline{\text{C}}$----	20–24
p59^{fyn}	^{1}MG$\underline{\text{C}}$V$\underline{\text{C}}$----	25
p56^{lck}	^{1}MG$\underline{\text{C}}$VQ$\underline{\text{C}}$	26–28
H-Ras	----G$\underline{\text{C}}$MS$\underline{\text{C}}$K**C**VLS-COOH	29
N-Ras	----G$\underline{\text{C}}$MGLP**C**VVM-COOH	29
$G_s\alpha$	^{1}MG$\underline{\text{C}}$----	22, 30
$G_q\alpha$	^{1}MTLESIMA$\underline{\text{CC}}$---	22, 30
$G_{12}\alpha$	^{1}MSGVVRTLSR$\underline{\text{C}}$LL---	31, 32
GAP-43 (neuromodulin)	^{1}ML$\underline{\text{CC}}$M---	33, 34
PSD-95	^{1}MD$\underline{\text{C}}$L$\underline{\text{C}}$ITT---	35
SNAP-25	---^{83}KF$\underline{\text{C}}$GL$\underline{\text{C}}$V$\underline{\text{C}}$P$\underline{\text{C}}$NKL^{95}---	36, 37

[a] Thioacylated cysteines are underlined ($\underline{\text{C}}$). Prenylated cysteines are indicated in boldface type (**C**). N-Myristoylated glycines are indicated in boldface type (**G**). The superscript numbers refer to the amino acid position in the protein sequence. A slash mark (/) signifies the border between the transmembrane and cytoplasmic domain of the polypeptide.

of viral and cellular proteins are recognized as substrates for thioester-linked fatty acylation. The term *palmitoylation* has come into general use to describe the covalent attachment of fatty acids to proteins at cysteine residues through a thioester linkage. However, protein thioacylation or protein S-fatty acylation are more appropriate names because of the diversity of fatty acids that have been detected on thioacylated proteins.

Protein thioacylation is a posttranslational event that is almost exclusively a property of membrane-associated proteins (reviewed in Refs. *6–8*). Thioacylated cysteine residues are found in a broad spectrum of sequence contexts [Table I (*9–37*)]. These characteristics distinguish protein thioacylation from the other mode of protein fatty acylation, N-myristoylation (*38*), and from isoprenylation (39). Myristic acid is added to proteins through an amide linkage at an amino-terminal glycine residue that is exposed on removal of the initiator methionine. Isoprenoid modification occurs at cysteine residues near the carboxyl terminus through a thioether linkage. N-Myristoylation and prenylation occur at residues within well-characterized sequence motifs that are described elsewhere in this volume. Both of these processes occur in the cytoplasm, cotranslationally for N-myristoylation, or shortly after protein synthesis for prenylation. In contrast, thioacylation takes place on cellular membranes and its substrates include proteins synthesized on both soluble and membrane-bound polysomes. Perhaps the most important distinction between protein thioacylation and N-myristoylation and prenylation is reversibility. N-Myristoylation and prenylation are stable modifications. However, for many thioacylated proteins, the half-life of thioester-linked fatty acid is much shorter than that of the polypeptide, indicating that proteins undergo cycles of acylation and deacylation. Thus, protein thioacylation has the potential to be regulated.

The dynamic nature of protein thioacylation dictates that mechanisms exist not only for the addition, but also for the removal, of fatty acids from proteins. Despite more than 20 years of research on protein thioacylation, our understanding of the enzymology of these processes is in its infancy. Significant progress has been made in purifying palmitoyl-protein thioesterase activities and cloning the genes corresponding to these activities, permitting our first glimpse of the physiological settings in which protein deacylation is functionally important (40, 41). The mechanism of fatty acid addition to proteins is controversial. Although several protein acyltransferase activities have been purified to apparent homogeneity, it has not been established that these enzymes mediate protein thioacylation in cells (42, 43). Spontaneous and efficient transfer of fatty acid from acyl-CoA to proteins has been observed in vitro, suggesting that thioacylation may be an autocatalytic process in vivo (44, 45).

This review surveys our current understanding of the mechanisms underlying protein thioacylation and deacylation. The structural requirements for this modification and the intracellular sites where proteins are thioacylated are reviewed. This is followed by a discussion of the evidence supporting enzymatic versus nonenzymatic mechanisms for the addition of fatty acids to proteins. Finally, the properties, localization, and function of palmitoyl-protein thioesterases are presented and discussed.

II. Structural Requirements for Protein Thioacylation

Thioacylated proteins can be classified according to their membrane topology. Representative proteins from each class are shown in Table I.

Polytopic membrane proteins that are thioacylated include CD36, an 88-kDa glycoprotein that is palmitoylated on both N- and C-terminal cytoplasmic tails (9), and many G protein-coupled receptors (GPCRs) (46). Rhodopsin was the first receptor to be characterized as a palmitoylated protein and one of the few proteins in which the modification has been rigorously characterized chemically. The fatty acid composition of bovine rhodopsin is predominantly palmitate (83%), with smaller amounts of oleic, stearic, and linoleic fatty acids (47). The sites of modification have been mapped to cysteines 322 and 323, using a combination of chemical and enzymatic cleavage in conjunction with tandem mass spectroscopy (12). These residues are found within the cytoplasmic carboxyl-terminal tail of the protein. Introduction of fluorescent fatty acid analogs into purified rhodopsin has allowed the membrane accessibility of the palmitoylation sites in rhodopsin to be assessed. These studies indicate that the fatty acyl chains are inserted in the membrane, thereby creating a fourth cytoplasmic loop of the protein (48). This structural motif appears to be conserved in a large number of GPCRs (46). However, a clear conserved functional role has not emerged from studies of thioacylation-defective receptor mutants (46, 49). For the β-adrenergic receptor, accessibility of the cytoplasmic tail to protein kinases may be regulated by reversible thioacylation. Mutation of the thioacylation site in the β-adrenergic receptor results in uncoupling of the receptor from G protein, presumably because of hyperphosphorylation of the C-terminal cytoplasmic tail and chronic desensitization of the receptor (50). In contrast, the corresponding mutation in the α_2-adrenergic receptor does not perturb G protein coupling (51). Clarification of the role of fatty acylation in receptor function will require studies of a larger number of receptors.

Monotopic membrane proteins constitute the second class of thioacylated proteins (Table I) that includes viral glycoproteins, the transferrin receptor, and CD4, a cell surface protein expressed in lymphocytes that acts as a coreceptor for the T cell receptor. These proteins are typically modified at one or more cysteine residues near the cytoplasm–membrane interface. Modification can occur within the predicted transmembrane span or within the cytoplasmic domain of the protein. The lack of a consensus sequence for thioacylation of monotopic proteins suggested that any free cysteine in the vicinity of the cytoplasmic membrane border might be a site for thioacylation. However, the small hydrophobic protein of human respiratory syncytial virus is not acylated and has a cysteine residue located in this context (52). Furthermore, studies of chimeric mutant proteins of the acylated influenza virus hemagglutinin protein and a nonacylated Sendai virus fusion protein reveal that cysteine residues in a juxtamembranous region are not sufficient for thioacylation (53). The signals for thioacylation appear to be located primarily within the transmembrane domain of the hemagglutinin protein, but also involve the cytoplasmic tail. Interestingly, the length of the cytoplasmic tail also appears to influence the fatty acid incorporated into viral glycoproteins (54).

The functional consequences of protein thioacylation are poorly understood for many monotopic membrane proteins (reviewed in Refs. 7 and 55). One potential function for thioacylation is the regulation of protein trafficking. Consistent with this possibility, mutation of the thioacylation sites in the transferrin receptor results in an increased rate of endocytosis of the receptor relative to the thioacylated form of the protein (56). The cation-dependent mannose 6-phosphate receptor (CD-MPR) is reversibly thioacylated at cysteine residues that are approximately 30 amino acids distant from the predicted end of the transmembrane domain (see Table I) (14). The cytoplasmic tail of the CD-MPR contains multiple signals that control receptor internalization and trafficking (57, 58). Mutation of residues 35–39 in the receptor impairs palmitoylation and results in abnormal receptor trafficking to lysosomes and the loss of lysosomal enzyme-sorting function. Anchoring of the cytoplasmic tail of the receptor at the plasma membrane interface may regulate its interactions with the cellular machinery that controls receptor trafficking (14).

The third major class of thioacylated proteins is made up of hydrophilic proteins that lack transmembrane spans (Table I) (8, 59). These proteins are synthesized on soluble ribosomes and become thioacylated when associated with a membrane compartment. Within this class are proteins that undergo sequential lipid modifications, either prenylation followed by thioacylation or N-myristoylation followed by thioacylation. All members of the Ras family of proteins are farnesylated at a cysteine residue four amino acids from the C terminus, followed by proteolytic cleavage of the C-terminal three amino acids, and carboxyl methylation of the farnesylated cysteine. H-Ras and N-Ras are subsequently fatty acylated at cysteine residues immediately upstream of the farnesylated C terminus. All members of the Src family of protein tyrosine kinases and the $G_i\alpha$ subfamily are N-myristoylated. The majority of these proteins are also thioacylated at one or more cysteine residues that are located adjacent to or near the myristoylated glycine. Thioacylation of dually modified proteins is dependent on prior modification with either the prenyl or N-myristoyl group. The first lipid modification facilitates membrane binding of the protein, bringing it to a compartment where thioacylation can take place. In the absence of the N-myristoylation or prenylation, other mechanisms that provide affinity for membranes can substitute for the addition of the first lipid. For example, myristoylation-defective $G_i\alpha$ coexpressed with $G\beta\gamma$ subunits is thioacylated (20). The prenyl group on the $G\gamma$ subunit serves as the membrane anchor for the heterotrimer, providing $G_i\alpha$ access to the palmitoylating machinery at the membrane.

A second group of hydrophilic thioacylated proteins is modified exclusively with thioester-linked fatty acids. Within this group are three of the major substrates for thioacylation in the nervous system: GAP-43/neuromodulin, an abundant protein in neuronal growth cones (33); SNAP-25, a protein essential for synaptic vesicle exocytosis (36); and PSD-95, a scaffolding protein that facilitates the localization of ion channels and receptors at postsynaptic sites (35). Most of the proteins within this group are modified at cysteine residues within the 20 N-terminal amino acids (Table I). SNAP-25 is atypical in that the thioacylated cysteine residues are in a central domain of the protein between two α helices that form coiled–coil interactions with other proteins at the presynaptic membrane (60).

Thioacylation provides a membrane anchor for class III proteins (Table I), presumably by insertion of the fatty acid into the lipid bilayer. In addition to contributing to the membrane avidity of some class III proteins (reviewed in Refs. 8 and 61), thioacylation also appears to be an important determinant of certain protein–protein interactions. This has been best studied with heterotrimeric G proteins and has been reviewed in detail elsewhere (8, 62). Both positive and negative regulatory roles for thioacylation of $G\alpha$ are indicated by reconstitution of various G-protein activities in vitro. Thioacylation of $G_s\alpha$ increases its affinity for $G\beta\gamma$ subunits (63). Comparison of palmitoylated $G_s\alpha$ to protein that had been enzymatically deacylated revealed a fivefold difference in the affinity between palmitoylated and nonpalmitoylated G protein. A negative regulatory role for thioacylation of $G\alpha$ has been demonstrated for its interactions with RGS (regulators of G-protein signaling) proteins. Members of the RGS family act as GTPase-activating proteins or GAPs for G-protein α subunits, thereby regulating the deactivation of G-protein pathways (reviewed in Ref. 64). Thioacylation of $G\alpha$ dramatically inhibits the GAP activity of RGS proteins in vitro, suggesting that thioacylation may be an important determinant of the duration of G protein-mediated responses (65). It will be important to address whether this mode of regulation is found in vivo.

The strength of these biochemical analyses is that the function of thioacylation is measured directly by comparison of stoichiometrically modified protein to unmodified protein. In vivo, the functional consequences of thioacylation are typically assessed by analyzing the phenotype of thioacylation-defective mutants. Interpretation of these experiments is difficult because it is not possible to distinguish whether a defect in protein function is due to lack of the lipid modification or to the change in amino acid residue. This problem underscores the need to identify enzymes that mediate protein thioacylation. Genetic and pharmacological manipulation of the enzymes that acylate and deacylate proteins will have enormous value for investigating the functional significance of protein thioacylation.

III. Intracellular Sites of Thioacylation

The posttranslational addition of palmitate to proteins occurs at multiple sites within the cell. Thioacylation of many transmembrane-spanning proteins occurs during transit of the newly synthesized proteins through the secretory pathway. Viral glycoproteins are thioacylated after translocation into the endoplasmic reticulum (ER) and completion of polypeptide synthesis, but prior to trimming of the high-mannose residues on the protein-bound oligosaccharide, a modification that occurs in the *cis*-Golgi (*66*). Thioacylation is blocked by incubation of cells at 15°C, a condition that also prevents transit from the ER to the Golgi (*67*). Proteins accumulate in structures that are distinct from ER and Golgi during the 15°C block, suggesting that the protein has moved into a compartment designated as "intermediate" between the ER and *cis*-Golgi. Release of the temperature block results in rapid thioacylation, followed by trimming of the oligosaccharides in the *cis*-Golgi. These data are consistent with thioacylation occurring in the intermediate compartment and suggest that protein acyltransferases are localized there. Interestingly, p63, a thioacylated transmembrane protein that is a resident of the ER, is only weakly labeled with tritiated palmitate at steady state (*68*). p63 thioacylation is dramatically increased in cells treated with brefeldin A (BFA) (*15, 69*), a fungal metabolite that blocks anterograde transport through the secretory pathway (*70*). The Golgi apparatus disassembles in cells treated with BFA, and Golgi resident proteins are redistributed into the ER. Protein acyltransferase activity associated with the intermediate compartment may have greater access to p63 when BFA induces the mixing of compartments in the early secretory pathway (*16*).

The majority of thioacylated proteins are associated with the plasma membrane. Analysis of total cellular proteins radiolabeled with tritiated palmitate reveals that most palmitoylated proteins cofractionate with plasma membrane markers, whereas N-myristoylated proteins are more widely dispersed among subcellular compartments (*71*). This finding is not surprising in that many substrates for thioacylation are participants in signal transduction cascades. The major substrates for thioacylation in brain are also associated with the plasma membrane (*33, 35, 36*). The concentration of thioacylated proteins at the plasma membrane and the dynamic nature of the modification strongly suggest that fatty acid addition must occur at this site. As discussed below, protein acyltransferase (PAT) activity has been detected in plasma membrane fractions derived from rat liver (72) and in erythrocyte membranes that contain only surface membranes (6).

Class III proteins are synthesized on soluble polysomes and must associate with a membrane compartment to be thioacylated. The posttranslational processing and trafficking of newly synthesized protein have been characterized for several class III proteins and reveal that there are distinct pathways for delivery of these proteins to the plasma membrane. These differences may be observed even within members of the same protein family. p56^{lck} and p59^{fyn} are tandemly modified with amide-linked myristate and thioester-linked palmitate (25, 26). High-affinity binding of the kinases to membranes requires both lipid modifications (25, 27). N-Myristoylation occurs in the cytoplasm, whereas thioacylation is tightly coupled to membrane association because mutant kinases that cannot be thioacylated fail to bind well to membranes. Newly synthesized p59^{fyn} is thioacylated and targeted to the plasma membrane rapidly, within 5 min of synthesis (73). Thioacylation is believed to occur at the plasma membrane as there is no evidence that newly synthesized p59^{fyn} is associated with another membrane compartment prior to its association with the plasma membrane. In contrast, the kinetics of membrane association of newly synthesized p56^{lck} are much slower than those reported for p59^{fyn} (74). p56^{lck} binds to the T cell coreceptors CD4 and CD8 shortly after associating with membranes. Intracellular CD4 can be distinguished from cell surface CD4 by the extent of oligosaccharide processing, providing a convenient assay for the localization p56^{lck} bound to CD4. Newly synthesized p56^{lck} is associated with an intracellular pool of CD4 early after synthesis and with cell surface CD4 at later times, suggesting that initial binding of p56^{lck} and its thioacylation must occur on intracellular membranes, followed by transport through the exocytic pathway to the plasma membrane (74). Yet a third pathway is apparent for newly synthesized G$_z\alpha$, another dually fatty acylated protein (Table I) (75). Newly synthesized G$_z\alpha$ associates rapidly with all cellular membranes shortly after synthesis, but accumulates with time in the plasma membrane. Thioacylation of G$_z\alpha$ is observed only with the plasma membrane-associated protein, suggesting that it acquires the second lipid modification at this site.

The results of these studies support the model that has emerged from the biophysical measurements of the relative affinities of lipidated peptides for model membranes. The biophysical studies demonstrate that myristate alone provides only a weak affinity for membranes (76), whereas tandem lipid modifications are sufficient for high-affinity interactions (77). The kinetic membrane-trapping model proposed by Shahinian and Silvius is based on their measurements of the rates of interbilayer transfer of singly and dually lipid-modified peptides (77). The model proposes that a protein modified with myristate will undergo transient associations with membranes until it encounters a membrane with an appropriate "membrane-targeting receptor." This interaction leads to the addition of the second lipid. Consequently the protein becomes a permanent resident of the membrane where the second lipid modification occurred. The protein interaction that permits thioacylation need only be transient. Thus, a protein acyltransferase is an excellent candidate for the membrane-targeting receptor. If the protein acyltransferase is restricted in its localization, this provides a mechanism for targeting a thioacylated protein to a specific site within the cell. This model is supported by studies of the localization of simple lipid-modified peptides in mammalian cells (78). Short N-myristoylated or farnesylated fluorescent peptides easily penetrate cells and equilibrate into all cellular membranes (78). Peptides containing a free cysteine thiol are acylated and are retained in the cell, presumably at the membrane where they are thioacylated. Peptides that cannot be thioacylated are washed out. Interestingly, most of the peptides tested were preferentially associated with the plasma membrane, including myristoylated Gly–Cys (myr-GC) and a farnesylated N-Ras peptide (78, 79).

Retention of newly synthesized p59^{fyn}, G$_z\alpha$, and the myr-GC and N-Ras peptides at the plasma membrane may be mediated by the action of a plasma membrane-associated PAT. Plasma membrane localization of p56^{lck} does not fit this paradigm as this protein appears to be thioacylated by a protein acyltransferase activity on intracellular membranes and then move to the plasma membrane by vesicular transport. The signals that direct newly synthesized p56^{lck} to intracellular membranes for thioacylation are unknown. Binding to CD4 is an obvious mechanism, but there must be others, as the membrane-binding kinetics of p56^{lck}, and therefore thioacylation, are similar in cells that express CD4 and those that do not (74).

Accessory proteins may facilitate thioacylation, either by directing substrates to specific membranes or facilitating their interaction with PAT. Whereas G$_z\alpha$ associates transiently with intracellular membranes prior to its thioacylation at the plasma membrane (75), p59^{fyn} appears to be transported there directly, perhaps by binding to a shuttle protein (73). The heat shock proteins pp50/pp90 are possible candidates for this role as they associate with newly synthesized pp60^{v-src} and p59^{fyn} (73). Binding of G$_z\alpha$ to G$\beta\gamma$ at the plasma membrane appears to stabilize its localization there and facilitate thioacylation (72, 75).

The site where newly synthesized SNAP-25 and GAP-43 are thioacylated has not been determined, but is suspected to be on intracellular membranes. BFA and other agents that perturb the integrity of intracellular membrane compartments inhibit thioacylation of these proteins (80). This property appears to be unique to SNAP-25 and GAP-43 (74, 75, 80), suggesting that the mechanism of fatty acid addition to these proteins may be distinct from other class III proteins. A centrally located 35-amino acid domain within SNAP-25 is sufficient for thioacylation and targets green fluorescent protein (GFP) to the plasma membrane (60). Thioacylation of SNAP-25 is dependent not only on the cluster of cysteine residues at the 5' end of the domain, but also on a five-amino acid motif (QPARV) at the C terminus of the domain. This motif may serve as part of a binding site for the protein acyltransferase or as an accessory factor that facilitates SNAP-25 thioacylation.

IV. Thioacylating Activities

Despite intensive efforts, no *bona fide* protein acyltransferase activity has been purified to homogeneity, nor has the gene associated with a protein acyltransferase activity been identified. Several reports have appeared of purified activities that thioacylate proteins *in vitro*, but their *in vivo* significance as protein acyltransferases remains to be determined. Liu and coworkers purified a palmitoyltransferase activity from the microsome-enriched fraction of rat liver, using H-Ras as a substrate (42). The sequence of the purified enzyme matched that of rat peroxisomal thiolase A (3-oxoacyl-CoA thiolase A) (81), an enzyme required for fatty acid β-oxidation. Given the proper assay conditions, thiolase A can catalyze the palmitoylation of Ras at the physiological cysteine residues, using palmitoyl-CoA as the fatty acid donor. Palmitoylation of H-Ras is dependent on the presence of a farnesyl group on H-Ras (42), similar to the requirement for H-Ras farnesylation *in vivo* (29). The unlikely possibility that thiolase A acts as a protein acyltransferase *in vivo* remains to be investigated.

The soluble proteins p260/270 from *Bombyx mori* can transfer palmitate from acyl-CoA to synthetic peptides of the C-terminal regions of *Drosophila ras1* and *ras2* sequences (82). The activity is not dependent on prenylation of the peptide, and thus it is unlikely to be physiologically relevant for Ras palmitoylation. The cDNA encoding p260/270 contains a domain that resembles rat fatty-acid synthase, which synthesizes palmitate from acetyl-CoA and malonyl-CoA. Both of these enzymes, which act as protein acyltransferases *in vitro*, are related to some aspect of acyl-CoA metabolism and illustrate the facile transfer of activated fatty acids from acyl-CoA donors to protein thiols.

A protein palmitoyltransferase has been purified to apparent homogeneity from red blood cells, using spectrin as a substrate (43). The enzyme activity was released from membranes with salt and required only 86-fold purification to yield a single 70-kDa band on sodium dodecyl sulfate–

polyacrylamide gel electrophoresis (SDS–PAGE). No amino acid sequence has been reported for this protein and it remains uncertain whether the 70-kDa protein is responsible for the reported palmitoyltransferase activity.

In contrast to the activities described in the preceding section, in which acyltransferase activity is either soluble or extracted from membranes without detergent, there are a number of reports of protein acyltransferase (PAT) activity that behaves as an integral membrane protein. Schmidt and co-workers were the first to describe such an activity using placental microsomal membranes as a source of enzyme and viral glycoproteins (83), or more recently, rhodopsin (84), as protein substrates. Partial purification of PAT using p59^{fyn} (85) or G-protein heterotrimer (72) as substrates has also been reported. At present it is not known whether these activities represent the same or distinct polypeptides. In each case, enzyme activity is associated with membranes and requires detergent for solubilization. Transfer of palmitate from an acyl-CoA donor to the protein substrate occurs at the physiological sites of thioacylation. The reaction is protein mediated on the basis of its susceptibility to proteases, denaturing detergents, and heat inactivation. Highly purified preparations of detergent-solubilized PAT have not been obtained. Activity is labile after several chromatography steps and recoveries at each step are poor (85) (J. T. Dunphy and M. E. Linder, unpublished observations, 1998). Thus, the properties described below for detergent-solubilized PAT are ascribed to partially purified activities.

PAT displays broad acyl-CoA substrate specificity, with a modest preference for palmitoyl-CoA over other long-chain fatty acyl-CoAs (72, 85). Both saturated and unsaturated acyl chains are transferred efficiently. The acyl-CoA substrate specificity reported for PAT is not surprising, given the heterogeneity of fatty acids that have been detected on proteins purified from tissues.

The protein substrate specificity of PAT described for Gα and p59^{fyn} is consistent with the properties of thioacylated G-protein α subunits and nonreceptor tyrosine kinases *in vivo* (72, 85). Substrates with the N-terminal sequence motif, N-myristoyl-G-C-, are thioacylated *in vitro* at Cys-3, a physiological site of thioacylation for many G-protein α subunits and nonreceptor tyrosine kinases (see Table I). PAT activity toward G$_i\alpha$ and p59^{fyn} is dependent on myristoylation (85). Myristoylation appears to increase hydrophobicity of the protein substrate so that it can access PAT in a detergent micelle (61, 72). In the absence of myristate, G$\beta\gamma$ subunits can serve as a hydrophobic anchor because PAT recognizes nonmyristoylated G$_i\alpha$ when it is bound to G$\beta\gamma$ subunits. The substrate requirements for PAT activity *in vitro* echo those of thioacylation *in vivo* (see Section III).

In addition to its role as a hydrophobic anchor, G$\beta\gamma$ may enhance the substrate affinity of $G_i\alpha$ through a conformational change in the N-terminal domain of the protein, which includes the thioacylation site, Cys-3. G-protein PAT activity prefers myristoylated $G\alpha$ bound to $\beta\gamma$ as a substrate to the monomeric form. However, prenylated or nonprenylated $\beta\gamma$ support *in vitro* thioacylation of myristoylated $G_i\alpha$ with equal efficiency (72). Non-prenylated $\beta\gamma$ does not bind to membranes (86). Therefore, it cannot be acting as a hydrophobic anchor. Instead, the conformational change that is induced in the amino-terminal region of $G_i\alpha$ by binding to $\beta\gamma$ (87) may make $G_i\alpha$ a better substrate.

At present, the subcellular localization of PAT can be determined only by assaying enzyme activity in fractionated cell membranes. The distribution of PAT in subcellular fractions of rat liver has been characterized with G-protein α subunits as substrate (72). PAT activity for $G_i\alpha$ is highly enriched in plasma membrane. This localization is consistent with the evidence that newly synthesized $G_z\alpha$ (a $G_i\alpha$ family member) is not palmitoylated until it reaches the plasma membrane (75). Using G protein as a substrate, little PAT activity is detected in Golgi, ER, or mitochondrial fractions. In contrast, PAT activity for N-Ras appears to cofractionate with Golgi markers in mouse fibroblasts, with little activity associated with the plasma membrane (88). Proteolytic cleavage and carboxyl methylation of pre-nylated Ras occur in the ER (89, 90). It is possible that subsequent processing by thioacylation also occurs in an intracellular compartment. However, mature Ras that is localized at the plasma membrane and undergoing palmitate turnover is likely to be modified by a plasma membrane-associated activity. Similarly, thioacylation of newly synthesized p56[lck] and the mature form may also occur at different subcellular locations (74) (see Section III). A farnesylated cysteinyl-containing peptide derived from the N-Ras sequence and myristoyl-GC are localized at the plasma membrane when introduced into intact cells, consistent with their thioacylation occurring at this site (78, 79). Interestingly, two nonphysiological peptide substrates (myristoyl-C and myristoyl-CG) are thioacylated and localized on intracellular membranes as well as the plasma membrane, suggesting that a protein acyltransferase with distinct substrate specificity is present on intracellular membranes (78). Resolution of the questions regarding the number and distribution of PAT activities *in vivo* awaits characterization of PAT activity at the molecular level.

V. Mechanisms of Thioacylation

A number of cellular proteins will undergo spontaneous thioacylation *in vitro* in the presence of long-chain acyl-CoAs. Nonenzymatic acylation (autoacylation) has been documented for a number of proteins that are known to be substrates for thioacylation *in vivo,* including rhodopsin (*47*), myelin proteolipid proteins, lipophilin (*44, 91*) and P_0 (*92*), and G-protein α subunits (*45*). The possibility that autoacylation is due to contamination with trace amounts of an enzyme has been ruled out by the findings that synthetic peptides (*92–94*) and bacterially derived recombinant protein (*45*) are effective substrates for this process. The cysteine acceptor sites for autoacylation of $G_i\alpha$, myelin proteins P_0 and lipophilin, and rhodopsin map to the same sites that are modified *in vivo.* Acyl transfer appears to be specific for cysteine thiols, as serine or threonine substitutions in peptide substrates were not acylated *in vitro* (*93*). Efficient, stoichiometric thioacylation of proteins and peptides occurs at physiological pH and temperature in the presence of lipid vesicles or detergent micelles containing long-chain acyl-CoAs. The only requirement for the reaction appears to be the suitable disposition of the cysteine residue at the aqueous–lipid bilayer interface (but see Ref. *94*). Thus, autoacylation of short cysteinyl-containing peptides (myristoyl-G-C-X, X = G, L, R, T, or V) (*93*) or the G-protein α subunit (*45*), $G_i\alpha$, is dependent on a hydrophobic lipid anchor that facilitates association with liposomes or detergent micelles. The anchor is either an *N*-myristoyl group, or in the case of $G_i\alpha$, prenylated $G\beta\gamma$ subunits will substitute for N-myristoylation. A synthetic peptide (IRYCWLRR) corresponding to the thioacylated sequence of myelin P_0 glycoprotein is also an efficient substrate for autoacylation (*92*). Although the peptide lacks a hydrophobic anchor, it binds to synthetic liposomes under conditions used in the autoacylation assay (*95*). Conditions that reduce the fraction of peptide bound to vesicles decrease the rate of autoacylation, consistent with the reaction occurring at the membrane surface (*95*).

The ability to observe facile nonenzymatic acylation of proteins in the absence of a source of enzyme and the lack of a viable molecular candidate for a protein acyltransferase have led to the proposal that thioacylation in cells does not require an enzyme (*44, 45, 94*). The properties of $G_i\alpha$ autoacylation are strikingly similar to those observed for PAT activity solubilized from bovine brain, i.e., dependence on an intact protein substrate, prior myristoylation, and increased rate of acylation in the presence of $\beta\gamma$ subunits (*45, 72*). However, an important distinction between the assay conditions for optimal enzymatic acylation and those for autoacylation of G-protein α subunits is the concentration of acyl-CoA. Autoacylation is performed in the presence of 10–20 μM palmitoyl-CoA (*45*), compared with 1 μM palmitoyl-CoA in PAT assays (*72*). The availability of free long-chain acyl-CoAs is an important consideration with regard to the feasibility of a nonenzymatic mechanism for protein thioacylation *in vivo* (*95*).

Long-chain acyl-CoAs bind reversibly and with high affinity to acyl-CoA-binding proteins (ACBPs) (*96, 97*). Estimates of ACBP concentrations in several cell types suggest that the proteins are present in molar excess over long-chain acyl-CoAs, suggesting that most cytoplasmic acyl-CoAs are sequestered in ACBP–acyl-CoA complexes and that free acyl-CoA concentrations are low. This provides a mechanism for minimizing otherwise uncontrolled effects of acyl-CoAs on cellular processes. Under physiological concentrations and ratios of ACBP, acyl-CoA, and membrane lipids, ACBP–acyl-CoA complexes inhibit nonenzymatic acylation of short cysteinyl-peptides *in vitro* (*95*). On the basis of the peptide data, the half-time of nonenzymatic palmitoylation under these conditions was calculated to be on the order of tens of hours. Thus, the rate of palmitoylation of proteins in the absence of enzyme is unlikely to be able to keep up with signaling events that trigger rapid palmitoylation and depalmitoylation. This study strongly argues against a nonenzymatic mechanism *in vivo* for proteins that undergo rapid turnover of palmitate, but does not exclude the possibility that long-chain acyl-CoAs contribute to modification of proteins that are thioacylated with slow kinetics (*95*).

One way to reconcile the argument regarding the mechanism of thioacylation is to invoke protein factors that make the process of autoacylation more efficient. Although a protein acyltransferase is the obvious candidate, there are other possibilities. A protein could facilitate autoacylation by increasing local concentrations of long-chain acyl-CoAs in membrane compartments that thioacylate proteins. This could occur, for example, through synthesis of new acyl-CoAs or by extraction of acyl-CoAs from ACBP. It is also possible that thioacylation of protein substrates at specific membranes is regulated strictly by targeting factors that localize the substrate at the site of thioacylation where local concentrations of acyl-CoA are sufficient to drive autoacylation. The discovery of a prenylated Ras receptor that is localized at the plasma membrane is an attractive candidate for such a role (*98*). As discussed in Section III, there are accessory proteins that facilitate targeting of $G\alpha$ subunits and nonreceptor tyrosine kinases to the plasma membrane. It is less obvious how a targeting mechanism would apply to monotopic and polytopic membrane proteins.

Resolving the relative contributions of enzymatic and nonenzymatic mechanism to protein thioacylation *in vivo* will require additional research in two areas. First, the protein (or proteins) that harbor PAT activity must be characterized at the molecular level. This will establish whether PAT is a protein acyltransferase or encodes an activity that facilitates protein thioacylation by another mechanism. With either outcome, the physiological function of PAT must be determined to conclude that its function *in vivo* is the regulation of protein thioacylation. Analysis of PAT in genetically tractable organisms may be particularly helpful in this regard. Second, more information on the biosynthesis and metabolism of long-chain acyl-CoAs is needed, particularly with respect to their local concentrations in membranes where thioacylation is occurring. The relationship between protein thioacylation and proteins that regulate acyl-CoA availability must be better characterized to understand how acyl-CoAs are utilized in protein thioacylation. Elucidation of the mechanism of thioacylation *in vivo* is critical for the advancement of our knowledge regarding the functions of protein thioacylation.

VI. Palmitoyl-protein Thioesterases

A. PPT1 AND PPT2: LYSOSOMAL PALMITOYL-PROTEIN THIOESTERASES

1. *Molecular Cloning*

The first palmitoyl-protein thioesterase characterized at the molecular level was PPT1 (*40*). Enzyme activity was identified by its ability to remove palmitate from radiolabeled H-Ras and G-protein α subunits and was purified 33,000-fold to apparent homogeneity from the soluble fraction of a bovine brain extract (*99*). Peptide sequence derived from the purified enzyme permitted cloning of bovine and rat cDNAs, which predicted synthesis of a 306-amino acid protein (*40*). A human cDNA has also been isolated (*100, 101*). Sequence conservation among species is high; human PPT1 is 91 and 85% identical to bovine and rat PPT1, respectively (*100*). Comparison of the NH_2-terminal sequence of the purified enzyme with the predicted sequence revealed that PPT1 has a cleavable leader peptide of 27 amino acids that precedes the sequence of the mature protein. This finding, coupled with the discovery that the enzyme purified from brain was N-glycosylated, suggested that PPT1 was not a cytoplasmic enzyme, but instead was synthesized on membrane-bound ribosomes and underwent processing during transit through the secretory pathway. Indeed, most of the enzyme activity was secreted when PPT1 was expressed in mammalian or insect cells (*40*). The unexpected subcellular distribution of the enzyme suggested that its physiological targets were not intracellular signaling proteins, but instead were either extracellular or within the lumen of intracellular organelles. As discussed below, PPT1 is localized in lysosomes and thus its substrates are likely to be thioacylated proteins or peptides undergoing degradation in lysosomes.

A second PPT was identified by a database search for homologs of PPT1 (*102*). Human PPT2 shares 18% identity with PPT1. The sequence motif Ψ-X-Ψ-X-G-X-S-X-G-G-X-X-Ψ, where Ψ is a hydrophobic amino acid, is characteristic of many thioesterases and lipases. PPT1 and PPT2 possess this sequence, although in PPT2 a cysteine residue is found in the position of the first glycine in the motif. A second motif found in several mammalian thioesterases, G-H-D, is found in PPT1, but not PPT2. Like PPT1, PPT2 has a cleavable leader sequence and multiple potential N-linked glycosylation sites.

Chromosomal mapping of PPT1 to human chromosome 1p32 led to the exciting discovery that defects in the enzyme were responsible for a severe neurodegenerative disorder, infantile neuronal ceroid lipofuscinosis (INCL) (103). As discussed in more detail below, patients with INCL harbor mutations in the PPT1 gene that result in the intracellular accumulation of the enzyme and a lack of detectable enzyme activity. The PPT2 gene maps to the major human histocompatibility complex on chromosome 6p21.3 (102). To date, no link to disease has been found for PPT2.

2. Lysosomal Targeting of PPT1 and PPT2

Soluble lysosomal enzymes are targeted to the lysosome through a well-characterized receptor-mediated pathway (reviewed in Ref. 104). N-Linked oligosaccharides on soluble lysosomal enzymes are modified with mannose 6-phosphate (M-6-P). This moiety is recognized by receptors that direct the vesicular transport of newly synthesized lysosomal enzymes from the Golgi to the lysosome. On arrival in the lysosome, the enzyme dissociates from the receptor and the targeting signal is rapidly removed. The M-6-P receptors recycle back to the Golgi to carry out another round of transport. Receptors also traffic from the Golgi to the plasma membrane and back through the endocytic pathway. At the cell surface, the receptors retrieve mannose 6-phosphorylated proteins that have been secreted into the extracellular milieu and transport them to the lysosome. This process is defective in I-cell disease, another lysosomal storage disorder. In I-cell disease, lysosomal enzymes lack the M-6-P recognition signal, thus escaping detection by receptors, and are secreted rather than targeted to lysosomes.

PPT1 has the characteristics of a lysosomal enzyme that is targeted to lysosomes through the M-6-P recognition pathway. First, endogenous PPT1 cosediments with dense lysosomes in Madin–Darby canine kidney cells (105). Second, recombinant PPT1 codistributes with lysosomal markers in COS cells by immunofluorescence, but not with markers of the Golgi apparatus or ER (101). Third, PPT1 is one of the major mannose 6-phosphorylated glycoproteins isolated from brain (106). Fourth, recombinant PPT1, when supplied exogenously to cells, is taken up in a manner that is blocked by M-6-P (105). Fifth, in normal human fibroblasts, PPT1 immunoreactivity and enzyme activity are cell associated. However, fibroblasts from patients with I-cell disease have markedly reduced intracellular levels of PPT, with increased levels of PPT in cell culture medium (107). Thus, in a physiological setting, perturbation of the M-6-P recognition system results in the mislocalization of endogenous PPT1.

PPT2 also appears to be localized in lysosomes (*102*). The protein is N-glycosylated and has a cleavable signal peptide that is not present in the mature protein, suggesting that its biosynthesis and posttranslational processing occur in the secretory pathway. Recombinant PPT2 expressed in COS cells codistributes with lysosomal markers on Percoll density gradients. Uptake of recombinant PPT2 from conditioned media is M-6-P dependent. Although the localization of endogenous PPT2 has not been reported, the behavior of the recombinant protein expressed in a heterologous system is consistent with it being a soluble lysosomal enzyme.

3. Enzymatic Properties of PPT1 and PPT2

a. *Substrates.* PPT1 is distinguished from mammalian thioesterases I and II by its palmitoyl-protein hydrolase activity (*99*). Whereas all three enzymes are acyl-CoA hydrolases, only PPT1 utilizes palmitoylated H-Ras as a substrate. In addition to H-Ras, PPT1 removes thioester-linked palmitate from G-protein α subunits (*99*) and albumin (*102*), as well as from palmitoylated peptides and palmitoyl-cysteine (*108*). PPT2 has distinct substrate specificity from PPT1, demonstrating palmitoyl-CoA hydrolase activity, but no activity toward the panel of protein and peptide substrates utilized by PPT1 (*102*). The fatty acyl-CoA specificity of PPT1 has been examined, with a preference for acyl chains of 14 to 18 carbons (*40*). The acyl chain length profile for PPT1 acyl-CoA hydrolase activity is similar to the profile of fatty acids that modify proteins, consistent with the suggestion that the primary role of PPT1 may be in the degradation of thioacylated proteins.

The localization of PPT1 and PPT2 in lysosomes suggests that the physiological role of these enzymes is in the catabolism of thioacylated proteins. This hypothesis is reinforced by the finding that lipidated thioesters derived from acylated proteins accumulate in cells from patients with INCL (*109*). At least five distinct species of [^{35}S]cysteine-labeled lipids are extracted from lymphoblasts and fibroblasts derived from patients with INCL, but not from normal controls. The accumulation of these species is dependent on protein synthesis, suggesting the lipidated thioesters are derived from thioacylated proteins. Recombinant PPT1, introduced into cells by uptake through the M-6-P receptor, reverses the accumulation of the lipid thioesters. These moieties are also substrates for PPT1 *in vitro*. Thus, PPT1 corrects the metabolic defect in INCL lymphoblasts. Interestingly, PPT2 does not complement the PPT1 deficiency, providing additional evidence that PPT1 and PPT2 have distinct substrate specificities (*102*). It is possible that PPT2 deacylates a distinct spectrum of proteins or may have a role in degrading other types of lipid thioesters (*102*).

b. *pH Activity Profile.* PPT1 and PPT2 are maximally active at pH 7, an unusual pH optimum for lysosomal enzymes that typically display a preference for acidic pH (*99, 102*). Two other lysosomal enzymes, aspartyl-glucosaminidase (*110*) and sialic acid-specific *O*-acetylesterase (*111*), also have neutral pH optima. These enzymes share the property of removing posttranslational modifications of proteins. It has been proposed that lysosomes may undergo cycles of acidification; thus, these enzymes may be active in a periodic fashion (*112*).

c. *Inhibitors.* Unlike most mammalian thioesterases, PPT1 and PPT2 are insensitive to inhibition by *N*-ethylmaleimide and phenylmethylsulfonyl fluoride (*99, 102*). The enzymes are sensitive to diethyl pyrocarbonate, suggesting that a histidine residue may be involved in the active site. The GDH motif near the C terminus of PPT1 is presumed to contain the active-site histidine residue. PPT2 does not have the GDH motif, but there are two histidine residues in the vicinity of the C terminus that may fulfill the role of active-site histidine. Characterization of site-directed mutant proteins of PPT1 and PPT2 will be informative in defining the active site of the enzyme.

PPT1 was discovered independently as a didemnin-binding protein (*113*). Didemnins are marine natural products with potent cytostatic and immunosuppressive effects. Thus, the physiological targets of didemnin family members are of great interest. PPT1 was purified from a brain extract with an affinity matrix of immobilized didemnin. The effects of didemnin binding to PPT1 have been characterized *in vitro* (*114*). Didemnin inhibits PPT1 activity for H-Ras with a median inhibitory concentration (IC_{50}) of 5 μM. The kinetics of human recombinant PPT1 inhibition by didemnins was studied with myristoyl-CoA as a substrate. This analysis revealed that didemnin family members act as uncompetitive inhibitors of PPT1, suggesting that didemnin binds to the enzyme–substrate complex. Indeed, PPT1 binding to didemnin A-immobilized resin is induced by palmitoyl-CoA, but not palmitate. It seems unlikely that the biological effects of didemnin family members are mediated by its inhibition of PPT1. Mice treated with lethal doses of didemnin B lack the characteristic lysosomal accumulation of lipofuscin associated with INCL. Furthermore, there is a lack of correlation between the potency of PPT1 inhibition by various didemnin analogs and their IC_{50} values for *in vivo* biological activities (*114*).

4. *Molecular Genetics of PPT1 Deficiency*

INCL is one of several severe neurodegenerative disorders of children that have an autosomal recessive pattern of inheritance (reviewed in Ref. *115*). Children with INCL undergo early visual loss and mental deterioration. By age 3 years, they are frequently in a vegetative state with death occuring at ages 8–11 years. There is a selective loss of cortical neurons in patients with INCL, with the brainstem and spinal cord remaining intact. A hallmark of all neuronal ceroid lipofuscinoses is the accumulation of autofluorescent material in all tissues, including the brain. The discovery that PPT1 was a lysosomal enzyme firmly established that INCL could be classified as a lysosomal storage disorder (*101, 105*).

The relationship between INCL and PPT1 was discovered in studies of patients in the Finnish population (*103*). A single missense mutation (R122W) in PPT1 appears to be responsible for most of these cases. Transfected cells expressing PPT1(R122W) display no detectable PPT1 activity (*101*). Furthermore, the enzyme is mislocalized, accumulating in the ER. This strongly suggests that the mutation results in a misfolded protein, which is retained in the ER and subsequently degraded. A recent study of the U.S. population documented 19 different gene mutations in PPT1 that manifest in a broader spectrum of clinical presentations (*116*). Nonsense or frameshift mutations are associated with the earliest onset of symptoms and most severe clinical disease, similar to that observed with the R122W mutation. Missense mutations were identified throughout the coding region. PPT1 enzyme activity could not be detected in lymphoblasts from the patients tested, even those with later onset disease. However, residual enzyme activity in the range of 2–4% of normal was associated with several missense alleles when expressed in a heterologous cell system. Whether missense mutations interfere with enzyme function or protein expression remains to be determined. However, analysis of lymphoblasts from 26 subjects with various missense mutations revealed that all had severely reduced levels of protein.

One of the most interesting questions to be addressed in the future is how the lack of PPT1 enzyme activity manifests itself in the accumulation of storage bodies and the selective loss of cortical neurons in INCL. Hofmann and colleagues have proposed a model whereby intracellular palmitoylated proteins are targeted for deacylation by PPT through the process of autophagy (*108*). The formation of an autophagosome occurs by the engulfment of large membrane fragments, organelles, and cytoplasm into a vacuole. This organelle subsequently fuses with lysosomes, generating an autophagolysosome in which PPT1 would have access to palmitoylated substrates. A defect in PPT1 leads to the accumulation of material derived from the autophagolysosomes. This could occur directly through the accumulation of PPT1 substrate. A second hypothesis is that the small thioesters that accumulate in lymphoblasts lacking PPT1 activity inhibit lysosomal proteases, thereby causing a more global defect in lysosomal function (*108*). It has also been suggested that a population of PPT1 may be secreted normally and acts on extracellular substrates (*101*). Depalmitoylation of a PPT1 substrate may be essential for the postnatal development or mainte-nance of cortical neurons (*101*). Identification of novel substrates of PPT1 and a more complete characterization of the lipid thioesters that accumulate in lymphoblasts from INCL patients should provide insight into the pathol-ogy associated with the lack of PPT1 activity.

B. ACYLPROTEIN THIOESTERASE 1: A CYTOPLASMIC
    PALMITOYL-PROTEIN THIOESTERASE

Although one functional role for palmitoyl-protein thioesterases is the catabolism of thioacylated proteins, the rapid turnover of palmitate on intracellular signaling proteins suggests that thioesterases with access to the cytoplasmic face of the plasma membrane are present in cells. Acylprotein thioesterase 1 (APT1) is a candidate for a physiological regulator of palmi-tate turnover on intracellular signaling proteins. Duncan and Gilman puri-fied APT1 from the soluble fraction of rat liver by using palmitoylated G-protein α subunit as substrate (*41*). Acylprotein thioesterase activity in rat liver could be accounted for by PPT1 and APT1, suggesting that these two enzymes represent all acylprotein thioesterase activity that can be detected using palmitoylated $G_i\alpha$ as a substrate.

1. *Properties of Enzyme*

Purified APT1 is a monomer of 25 kDa (*41*). Amino acid sequence derived from the purified enzyme matched that of a previously identified lysophospholipase (*117, 118*). APT1 displays both acylprotein thioesterase activity and lysophospholipase activity, whether purified from rat liver or after expression in *Escherichia coli* (*41*). Therefore, activity toward both substrates is intrinsic to a single entity and not the result of copurification of distinct enzyme activities. Similar to PPT1 and PPT2, APT1 also acts as an acyl-CoA hydrolase. Thus, the enzyme is capable of hydrolyzing thioes-ters in the context of a protein or small molecule, as well as hydrolyzing oxyesters on lysolipids. Comparison of the kinetic parameters of APT1 for palmitoyl-$G_i\alpha$, palmitoyl-CoA, and lysophosphocholine (lyso-PC) suggests that palmitoyl-$G_i\alpha$ is the preferred substrate for APT1 (see Table II). The apparent K_m of APT1 for palmitoyl-$G_i\alpha$ was 25- and 250-fold lower than the values observed for palmitoyl-CoA and lyso-PC, respectively. Furthermore, the catalytic efficiency (V_{max}/K_m) for thioesters in acyl-CoA and acylprotein was 200-fold higher than that for the oxyester-containing lyso-PC. APT1 also recognizes other protein substrates, deacylating palmitoyl-H-Ras with a catalytic efficiency that is approximately threefold lower than the value estimated for palmitoyl-$G_i\alpha$.

TABLE II

KINETIC PARAMETERS OF APT1 FOR ITS SUBSTRATES[a]

Substrate	K_m (mM)	V_{max} (μmol/min/mg)
Lysophosphocholine	5.1	4.6
Palmitoyl-CoA	0.51	180
Palmitoyl-$G_i\alpha_1$	0.02	40

[a] From Ref. *41*.

2. Acylprotein Thioesterase 1 Regulation of Palmitate Turnover on $G_s\alpha$

$G_s\alpha$ is one of the best-studied proteins that undergo dynamic fatty acyla-tion *in vivo* (*21, 119, 120*). The basal rate of turnover of thioester-linked fatty acid on $G_s\alpha$ occurs with a $t_{1/2}$ of 20–90 min. Activation of $G_s\alpha$ by β-adrenergic agonists, cholera toxin, or mutations that prevent GTP hydroly-sis causes a greater than 10-fold increase in the rate of palmitate turnover. Reacylation of the protein may be rapid, because receptor stimulation does not result in a significant change in the stoichiometry of thioacylation of $G_s\alpha$ (*121*). The characteristics of the thioacylation cycle of $G_s\alpha$ in intact cells suggest that $G_s\alpha$ is thioacylated when bound to Gβγ and becomes a substrate for a thioesterase on activation and dissociation from Gβγ. The subcellular distribution of depalmitoylated $G_s\alpha$ has been a matter of debate. Translocation of $G_s\alpha$ to the cytoplasm was detected by immunofluorescence of cells treated with β-adrenergic agonists (*122*). However, one study sug-gests that G-protein α subunits, including $G_s\alpha$, activated with GTPγS and/ or depalmitoylated by APT1, remain associated with the plasma mem-brane (*123*).

The protective effect of Gβγ on depalmitoylation of $G_s\alpha$ was suggested first by experiments in broken cell preparations (*119*) and later confirmed by reconstitution of purified APT1 and G_s heterotrimer (*41*). Free $G_s\alpha$ is depalmitoylated at a faster rate than that bound to Gβγ. APT1 is indifferent to the activation state of free $G_s\alpha$ because GDP-bound $G_s\alpha$ is deacylated at the same rate as the GTPγS-bound form. Thus, subunit dissociation is associated with the enhanced rate of depalmitoylation and not the activated conformation of $G_s\alpha$. The cycle of subunit association and dissociation appears to account for the regulation of thioacylation of $G_s\alpha$ *in vivo*.

Support for the role of APT1 as a physiological regulator of G-protein thioacylation comes from studies of APT1 expression in mammalian cells (*41*). In stable cell lines expressing APT1, the rate of palmitate removal from $G_s\alpha$ is faster than that of control lines, suggesting that G-protein α subunits are substrates for APT1 *in vivo*, as well as *in vitro*. Although APT1 has acyl-CoA hydrolase activity, its expression did not significantly increase acyl-CoA hydrolase activity in cell extracts nor did it perturb incorporation of [^{3}H]palmitate into cellular lipids in cells overexpressing APT1. Thus, the effects of APT1 on palmitate turnover on $G_s\alpha$ are not due to effects on the rate of turnover of palmitoyl-CoA.

The discovery of APT1 provides an important tool for deciphering the functional significance of cycles of protein thioacylation. It will be important to establish whether APT1 represents all the acylprotein thioesterase activity with access to the cytoplasmic face of the plasma membrane or whether this is the first member of a protein family. The identification of orthologs of APT1 in *Saccharomyces cerevisiae* and *Caenorhabditis elegans* means that this question will be addressed, at least for these organisms, in the near future.

VII. Future Prospects

In contrast to protein N-myristoylation and prenylation, the enzymology of protein thioacylation is in its infancy. Molecular characterization of palmitoyl-protein thioesterases is a significant step forward and has already yielded an unanticipated link between protein depalmitoylation and INCL (*103*). This represents the second example of a genetic disorder related to protein lipidation. The first is choroideremia, an X-linked form of retinal degeneration, that is caused by defects in Rab escort protein, a component of the Rab geranylgeranyltransferase (*124*). These findings underscore the importance of elucidating the molecular mechanisms that underlie protein modifications. The ability to produce soluble recombinant PPT1 (*40*) and APT1 (*41*) in quantity portends that structural information will be available for these enzymes in the near future. In combination with *in vivo* characterization of palmitoyl-protein thioesterases, these studies will likely reveal novel insights into the biology of protein thioacylation. The mechanism that underlies the addition of thioester-linked fatty acids has remained elusive. Traditional biochemical methods have yielded limited information to date and must be combined with *in vivo* approaches to define the gene products that mediate this process. With the sequences of the yeast and worm genomes complete and the human genome nearing completion, the information is available and strategies must be developed to recognize it.

ACKNOWLEDGMENTS

The author thanks Susana Gonzalo, Carol Manahan, and Serge Moffett for comments on the manuscript. Work in the author's laboratory is supported by National Institutes of Health Grant GM51466 and the Monsanto-Searle/Washington University Biomedical Research Program.

REFERENCES

1. Braun, P. E., and Radin, N. S. (1969). *Biochemistry* **8,** 4310.
2. Stoffyn, P., and Folch, J. (1971). *Biochem. Biophys. Res. Commun.* **44,** 157.
3. Gagnon, J., Finch, P. R., Wood, D. D., and Moscarello, M. A. (1971). *Biochemistry* **10,** 4576.
4. Schmidt, M. F. G., and Schlesinger, M. J. (1979). *Cell* **17,** 813.
5. Schlesinger, M. J., Magee, A. I., and Schmidt, M. F. G. (1980). *J. Biol. Chem.* **255,** 10021.
6. Schlesinger, M. J., Veit, M., and Schmidt, M. F. G. (1993). *In* "Lipid Modifications of Proteins" (M. J. Schlesinger, ed.), pp. 2. CRC Press, Boca Raton, Florida.
7. Bizzozero, O. A., Tetzloff, S. U., and Bharadwaj, M. (1994). *Neurochem. Res.* **19,** 923.
8. Dunphy, J. T., and Linder, M. E. (1998). *Biochim. Biophys. Acta* **1436,** 245.
9. Tao, N., Wagner, S. J., and Lublin, D. M. (1996). *J. Biol. Chem.* **271,** 22315.
10. O'Dowd, B. F., Hnatowich, M., Caron, M. G., Lefkowitz, R. J., and Bouvier, M. (1989). *J. Biol. Chem.* **264,** 7564.
11. Kennedy, M. E., and Limbird, L. E. (1993). *J. Biol. Chem.* **268,** 8003.
12. Papac, D. I., Thornburg, K. R., Bullesbach, E. E., Crouch, R. K., and Knapp, D. R. (1992). *J. Biol. Chem.* **267,** 16889.
13. Crise, B., and Rose, J. K. (1992). *J. Biol. Chem.* **267,** 13593.
14. Schweizer, A., Kornfeld, S., and Rohrer, J. (1996). *J. Cell Biol.* **132,** 577.
15. Schweizer, A., Rohrer, J., Jeno, P., DeMaio, A., Buchman, T. G., and Hauri, H. P. (1993). *J. Cell Sci.* **104,** 685.
16. Schweizer, A., Rohrer, J., and Kornfeld, S. (1995). *J. Biol. Chem.* **1995,** 9638.
17. Omary, M. B., and Trowbridge, I. S. (1981). *J. Biol. Chem.* **256,** 4715.
18. Jing, S. Q., and Trowbridge, I. S. (1987). *EMBO J.* **6,** 327.
19. Rose, J. K., Adams, G. A., and Gallione, C. J. (1984). *Proc. Natl. Acad. Sci. U.S.A.* **81,** 2050.
20. Degtyarev, M. Y., Spiegel, A. M., and Jones, T. L. Z. (1994). *J. Biol. Chem.* **269,** 30898.
21. Mumby, S. M., Kleuss, C., and Gilman, A. G. (1994). *Proc. Natl. Acad. Sci. U.S.A.* **91,** 2800.
22. Linder, M. E., Middleton, P., Hepler, J. R., Taussig, R., Gilman, A. G., and Mumby, S. M. (1993). *Proc. Natl. Acad. Sci. U.S.A.* **90,** 3675.
23. Wilson, P., and Bourne, H. (1995). *J. Biol. Chem.* **270,** 9667.
24. Parenti, M., Vigano, M. A., Newman, C. M., Milligan, G., and Magee, A. I. (1993). *Biochem. J.* **291,** 349.
25. Alland, L., Peseckis, S. M., Atherton, R. E., Berthiaume, L., and Resh, M. D. (1994). *J. Biol. Chem.* **269,** 16701.
26. Paige, L. A., Nadler, M. J. S., Harrison, M. L., Cassady, J. M., and Geahlen, R. L. (1993). *J. Biol. Chem.* **268,** 8669.

27. Kwong, J., and Lublin, D. M. (1995). *Biochem. Biophys. Res. Commun.* **207**, 868.
28. Rodgers, W., Crise, B., and Rose, J. K. (1994). *Mol. Cell. Biol.* **14**, 5384.
29. Hancock, J. F., Magee, J. I., Childs, J. E., and Marshall, C. J. (1989). *Cell* **57**, 1167.
30. Wedegaertner, P. B., Chu, D. H., Wilson, P. T., Levis, M. J., and Bourne, H. R. (1993). *J. Biol. Chem.* **268**, 25001.
31. Veit, M., Nurnberg, B., Spicher, K., Harteneck, C., Ponimaskin, E., Schultz, G., and Schmidt, M. F. G. (1994). *FEBS Lett.* **339**, 160.
32. Jones, T. L. Z., and Gutkind, J. S. (1998). *Biochemistry* **37**, 3196.
33. Skene, J. H. P., and Virag, I. (1989). *J. Cell Biol.* **108**, 613.
34. Liu, Y., Fisher, D. A., and Storm, D. R. (1993). *Biochemistry* **32**, 10714.
35. Topinka, J. R., and Bredt, D. S. (1998). *Neuron* **20**, 125.
36. Hess, D. T., Slater, T. M., Wilson, M. C., and Skene, J. H. P. (1992). *J. Neurosci.* **12**, 4634.
37. Lane, S. R., and Liu, Y. (1997). *J. Neurochem.* **69**, 1864.
38. Johnson, D., Bhatnagar, R., Knoll, L., and Gordon, J. (1994). *Annu. Rev. Biochem.* **63**, 869.
39. Glomset, J. A., and Farnsworth, C. C. (1994). *Annu. Rev. Cell Biol.* **10**, 181.
40. Camp, L. A., Verkruyse, L. A., Afendis, S. J., Slaughter, C. A., and Hofmann, S. L. (1994). *J. Biol. Chem.* **269**, 23212.
41. Duncan, J. A., and Gilman, A. G. (1998). *J. Biol. Chem.* **273**, 15830.
42. Liu, L., Dudler, T., and Gelb, M. (1996). *J. Biol. Chem.* **271**, 23269.
43. Das, A. K., Dasgupta, B., Bhattacharya, R., and Basu, J. (1997). *J. Biol. Chem.* **272**, 11021.
44. Bizzozero, O., McGarry, J., and Lees, M. (1987). *J. Biol. Chem.* **262**, 13550.
45. Duncan, J. A., and Gilman, A. G. (1996). *J. Biol. Chem.* **271**, 23594.
46. Morello, J.-P., and Bouvier, M. (1996). *Biochem. Cell Biol.* **74**, 449.
47. O'Brien, P., St. Jules, R., Reedy, T., Bazan, N., and Zatz, M. (1987). *J. Biol. Chem.* **262**, 5210.
48. Moench, S. J., Moreland, J., Stewart, D. H., and Dewey, T. G. (1994). *Biochemistry* **33**, 5791.
49. Milligan, G., Parenti, M., and Magee, A. (1995). *Trends Biochem. Sci.* **20**, 181.
50. Moffett, S., Mouillac, B., Bonin, H., and Bouvier, M. (1993). *EMBO J.* **12**, 349.
51. Kennedy, M. E., and Limbird, L. E. (1994). *J. Biol. Chem.* **269**, 31915.
52. Collins, P. L., and Mottet, G. (1993). *J. Gen. Virol.* **74**, 1445.
53. Ponimaskin, E., and Schmidt, M. F. G. (1998). *Virology* **249**, 325.
54. Veit, M., Reverey, H., and Schmidt, M. F. G. (1996). *Biochem. J.* **318**, 163.
55. Schlesinger, M. (ed.). (1993). "Lipid Modifications of Proteins." CRC Press, Boca Raton, Florida.
56. Alvarez, E., Girones, N., and Davis, R. J. (1990). *J. Biol. Chem.* **265**, 16644.
57. Rohrer, J., Schweizer, A., Johnson, K. F., and Kornfeld, S. (1995). *J. Cell Biol.* **130**, 1297.
58. Hille-Rehfeld, A. (1995). *Biochim. Biophys. Acta* **1241**, 177.
59. Casey, P. (1995). *Science* **268**, 221.
60. Gonzalo, S., Greentree, W. K., and Linder, M. E. (1999). *J. Biol. Chem.* **274**, 21313.
61. Resh, M. D. (1996). *Cell. Signal.* **8**, 403.
62. Mumby, S. M. (1997). *Curr. Opin. Cell Biol.* **9**, 148.
63. Iiri, T., Backlund, P. S., Jr., Jones, T. L. Z., Wedegaertner, P. B., and Bourne, H. R. (1996). *Proc. Natl. Acad. Sci. U.S.A.* **93**, 14592.
64. Berman, D. M., and Gilman, A. G. (1998). *J. Biol. Chem.* **273**, 1269.
65. Tu, Y., Wang, J., and Ross, E. M. (1997). *Science* **278**, 1132.

66. Schmidt, M. F. G., and Schlesinger, M. J. (1980). *J. Biol. Chem.* **255,** 3334.
67. Bonatti, S., Migliaccio, G., and Simons, K. (1989). *J. Biol. Chem.* **264,** 12590.
68. Schweizer, A., Rohrer, J., Slot, J., Geuze, H., and Kornfeld, S. (1995). *J. Cell Sci.* **108,** 2477.
69. Mundy, D. I., and Warren, G. (1992). *J. Cell Biol.* **116,** 135.
70. Klausner, R., Donaldson, J., and Lippincott-Schwartz, J. (1992). *J. Cell Biol.* **116,** 1071.
71. Wilcox, C. A., and Olson, E. N. (1987). *Biochemistry* **26,** 1029.
72. Dunphy, J. T., Greentree, W. K., Manahan, C. L., and Linder, M. E. (1996). *J. Biol. Chem.* **271,** 7154.
73. van't Hof, W., and Resh, M. D. (1997). *J. Cell Biol.* **136,** 1023.
74. Bijlmakers, M.-J., and Marsh, M. (1999). *J. Cell Biol.* **145,** 457.
75. Fishburn, C. S., Herzmark, P., Morales, J., and Bourne, H. R. (1999). *J. Biol. Chem.* **274,** 18793.
76. Peitzsch, R. M., and McLaughlin, S. (1993). *Biochemistry* **32,** 10436.
77. Shahinian, S., and Silvius, J. R. (1995). *Biochemistry* **34,** 3813.
78. Schroeder, H., Leventis, R., Shahinian, S., Walton, P. A., and Silvius, J. R. (1996). *J. Cell Biol.* **134,** 647.
79. Schroeder, H., Leventis, R., Rex, S., Schelhaas, M., Nagele, E., Waldmann, H., and Silvius, J. R. (1997). *Biochemistry* **36,** 13102.
80. Gonzalo, S., and Linder, M. E. (1998). *Mol. Biol. Cell* **9,** 585.
81. Liu, L., Dudler, T., and Gelb, M. H. (1999). *J. Biol. Chem.* **274,** 3252a.
82. Ueno, K., and Suzuki, Y. (1997). *J. Biol. Chem.* **272,** 13519.
83. Schmidt, M. F. G., and Burns, G. R. (1989). *Biochem. Soc. Trans.* **17,** 625.
84. Veit, M., Sachs, K., Heckelmann, M., Maretzki, D., Hofmann, K. P., and Schmidt, M. F. G. (1998). *Biochim. Biophys. Acta* **1394,** 90.
85. Berthiaume, L., and Resh, M. (1995). *J. Biol. Chem.* **270,** 22399.
86. Muntz, K. H., Sternweis, P. C., Gilman, A. G., and Mumby, S. H. (1992). *Mol. Biol. Cell* **3,** 49.
87. Wall, M., Coleman, D., Lee, E., Iniguez-Lluhi, J., Posner, B., Gilman, A., and Sprang, S. (1995). *Cell* **83,** 1047.
88. Gutierrez, L., and Magee, A. I. (1991). *Biochim. Biophys. Acta* **1078,** 147.
89. Dai, Q., Choy, E., Chiu, V., Romano, J., Slivka, S. R., Steitz, S. A., Michaelis, S., and Philips, M. R. (1998). *J. Biol. Chem.* **273,** 15030.
90. Schmidt, W. K., Tam, A., Fujimura-Kamada, K., and Michaelis, S. (1998). *Proc. Natl. Acad. Sci. U.S.A.* **95,** 11175.
91. Ross, N. W., and Braun, P. E. (1988). *J. Neurosci. Res.* **21,** 35.
92. Bharadwaj, M., and Bizzozero, O. A. (1995). *J. Neurochem.* **65,** 1805.
93. Quesnel, S., and Silvius, J. R. (1994). *Biochemistry* **33,** 13340.
94. Bano, M. C., Jackson, C. S., and Magee, A. I. (1998). *Biochem. J.* **330,** 723.
95. Leventis, R., Juel, G., Knudsen, J. K., and Silvius, J. R. (1997). *Biochemistry* **36,** 5546.
96. Faergeman, N. J., and Knudsen, J. (1997). *Biochem. J.* **323,** 1.
97. Frolov, A., and Schroeder, F. (1998). *J. Biol. Chem.* **273,** 11049.
98. Siddiqui, A. A., Garland, J. R., Dalton, M. B., and Sinensky, M. (1998). *J. Biol. Chem.* **273,** 3712.
99. Camp, L., and Hofmann, S. L. (1993). *J. Biol. Chem.* **268,** 22566.
100. Schriner, J. E., Yi, W., and Hofmann, S. L. (1996). *Genomics* **34,** 317.
101. Hellsten, E., Vesa, J., Olkkonen, V. M., Jalanko, A., and Peltonen, L. (1996). *EMBO J.* **15,** 5240.
102. Soyombo, A. A., and Hofmann, S. L. (1997). *J. Biol. Chem.* **272,** 27456.

103. Vesa, J., Hellsten, E., Verkruyse, L., Camp, L., Rapola, J., Santavuori, P., Hofmann, S., and Peltonen, L. (1995). *Nature* (*London*) **376**, 584.
104. Kornfeld, S. (1992). *Annu. Rev. Biochem.* **61**, 307
105. Verkruyse, L., and Hofmann, S. (1996). *J. Biol. Chem.* **271**, 15831.
106. Sleat, D. E., Sohar, I., Lackland, H., Majercak, J., and Lobel, P. (1996). *J. Biol. Chem.* **271**, 19191.
107. Verkruyse, L. A., Natowicz, M. R., and Hofmann, S. L. (1997). *Biochim. Biophys. Acta* **1361**, 1.
108. Hofmann, S. L., Lee, L. A., Lu, J.-Y., and Verkruyse, L. A. (1997). *Neuropediatrics* **28**, 27.
109. Lu, J.-Y., Verkruyse, L. A., and Hofmann, S. L. (1996). *Proc. Natl. Acad. Sci. U.S.A.* **93**, 10046.
110. Tollersrud, O. K., and Aronson, N. N. J. (1989). *Biochem. J.* **260**, 101.
111. Higa, H. H., Manzi, A., and Varki, A. (1989). *J. Biol. Chem.* **264**, 19435.
112. Butor, C., Griffiths, G., Aronson, N. N. J., and Varki, A. (1995). *J. Cell Sci.* **108**, 2213.
113. Crews, C. M., Lane, W. S., and Schreiber, S. L. (1996). *Proc. Natl. Acad. Sci. U.S.A.* **93**, 4316.
114. Meng, L., Ny, S., and Crews, C. M. (1998). *Biochemistry* **37**, 10488.
115. Goebel, H. H. (1995). *J. Child Neurol.* **10**, 424.
116. Das, A. K., Becerra, C. H. R., Yi, W., Lu, J.-Y., Siakotos, A. N., Wisniewski, K. E., and Hofmann, S. L. (1998). *J. Clin. Invest.* **102**, 361.
117. Sugimoto, H., Hayashi, H., and Yamashita, S. (1996). *J. Biol. Chem.* **271**, 7705.
118. Wang, A., Deems, R. A., and Dennis, E. A. (1997). *J. Biol. Chem.* **272**, 12723.
119. Wedegaertner, P. B., and Bourne, H. R. (1994). *Cell* **77**, 1063.
120. Degtyarev, M. Y., Spiegel, A. M., and Jones, T. L. Z. (1993). *J. Biol. Chem.* **268**, 23769.
121. Jones, T. L., Degtyarev, M. Y., and Backlund, P. S., Jr. (1997). *Biochemistry* **36**, 7185.
122. Wedegaertner, P., Bourne, H., and von Zastrow, M. (1996). *Mol. Biol. Cell* **8**, 1225.
123. Huang, C., Duncan, J. A., Gilman, A. G., and Mumby, S. M. (1999). *Proc. Natl. Acad. Sci. U.S.A.* **96**, 412.
124. Seabra, M. C., Brown, M. S., and Goldstein, J. L. (1993). *Science* **259**, 377.

9

Biology and Enzymology of Protein N-Myristoylation

RAJIV S. BHATNAGAR* · KAVEH ASHRAFI* ·
KLAUS FÜTTERER† · GABRIEL WAKSMAN† ·
JEFFREY I. GORDON*

*Department of Molecular Biology and Pharmacology, and
†Department of Biochemistry and Molecular Biophysics
Washington University School of Medicine
St. Louis, Missouri 63110

I. Introduction

Protein N-myristoylation refers to the cotranslational covalent attachment of myristate, a rare 14-carbon saturated fatty acid (C14:0), to the N-

241

THE ENZYMES, Vol. XXI

terminal glycine of a subset of eukaryotic and viral proteins. This review summarizes a number of studies that have advanced our understanding of how proteins are N-myristoylated and how *N*-myristoyl proteins function. Thermodynamic analyses of the interactions of acyl peptides and acyl proteins with model membranes have confirmed the importance of the limited hydrophobicity of myristate in regulating dynamic interactions between *N*-myristoyl proteins and their membrane or protein partners. Genetic studies have shown that protein N-myristoylation is essential for the survival of *Saccharomyces cerevisiae* during periods when nutrients are unlimited as well as during periods when nutrients are scarce. Genetic studies have also shown that protein N-myristoylation is required for the survival of the common human fungal pathogens *Candida albicans* and *Cryptococcus neoformans.* Analyses of the enzyme responsible for catalyzing protein N-myristoylation—myristoyl-CoA : protein *N*-myristoyltransferase (Nmt)—have disclosed that orthologous Nmts have subtle differences in their peptide substrate specificities and that these differences can be exploited to generate species-selective inhibitors of fungal Nmts that are fungicidal. X-Ray crystallographic determination of the structure of Nmt has revealed how this enzyme interacts with its substrates, and how catalysis occurs through a novel mechanism. These latter results have implications for understanding how a large number of CoA-dependent acyltransferases, belonging to the GCN5-related *N*-acetyltransferase superfamily, operate.

II. Biology

A. How Myristate Is Used to Regulate Protein Function

The first *N*-myristoyl proteins were described in 1982 (*1, 2*). Since that time a large number of proteins containing an N-terminal tetradecanoyl group have been identified (reviewed in Ref. *3*). *N*-Myristoyl proteins have diverse functions and cellular locations. Many of these proteins are components of signal transduction cascades and include kinases, kinase substrates, phosphatases, and the α subunits of heterotrimeric G proteins. A number of structural and nonstructural proteins encoded by enveloped and nonenveloped viruses (retroviruses, hepadnaviruses, papovaviruses, picornaviruses, and reoviruses) are also N-myristoylated.

N-Myristoylation is known to be involved in the targeting of some proteins to cellular membranes, although not all *N*-myristoyl proteins are membrane associated. Moreover, myristate is not used as a universal signal to promote protein trafficking to a single cellular membrane: some *N*-myristoyl proteins affiliate with the plasma membrane, others with the endoplasmic reticulum and Golgi, and still others with specialized membrane subdomains such as caveolae (reviewed in Refs. 4 and 5). The mechanisms that direct N-myristoyl proteins to specific membrane targets remain poorly understood.

The diversity in N-myristoyl protein function and subcellular location has led to a number of studies that address two interrelated questions. First, why is myristate selected for this covalent modification of proteins as opposed to other, more abundant cellular acyl chains? Second, are there common paradigms by which myristate contributes to protein function?

In the few instances in which acyl-CoAs have been measured in cell lineages, myristoyl-CoA (C14:0-CoA) has been found to be rare, comprising less than a few percent of the total acyl-CoA pool (e.g., Refs. 6 and 7). The selection of myristate for covalent modification of a subset of cellular proteins appears to fulfill a need for weak, readily reversible interactions with lipids and/or proteins—interactions that can be severed at low thermodynamic cost. When Pietzsch and McLaughlin (8) measured the interactions of peptides containing acyl chains of varying lengths with model membranes of defined composition, they found that myristate provided barely enough hydrophobicity to result in stable association. Because a myristoyl moiety is typically insufficient to produce stable affiliation, a second source of affinity must be employed to achieve stable interaction with a membrane or protein partner. In this conceptualization, myristate can be viewed as providing a constitutive, albeit weak source of affinity. The challenge to an N-myristoyl protein is to deploy an additional regulated source of affinity that can be exploited to "switch on" or "switch off" its relationship to a potential partner or partners. A major advance in the field of protein N-myristoylation has been the identification and characterization of such switches.

1. *Myristoyl Electrostatic Switches: Myristate as Constitutive Source of Affinity*

Significant membrane affinity can be garnered by utilizing hydrophobic interactions provided by the myristoyl moiety together with electrostatic interactions provided by the positively charged side chains of a protein and the negatively charged phospholipid head groups of a membrane. Furthermore, a reversible covalent modification of the protein, such as phosphorylation, can be used to alter the character of its electrostatic interactions, thereby disrupting stable membrane association (see Refs. 9 and 10 for general theoretical discussions of myristoyl electrostatic switches).

The prototypic myristoyl electrostatic switch protein is MARCKS (myristoylated alanine-rich C-kinase substrate). This protein integrates protein kinase C (PKC) and Ca^{2+}–calmodulin signals that affect actin–cytoskeleton and actin–membrane interactions (*11*). Myristoyl-MARCKS translocates from the plasma membrane to the cytosol when it undergoes phosphorylation of serine residues located in its polybasic region. This switch can be operated by addition or subtraction of phosphate because the *N*-myristoyl group of the protein is not sufficient, in and of itself, for stable membrane association (*9*).

The tyrosine kinase, Src, can also be viewed as a myristoyl-electrostatic switch protein. All known members of the Src family are N-myristoylated. All family members, except Src and Blk, contain a cysteine residue near their N termini that can serve as a site for covalent attachment of palmitate. Dual acylation with an amide-linked myristate and a thioester-linked palmitate promotes stable association with cellular membranes (e.g., Refs. *8, 12;* this volume (*12a*); and below). Instead of palmitoylation, Src uses an N-terminal polybasic domain to provide a second source of membrane affinity (*13, 14*). As in MARCKS, the polybasic region of Src contains sites for phosphorylation. The existence of these sites suggests that Src may be subject to regulated dissociation from its membrane niches. Although phosphorylation-dependent switching has not been demonstrated in the same definitive manner as MARCKS, linkage of the N terminus of Src to β-galactosidase yields a chimeric protein whose distribution is shifted from membranes to the cytosol of COS cells when it is hyperphosphorylated (*10*).

2.   *Myristoyl Conformational Switches: Myristate as Regulated Source of Affinity*

A second type of switch has been identified, in which myristate functions as a regulated source of affinity rather than a constitutive source. In this switch, noncovalent association of an *N*-myristoyl protein with various ligands, or the acquisition or expulsion of a given ligand, results in a conformational change that either exposes or sequesters the acyl chain. When exposed, the myristoyl moiety is able to promote interactions with membrane (or protein) partners. The stability of such interactions may be defined further by other protein domains or attachments (modifications).

The prototypic myristoyl conformational switch protein is recoverin (*15*). This member of the EF-hand superfamily of Ca^{2+}-binding proteins is found in vertebrate photoreceptor cells. Ca^{2+}-bound recoverin appears to prolong the lifetime of photoexcited rhodopsin by inhibiting rhodopsin kinase (*16, 17*). In the absence of its Ca^{2+} ligand, the myristoylated N terminus of recoverin is sequestered in a hydrophobic binding site located on the protein. This unliganded form of recoverin is cytosolic. When Ca^{2+} is bound, recoverin undergoes a conformational change that extrudes and exposes its myristoyl moiety, allowing the protein to associate with membranes (*18*). Nuclear magnetic resonance (NMR) studies have shown that the conformational change involves large ligation-induced rotations of two glycines in the all α-helical protein (*19*).

The myristoyl binding site and glycine "swivels" are highly conserved in recoverin homologs present in organisms as diverse as yeast and humans, emphasizing the ancient nature of this type of calcium-sensitive switch (*19*). Myristoyl conformational switches may also define the behavior of a large family of intracellular neuronal calcium sensor proteins that participate in many signal transduction cascades (*20*).

The vertebrate photoreceptor guanylyl cyclase activating protein 2 (GCAP-2) is an N-myristoylated recoverin family member that regulates the activity of retinal guanylyl cyclase (RetGC), an integral membrane protein (*21*). Like recoverin, GCAP-2 is regulated by intracellular calcium. However, the effects of calcium binding on membrane partitioning of GCAP-2 are just the opposite of those observed for recoverin: when Ca^{2+} concentrations are low, GCAP-2 is membrane associated and stimulates RetGC activity; when Ca^{2+} concentrations are high, a greater fraction of the protein is cytosolic.

NMR spectroscopic studies indicate that unlike recoverin, the myristoyl moiety of GCAP-2 is not tightly bound in either its calcium-free or calcium-bound states (*22*). Even though myristate appears to be partially solvent exposed, the N terminus of GCAP-2 remains resistant to proteolysis whether or not calcium is present. Interaction of GCAP-2 with its RetGC partner is promoted by a calcium-induced conformational change that does not obviously affect or effect acyl chain presentation. Moreover, the intracellular distribution of GCAP-2 and its stimulation of RetGC do not require N-myristoylation (*23*). These findings raise the question of whether following calcium-regulated partnering of GCAP-2 and RetGC myristate affects the complex, or whether myristate is simply an evolutionary vestige of the ancient calcium myristoyl switch.

ADP-ribosylation factors (Arfs) are well known as regulators of coatmer-coated vesicle transport between components of the Golgi apparatus (reviewed in Ref. *24*). The Arf myristoyl-conformation switch is conceptually and functionally similar to the recoverin switch, but it operates through a different structural mechanism. GDP:Arf sequesters its myristoyl group within a hydrophobic pocket. When GDP is replaced by GTP, the protein undergoes a conformational change that exposes its myristoyl moiety, allowing it to associate with cellular membranes. Goldberg (*25*) has shown that the seven-stranded β sheet of GDP:Arf is transformed to a six-stranded sheet when GTP is acquired. In addition to the loss of the β strand in GTP:Arf, two additional strands shift in register relative to the remaining strands so as to eliminate the myristate-binding pocket and force exposure of the acyl chain.

3. Myristoyl–Palmitoyl Switch

Many Src family members (see above), several α subunits of heterotrimeric G proteins, and the constitutive form of endothelial nitric oxide synthase (eNOS) are dually lipidated with both myristoyl and palmitoyl groups. Palmitoylation of the cysteine present in proteins with an N-terminal glycine–cysteine sequence is inhibited when N-myristoylation is blocked by Gly-1 → Ala mutagenesis (reviewed in Ref. 5).

Although the factors that determine how most *N*-myristoyl proteins are targeted to specific cellular membranes remain largely unknown, an N-terminal glycine–cysteine motif containing N-myristoylation and S-palmitoylation sites appears to function as a signal for caveolar localization of at least two Src family members, Fyn and Hck (*12, 26, 27*).

Reversible palmitoylation of some *N*-myristoyl proteins may represent another type of switching mechanism. When eNOS is activated and translocates to the cytosol from membranes, palmitate turnover is increased (*28, 29*). However, it not yet certain that the state of palmitoylation of eNOS is the principal determinant of its membrane:cytosolic partitioning. For example, partitioning is also affected by competing interactions involving eNOS and membrane-associated calveolin and eNOS and cytosolic calmodulin (*30, 31*). Nonetheless, the ability of eNOS to be palmitoylated affects the kinetics of its reassociation with cell membranes and its targeting to caveolae (*30, 31*). The kinetic effect of palmitoylation of eNOS is similar to the effect of palmitoylation on partitioning of newly synthesized Fyn (*32*).

4. Switches Involving N-Myristoyl Protein–Protein Interactions

Some *N*-myristoyl proteins switch from a membrane-bound to cytosolic state based on their regulated interactions with other proteins. This phenomenon is exemplified by the calmodulin–eNOS interaction noted above. In the presence of high levels of intracellular calcium, the EF-hand-containing protein calmodulin undergoes a conformational change that allows it to interact with eNOS. This interaction permits eNOS to be extracted from caveolae (*30, 31*). MARCKS can alter its intracellular location through a similar calcium/calmodulin-dependent interaction (*33*).

In summary, a series of myristate-dependent protein switches have been identified. These switches are able to function because the limited hydrophobicity of myristate permits weak, readily reversible protein–membrane and/or protein–protein interactions.

B. PROTEIN N-MYRISTOYLATION CATALYZED BY
    MYRISTOYL-CO A : PROTEIN N-MYRISTOYLTRANSFERASE

Protein N-myristoylation appears to be a ubiquitous eukaryotic modification. The first myristoylCoA : protein N-myristoyltransferase to be identified, purified to apparent homogeneity, and characterized was from *Saccharomyces cerevisiae* (*34*). The yeast enzyme (Nmt1p) is monomeric and has
no cofactor requirements (*35*). The purified protein lacks methionylamino-
peptidase activity (*34*). Thus, removal of the initiator methionine from
nascent peptide substrates is a task relegated to cellular methionylamino-
peptidases (Map1p and Map2p in *S.cerevisiae;* Rcfs. *28, 36,* and *37*).
    Nmt1p has an ordered Bi–Bi kinetic mechanism (*38*). Apo-Nmt1p binds
myristoyl-CoA first. Assembly of this myristoyl-CoA : Nmt1p binary complex permits a functional peptide-binding site to be created. Once a myristoyl-CoA : Nmt1p : peptide ternary complex has formed, there is catalytic
transfer of myristate from CoA to the glycine amine of a substrate peptide.
The CoA product is then released followed by myristoylpeptide.
    Calorimetric and kinetic studies (*39, 40*) have shown that Nmt1p has a
high degree of specificity for transfer of myristate. Palmitoyl-CoA, the
much more abundant cellular acyl-CoA in yeast (*7*), is a poor substrate:
the N-palmitoyltransferase activity of Nmt1p is <1% of its N-myristoyl-
transferase activity (*40, 41*). *In vitro* kinetic analyses of >300 myristoyl-
CoA analogs indicated that Nmt1p requires the bound myristoyl chain to
adopt a bent conformation at C-5–C-6 (C-1 = carboxyl). The enzyme is
sensitive to the distance between this bend and C-1 as well as the distance
between the bend and the acyl chain ω terminus (*41–43*).
    Nmt1p peptide substrate specificities have been determined from surveys
of synthetic octapeptides representing systematic variations of the N-terminal sequences of known N-myristoyl proteins (*34, 44, 45*). There is an
absolute requirement for an N-terminal glycine. Positioning of the primary
amine is critical: substitution of β-alanine for glycine blocks peptide binding
to the binary myristoyl-CoA : Nmt1p complex. Uncharged residues are preferred at position 2 (e.g., cysteine, alanine, leucine). A spectrum of amino
acids can be accommodated at positions 3 and 4 although neutral residues
are preferred over basic, which in turn are favored over acidic side chains.
Serine is present at position 5 in all known yeast N-myristoyl proteins.
Substitution of Ser-[5] with alanine (replacement of hydroxyl with a proton)
decreases affinity (K_m) for Nmt1p by two to three orders of magnitude.
Lysine is favored at position 6. As described below, X-ray crystallographic
studies of Nmt1p with bound substrate analogs have revealed the structural
basis for many of these observed requirements for acyl-CoA and peptide
recognition.

The primary structures of at least 11 orthologous Nmts are known. These include *N*-myristoyltransferases from *S. cerevisiae* (*46*), *Candida albicans* (*47*), *Cryptococcus neoformans* (*48*), *Histoplasma capsulatum* (*48*), *Drosophila melanogaster* (*49*), *Caenorhabditis elegans* (*50*), *Schizosaccharomyces pombe* (locus SPBC2G2, accession number AL022103), *Homo sapiens* (*51–53*) and *Mus musculus* (*53*).

In vitro studies of human, *C. albicans*, and *C. neoformans* Nmts, purified from *Escherichia coli* (a bacterium with no endogenous Nmt activity; Ref. *54*), have established that their ordered reaction mechanism and acyl-CoA substrate specificities are conserved. However, some features of their peptide recognition have diverged (*55–64*). This divergence has therapeutic significance (see below).

C. PROTEIN N-MYRISTOYLATION REQUIRED FOR SURVIVAL OF *Saccharomyces cerevisiae*

Nmt1p is an essential enzyme in *S. cerevisiae:* an *NMT1* null allele (*nmt1Δ*) produces recessive lethality (*46*). The contributions of *N*-myristoyl proteins to cell survival have been explored using a conditional lethal allele (*nmt1-451D*). nmt451Dp contains a single amino acid substitution: aspartate replaces a highly conserved glycine located five residues from its C terminus. This substitution results in reduced affinity for myristoyl-CoA without changing the acyl chain length specificity of the enzyme (*50, 65*). *nmt1-451D* produces temperature-sensitive myristic acid auxotrophy. When *nmt1-451D* cells undergoing logarithmic growth in standard rich medium are shifted to a nonpermissive temperature ($\geq 30°C$), they stop dividing within 1 hr and die within 8–12 hr (*66*). Growth can be fully rescued by supplementing the medium with myristate but not with other naturally occurring fatty acids (*65, 67*).

The level of reduction in cellular protein N-myristoylation associated with the loss of viability of *nmt1-451D* cells has been defined using two functionally interchangeable but essential yeast Arfs. Both Arfs depend on N-myristoylation for expression of their biological functions (e.g., Refs. *68–70*). Remarkably, N-myristoylation of Arf1p and Arf2p produces a pronounced change in their mobilities during sodium dodecyl sulfate–polyacrylamide gel electrophoresis (SDS–PAGE): the N-myristoylated species (isoforms) migrate more rapidly than the nonmyristoylated isoforms (*48, 58*). By noting the ratio of N-myristoylated to nonmyristoylated isoforms in Western blots of total cellular proteins, the extent of protein N-myristoylation can be assessed under various growth conditions and as a function of wild-type or mutant *NMT1* alleles.

When Western blots are probed with antibodies specific for Arf1p or

Arf2p, only myristoyl-Arfs are detectable in wild-type *NMT1* cells cultured in standard rich medium at 24–37°C (*58*). When *nmt1-451D* cells are grown at the permissive temperature (24°C), these Arfs are also fully N-myristoylated. However, 2 hr after shifting to the nonpermissive temperature, 50% of Arf1p and Arf2p are not N-myristoylated (*58*). This reduced N-myristoylation and its associated lethality can be fully rescued by supplementing the medium with myristate or by introducing an episome containing *NMT1*. These results suggest that an increase in the representation of nonmyristoylated Arf1p or Arf2p to ≥50% of total Arf1p or Arf2p is a biochemical marker of a jeopardized yeast cell. However, reduced Arf N-myristoylation is not entirely responsible for the demise of these cells. For example, Arf1p accounts for ~90% of total cellular Arfs (*68*). *NMT1 arf1Δ ARF2* cells are viable even though their level of Arf is ~10% that of wild-type cells. Moreover, an *nmt1-451DARF1ARF2* cell has more N-myristoylated Arf1p and Arf2p 2–4 hr after growth arrest than does a proliferating *NMT1 arf1ΔARF2* cell (*58*).

Several factors make the Arf proteins attractive reporters for monitoring *in vivo* Nmt activity. First, the appearance of nonmyristoylated Arfs seems to be directly attributable to a reduction in Nmt1p activity because *S. cerevisiae* does not contain a detectable demyristoylation activity (*71*). Second, it is easy to identify changes in Arf acylation: (1) the "baseline" value for Arf1p and Arf2p N-myristoylation in wild-type cells is 100%; and (2) the N-myristoylation-dependent mobility shifts are readily detected by Western blotting because Arfs have low molecular weights and both N-myristoylated and nonmyristoylated isoforms have equivalent immunoreactivities (*58*). Finally, because Arfs are ubiquitous and highly conserved, the mobility shift assay can be applied to a wide variety of eukaryotes.

1. *Cellular Adaptations to Threat of Reduced Protein N-Myristoylation*

Comparisons of isogenic *NMT1* and *nmt1-451D* strains have revealed some of the cellular responses that are mounted to compensate for the reduced catalytic efficiency of the mutant acyltransferase. Log-phase *nmt1-451D* cells contain fourfold higher levels of Nmt mRNA compared with wild-type cells, even at the permissive temperature (*72*). The signaling pathways that produce this induction of *nmt1-451D* expression have not been identified: e.g., it is not known whether the state of N-myristoylation of one or more cellular proteins serves as a feedback signal to modulate *NMT1* transcription.

nmt1-451D also affects myristoyl-CoA production. There are two principal metabolic pathways for generating myristoyl-CoA while cells are undergoing logarithmic growth in standard rich medium: *de novo* synthesis by the fatty acid synthetase (Fas) complex and activation of myristate by

A

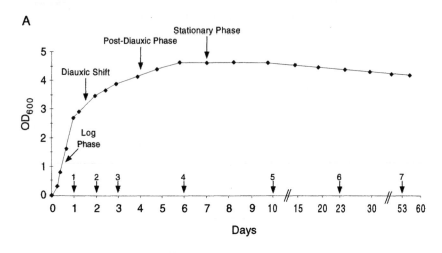

B

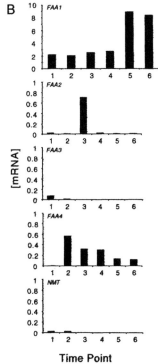

cellular acyl-CoA synthetases (fatty acid activation proteins, Faaps). Myristoyl-CoA represents ~5% of the acyl-CoAs produced by Fas (73, 74). *FAS1* encodes the β subunit of the $\alpha_6\beta_6$ Fas complex. Changes in *FAS1* expression affect expression of *FAS2* but not vice versa (75). At permissive temperatures, steady state Fas1p mRNA levels are twofold higher in *nmt1-451D* compared with isogenic *NMT1* strains (72). Promoter activity assays have shown that this increase is due to augmented *FAS1* transcription (76). The enhancement of *FAS1* transcription is associated with, and mediated in part by, alterations in the transcription of genes encoding Ino2p and Ino4p (76). These two basic helix–loop–helix transcription factors form heterodimeric complexes that bind to inositol-choline response elements present in *FAS1*.

Augmented production of myristoyl-CoA undoubtedly benefits nmt451Dp, which has reduced affinity for this acyl-CoA. However, increased production of myristoyl-CoA is not the exclusive responsibility of Fas: substituting *NMT1* with *nmt1-451D* has selective effects on expression of the four known and unlinked *FAA* genes of *S. cerevisiae* (72).

Genetic studies have shown that Faa1p and Faa4p are both able to activate myristate imported into the cell (77, 78). Although these two Faas have similar myristoyl-CoA synthetase activities *in vitro* (79, 80), they have distinct subcellular locations: an Faa1p–green fluorescent protein (GFP) chimera is associated with the plasma membrane in log-phase cells while Faa4p–GFP is present in an as yet undefined intracellular compartment (K. Ashrafi, C. G. Martin, and J. I. Gordon, unpublished observations, 1999). Faa1p and Faa4p also have distinct growth phase patterns of expression. As nutrients are depleted during log phase, *S. cerevisiae* undergoes a diauxic transition and begins a period of slow growth known as the postdiauxic phase. When nutrients are exhausted, cells enter a stationary phase during which proliferation ceases (Fig. 1A). Entry into stationary phase is

FIG. 1. Growth phase-specific patterns of *NMT1* and *FAA* expression. (A) Representative growth curve of a wild-type strain of *S. cerevisiae* at 24°C in standard rich medium (YPD). (B) Quantitation of Nmt1p and Faap mRNA levels in the wild-type strain as a function of growth phase. The time points surveyed are plotted in (A). Northern blots containing total cellular RNAs were probed with [^{32}P]DNAs having identical specific activities. mRNA levels have been normalized to the sum of the signals produced by 18S and 25S rRNA in cells at each of the time points surveyed. Mean values, expressed in arbitrary units, are from two independent experiments, each performed in duplicate. Values for a given mRNA varied <20% between experiments. Note that the *in vitro* myristoyl-CoA synthetase activity of Faa2p (a protein affiliated with peroxisomes: Ref. 82a) is equivalent to that of Faa1p and Faa4p, and considerably greater than that of Faa3p. Faa3p prefers monoenoic fatty acids [palmitoleic acid (C16 : 1$^{\Delta 9,\,10}$) and oleic acid (C18 : 1$^{\Delta 9,10}$)] as substrates and can also accommodate longer acyl chains than the other Faaps (79, 80).

accompanied by global changes in gene expression and metabolism (*81, 82b*). In *NMT1* cells, Faa1p mRNA is the most abundant Faap mRNA. This is true during log phase, the diauxic transition, the postdiauxic phase, and stationary phase (*72*) (Fig. 1B). Faa4p mRNA is barely detectable in log phase (0.02% of Faa1p mRNA levels) but undergoes a 100-fold increase in concentration as cells enter the diauxic/early postdiauxic phase. Levels are sustained during stationary phase. This pattern of expression is unique among the *FAA*s (Fig. 1B).

nmt1-451D exerts its most significant effect on *FAA4* expression (*72*). A precocious induction of Faa4p mRNA accumulation occurs in log phase, resulting in a 55-fold elevation in steady state levels compared with *NMT1* cells. The difference in Faa4p mRNA concentrations between log-phase wild-type and *nmt1-451D* strains diminishes as cells pass through the diauxic transition and enter stationary phase.

The change in *FAA4* expression has a great impact on the ability of *nmt1-451D* cells to maintain their pools of N-myristoylated proteins during stationary phase. As noted above, *FAA4* is the only *FAA* that undergoes an induction of expression during the diauxic/early postdiauxic phase. Northern and Western blot analyses plus N-myristoyltransferase assays of isogenic *NMT1* and *nmt1-451D* strains have shown that Nmt1p is present only during the log and diauxic/postdiauxic periods (Ref. *72* and Fig. 1B). This means that N-myristoyl proteins present in stationary phase are "inherited" from these earlier phases (*72*). Analysis of a panel of 10 isogenic strains containing all possible combinations of *NMT1* or *nmt1-451D,* plus a wild type of null allele of a given *FAA,* revealed that at 24°C, only *nmt1-451faa4Δ* cells exhibit defects in protein N-myristoylation during both log phase and the diauxic/early-postdiauxic transition. The deficiency, as defined by the Arf mobility shift assay, first appears during logarithmic growth, worsens through the postdiauxic phase, and becomes extreme in stationary phase. The deficiency in protein N-myristoylation is accompanied by a marked and progressive reduction in proliferative potential as cells spend increasing time in stationary phase. The reduction in proliferative potential is manifested by an ever-diminishing ability of cells, recovered from stationary-phase cultures, to form colonies when plated on standard rich medium. This decrease in colony-forming units (CFU) reaches 1 million-fold after 45 days in culture, and is not due to cell lysis, nor is it accompanied by a perceptible loss in the ability to synthesize complex lipids (*72*). The stationary-phase phenotype of *nmt1-451Dfaa4Δ* cells can be fully rescued by introducing a high copy centromeric plasmid containing *NMT1* under the control of its own promoter (*72*).

Additional genetic and biochemical studies have indicated that cells with *nmt1-451D* and wild-type *FAA* alleles have a reduced capacity for protein

N-myristoylation but compensate for the presence of the mutant N-myristoyltransferase by augmenting expression of *FAA4* as well as *nmt1-451D* itself (72). Loss of Faa4p from *nmt1-451D* cells is devastating: they are not able to support the required augmentation in *nmt1-451D* expression and the other Faaps are unable to overcome the loss of Faa4p. The result is that insufficient pools of N-myristoyl proteins are generated during log phase and the diauxic transition to execute their various functions during periods of gradual, progressive, and persistent, nutrient deprivation.

2. Identification of N-Myristoyl Proteins That Are Required for Surviving Periods of Gradual or Acute Nutrient Deprivation

To identify the N-myristoyl proteins involved in regulating cellular responses to gradual nutrient deprivation, the translation products of all standard open reading frames (ORFs) in the yeast genome database were scanned for the presence of $M^1G^2X^3X^4X^5X^6X^7$ at their N terminus (72). E, D, R, K, H, P, F, Y, and W were not allowed at position 3 (X^3). All possible amino acids were allowed at X^4 and X^5 while only S, T, A, G, C, or N were permitted at X^6. All residues except P were allowed at X^7. Among the 6220 ORFs surveyed, 70 (1.1%) were identified as known or putative N-myristoyl proteins: **YAL004W** Saccharomyces Genome database ((SGD) designation for the gene/open reading frame), YAL002W, **YBL056W, YBL049W, YBR097W, YBR125C,** YBR164C, YBR286W, **YDL192W, YDL160C, YDL137W,** YDL007W, **YDR079W, YDR105C, YDR181C, YDR364C,** YDR373W, **YDR436W, YDR440W,** YEL013W, **YER020W, YER038C, YER077C, YER089C,** YER144C, **YFR011C,** YGL246C, **YGL208W,** YGL108C, YGL022W, **YGR259C,** YGR280C, YHR005C, **YIL170W, YIR043C, YJL218W, YJL017W, YJR114W,** YKL053C-A, **YKL159C, YKL069W, YKL190W, YKR007W, YLR111W, YLR112W,** YLR232W, **YLR255C, YLR394W, YLR436C, YML115C, YML016C, YMR040W, YMR077C,** YMR140W, **YMR151W, YMR162C, YMR216C, YMR239C, YMR317W, YNL301C,** YNL048W, **YNL032W,** YOL120C, YOR070C, YOR094W, **YOR118W, YOR181W, YOR311C, YPL109C,** and YPR116W.

This value of 1.1% represented the first estimate of the number of N-myristoyl proteins produced by a eukaryote. A comparable search of the *C. elegans* genome (*www.sanger.ac.uk*) using the same octapeptide sequence disclosed 266 candidate N-myristoyl proteins among the 19,099 ORFs (1.4%) listed on this web site on February 23, 1999. However, unlike *S. cerevisiae* Nmt1p, a detailed analysis of the peptide substrate specificity of *C. elegans* N-myristoyltransferase has not been reported.

Fifty of the 70 *S. cerevisiae* genes encoding known or putative myristoyl proteins were successfully disrupted in *NMT1* cells (Ref. 72 plus K. Ashrafi, C. Martin, and J. I. Gordon, unpublished observations, 1999). These genes

TABLE I

GENE DELETIONS THAT PRODUCE REDUCTION IN COLONY-FORMING POTENTIAL OF NMT1
CELLS DURING ACUTE AND GRADUAL NUTRIENT DEPRIVATION

SGD gene designation	N-Terminal sequence[a]	CFU	
		Fold reduction[b]	Complete loss[c]
YBL049W	GLRYSIYI	10^4	+
YDL192W (ARF1)	GLFASKLF	10^2	±
YDL137W (ARF2)	GLYASKLF	10^4	+
YER089C (PTC2)	GQILSNPV	10^4	+
YGL208W (SIP2)	GTTTSHPA	10^4	−
YJR114W	GTDFSASH	10^4	+
YKR007W	GAVLSCCR	10^3	+
YML115C (VAN1)	GMFFNLRS	10^4	−
YMR077C	GQKSSKVH	10^4	+

[a] Residues 2–9 of the predicted or known product of the ORF are shown. The initiator methionine residue has been deleted.
[b] After 25 days in water. Reference control is the isogenic wild-type parent.
[c] After 7 days in 1% potassium acetate. +, Absence of any colonies after aliquots were removed after 7 days and incubated for 3 days at 24°C on YPD–agar.

are shown in boldface type in the preceding paragraph. Surveys of the resulting 50 isogenic strains disclosed 9 Nmt substrates whose loss mimicked the starvation-sensitive stationary-phase phenotype of $nmt1$-$451Dfaa4\Delta$ cells (Table I). These substrates include Arf1p, Arf2p, Ptc2p (a protein with homology to an S. pombe serine/threonine phosphatase 2c; Refs. 83 and 84), Sip2p (a member of an evolutionarily conserved group of homologous proteins that bind with high affinity to the Snf1p serine/threonine kinase required for expression of glucose-repressed genes during periods of glucose deprivation and for resistance to other forms of environmental stress; Refs. 85 and 86), Van1p (implicated in producing arrest in G_1 during stationary phase and, when removed, is associated with altered protein phosphorylation; Ref. 87), YBL049W [homology to Snf7p, a protein involved in derepression of SUC2 (invertase) during periods of glucose deprivation; Refs. 88 and 89], plus three proteins with no obvious homologies to known proteins (YJR114W, YKR007W, and YMR077C).

N-Myristoyl proteins not only play a role in maintaining the proliferative potential of S. cerevisiae during periods of gradual nutrient deprivation, but also affect the ability of cells to survive abrupt withdrawal of nutrients. When members of the panel of 10 isogenic strains with various combinations of NMT1, nmt1-451D, FAA, and $faa\Delta$ alleles were recovered during logarithmic growth at 24°C in standard rich medium, resuspended in minimal sporulation medium (1% potassium acetate; Ref. *90*), and CFU measured 7 days later, only *nmt1-451D* and *nmt1-451Dfaa4Δ* cells exhibited significant loss in their colony-forming ability (Fig. 2). As was the case with the *nmt1-451Dfaa4Δ* strain during stationary phase, *nmt1-451D* and *nmt1-451Dfaa4Δ* cells do not undergo lysis during acute nutrient deprivation and >90% retain their ability to reduce methylene blue (K. Ashrafi and J. I. Gordon, unpublished observations, 1999). A centromeric plasmid containing *NMT1* rescues the ability of *nmt1-451D* and *nmt1-451Dfaa4Δ* cells to maintain their colony-forming potential during the period of nutrient withdrawal (Fig. 2).

The difference in the colony-forming responses of *nmt1-451D* and *nmt1-451Dfaa4Δ* cells to gradual nutrient deprivation, as opposed to the similarity

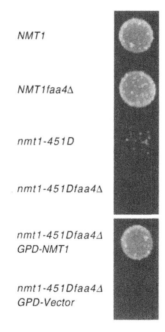

FIG. 2. Protein N-myristoylation is required to survive the stress of acute nutrient deprivation. Equal numbers of cells (10^7) from each of the indicated isogenic strains were removed during mid-log-phase growth in YPD medium and placed in 1% potassium acetate for 7 days at 24°C. Equal size aliquots of cells from each culture were then plated on YPD/agar. When cells containing *nmt1-451D* are acutely deprived of nutrients, they rapidly lose their ability to form colonies. This loss of CFU is rescued by an episome containing the *NMT1* open reading frame under the control of the glyceraldehyde-3-phosphate dehydrogenase promoter (*GPD*; Ref. *91*).

in their responses to acute nutrient deprivation, emphasizes that the format and timing of nutrient deprivation have an important impact on their capacity to return to the cell cycle. Removal of any 1 of 6 of the 50 known or putative Nmt1p substrates from *NMT1* cells reproduced the acute starvation-sensitive phenotype of *nmt1-451D* and *nmt1-451Dfaa4Δ* strains (Table I). Loss of each of these proteins in *NMT1* cells also reproduced the loss of proliferative potential manifested by *nmt1-451Dfaa4Δ* cells during stationary phase (Table I).

N-Myristoyl proteins appear to exert both positive and negative effects on maintenance of proliferative potential during periods of nutrient deprivation. Forty-five of the 50 genes that were deleted in the *NMT1* strain have been deleted successfully in an isogenic *nmt1-451D* strain. The deletional mutagenesis analysis disclosed that removal of any one of three genes encoding Nmt1p substrates fully rescued the loss of *nmt1-451D* CFU after acute nutrient deprivation (K. Ashrafi and J. I. Gordon, unpublished observations, 1999). These three Nmt1p substrates included (1) Gpa2p, the α subunit of a heterotrimeric G protein (*92*) that regulates growth and pseudohyphal development through a cAMP-dependent mechanism (*93*); (2) YBR125C, a protein with homology to a yeast protein phosphatase 2C (PP2Cp, Ref. *84*) and to a mammalian PP2C repressed in immortalized compared with senescent human diploid fibroblasts (*94*); and (3) YDR105C, a protein with weak homology to members of the ABC family of transporters. Moreover, when any of these three genes was deleted from *nmt1-451Dfaa4Δ* cells, there was partial or full rescue of their loss of proliferative potential during stationary phase (K. Ashrafi and J. I. Gordon, unpublished observations, 1999).

An implicit assumption underlying these analyses is that removal of an *N*-myristoyl protein via gene deletion/disruption is functionally equivalent to its hypo- or absent acylation. This assumption has not been formally tested for all of the *N*-myristoyl proteins described above, nor have the specific roles of these proteins in regulating proliferative potential during nutrient deprivation been elucidated. Nonetheless, when considered together with the phenotypes of the *nmt1-451D*-containing strains, the results establish that protein N-myristoylation is required for cell survival.

D. NMT: Target for Development of New Class of Fungicidal Drugs

Drug resistance is a troubling problem in the treatment of fungal pathogens that cause systemic infection in immunocompromised humans. The number of fungicidal compounds currently available is quite limited. Azoles inhibit lanosterol (C_{14}) demethylase (ERG11) in the ergosterol biosynthetic pathway and are the preferred class of drugs given to patients with acquired immune deficiency syndrome (AIDS) for chemoprophylaxis and/or treatment of their frequent systemic fungal infections (95, 96). Unfortunately, drug-resistant isolates are an increasingly common clinical problem (Refs. 97–99 for a review of the clinical, cellular, and molecular factors underlying development of resistance).

An ideal new target for antifungal therapy should fulfill several criteria. The gene product should be expressed under conditions of infection and should be essential for the survival of the pathogen. The metabolic pathway and/or substrate specificities of the target protein should be distinguishable from those represented in the human host but should be common to many fungal pathogens. As described below, Nmt satisfies several of these criteria. Genetic studies have established that NMT is essential for maintaining the viability of Candida albicans and Cryptococcus neoformans, the two principal causes of systemic fungal infections in immunodeficient humans. Moreover, biochemical studies have shown that the divergent peptide substrate specificities of fungal and human Nmts can be exploited to design biologically active inhibitors that are selective for fungal compared to human Nmts.

1. Genetic Proof That NMT Is Essential Gene in Candida albicans and Cryptococcus neoformans

Candida albicans nmt447D (Gly-447 → Asp) is analogous to S. cerevisiae nmt1-451D (100). Among isogenic NMT/NMT, NMT/Δnmt, and nmtΔ/nmt447D strains of C. albicans, only nmtΔ/nmt447D cells require supplemental myristate for growth on standard rich medium (yeast–peptone–dextrose; YPD) at 24 or 37°C. When switched from YPD supplemented with 500 μM myristate to YPD alone, 60% of the organisms die within 4 hr, even at 24°C, and all are dead within 24 hr (100). In the NMT/nmtΔ strain, 100% of Arf is N-myristoylated at 24 and 37°C. When nmtΔ/nmt447D cells are incubated at 24°C in YPD–myristate, <25% of a cellular Arf is nonmyristoylated, but 4 hr after withdrawal of myristate, ≥50% of the Arf lacks myristate (58).

Cryptococcus neoformans is a haploid yeast. Cryptococcus neoformans has tropism for the central nervous system, where it produces chronic meningitis. The incidence of infection in patients with AIDS is ~10% (101). Cryptococcus neoformans nmt487D (Gly-487 → Asp) is analogous to S. cerevisiae nmt1-451D (57). Like the nmtΔ/nmt447D strain of C. albicans, a strain of C. neoformans containing a single copy of nmt487D at the endogenous locus (59) is unable to grow at 24°C in the absence of myristate. Moreover, switching mid-log-phase cells from YPD–myristate to YPD produces lethality within 4hr (at 24°C). The nmt487D phenotype is not only distinctive but stable (reversion frequency $<10^{-8}$; Ref. *59*). Only the N-myristoylated Arf isoform is present in wild-type *C. neoformans* cells grown in YPD at 24 or 37°C. In contrast, the nonmyristoylated isoform represents >75% of the total 2 hr after withdrawal of myristate from *nmt487D* cells (*57*). Isogenic *NMT* and *nmt487D* strains and an immunosuppressed rabbit model of cryptococcal meningitis were used to show that genetic attenuation of Nmt activity allows the host to rid itself of an otherwise fatal infection (*57*).

These genetic studies provided strong evidence that inhibition of Nmt activity would be fungicidal and set the stage for a series of experiments designed to identify biologically active, species-selective inhibitors of Nmt.

2. Generating Species-Selective Inhibitors of Fungal Nmts That Are Fungicidal

Alanine-scanning mutagenesis of an octapeptide substrate derived from the N-terminal sequence of an Arf (GLYASKLS-NH$_2$) revealed that Gly-^{1}, Ser-^{5}, and Lys-^{6} play the most important role in recognition by the peptide-binding site of *C. albicans* Nmt (*102*). Replacement of Gly-1 with alanine produced an inhibitor (ALYASKLS-NH$_2$) that was competitive for peptide and noncompetitive for myristoyl-CoA. Remarkably, replacement of the N-terminal ALYA tetrapeptide with an 11-aminoundecanoyl group yielded an inhibitor (11-aminoundecanoylSKLS-NH$_2$) that was ~40-fold more potent than ALYASKLS-NH$_2$ (K_i = 0.4 ± 0.03 versus 15.3 ± 6.4 μM; Ref. *102*). An extensive structure–activity analysis of this inhibitor (*58, 60, 61, 63, 64, 102*) disclosed that increased potency and selectivity for *C. albicans* versus human Nmt could be obtained by adding an N-terminal 2-methylimidazole, rigidifying the flexible undecanoyl chain with a phenylacetyl group and replacing the C-terminal Leu–Ser with *N*-cyclohexylethyl. The product of this exercise was a dipeptide competitive inhibitor (SC-58272 in Fig. 3) that was potent (K_i = 56 nM) and 250-fold selective for purified *C. albicans* over human Nmt (*60*). Unfortunately, SC-58272 has no growth inhibitory effects when added to cultures of *C. albicans* [50% effective concentration (EC$_{50}$) >200 μM], no detectable effect on Arf N-myristoylation, and does not appear to gain access to Nmt within cells.

Introduction of a carboxyl group geminal to the lysine amide nitrogen in SC-58272 yielded a tripeptide inhibitor containing L-serine, L-lysine, and a cyclohexyl-L-alanine at its C terminus (see SC-59383 in Fig. 3). This compound is less potent than SC-58272 (IC$_{50}$ = 1.45 ± 0.08 μM for *C. albicans* Nmt) but is more selective for the fungal versus human Nmt (560-fold). Despite its more modest inhibitory activity *in vitro*, SC-59383 suppresses growth of *C. albicans* (EC$_{50}$ = 51 ± 17 μM). The effect is fungistatic rather than fungicidal (*58*). This latter observation is consistent

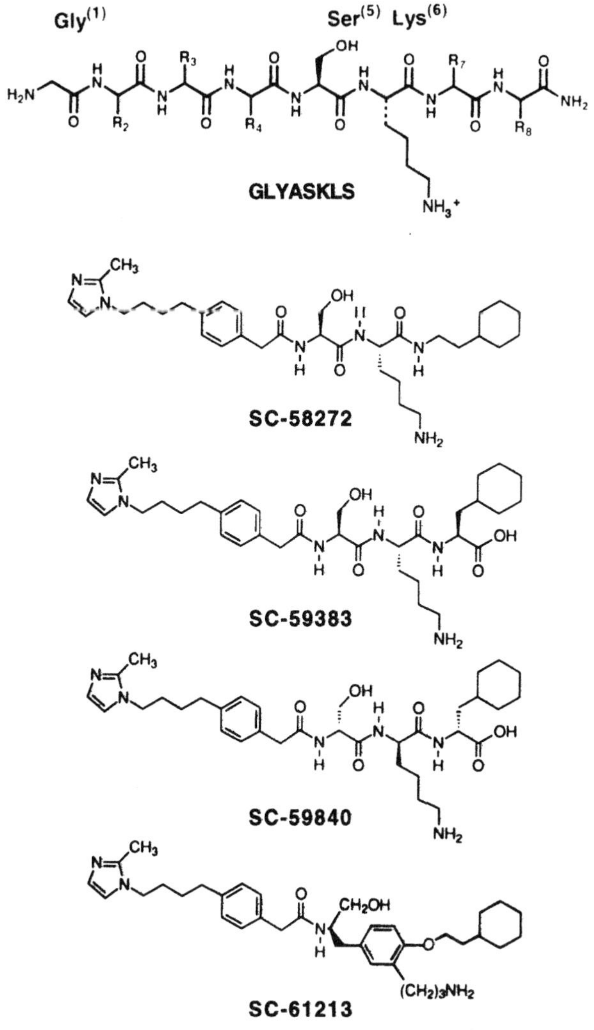

FIG. 3. Development of SC-61213, a fully depeptidized Nmt inhibitor derived from an octapeptide substrate. See text for further discussion.

with the finding that a single dose of 200 μM produces only a modest (10–15%) reduction in Arf N-myristoylation 2 to 4 hr after administration (58). Further evidence that the fungistatic effect of SC-59383 is due to its Nmt inhibitory activity came from an analysis of its enantiomer (SC-59840 in Fig. 3). Unlike SC-59383, SC-59840 contains D-serine, D-lysine, and a C-terminal cyclohexyl-D-alanine. Substitution of these D- for L-amino acids completely abolishes Nmt inhibitory activity *in vitro*. SC-59840 has no effect on growth of *C. albicans* after administering doses as high as 1000 μM. Moreover, exposing *NMT/nmtΔ* cells to SC-59840 results in no detectable inhibition of Arf N-myristoylation (*58*).

SC-59383 is cleaved by cellular carboxylpeptidases, likely limiting its biological activity. To achieve greater metabolic stability, another compound was produced with no peptide bonds and a single chiral center (SC-61213; Ref. *62*). The *p*-[(2-methylimidazol-l-yl)butylphenyl]acetyl moiety in SC-58272 and SC-59383 was retained in SC-61213 (Fig. 3). The C-terminal SKLS present in the parental inhibitor ALYASKLS-NH$_2$ was represented by a chiral tyrosinol scaffold that contained the serine alcohol and presented the lysine amine as a 3-aminobutyl side chain. The hydrophobic interactions mediated by Leu-7 were provided by the 4-cyclohexylethyl ether in SC-61213. This compound has modest potency against *C. albicans* and *C. neoformans* Nmts and is about fivefold selective for the fungal compared with human enzymes [K_i = 1.8 ± 1 μM (*C. neoformans*) vs 9 ± 2.4 μM (human); Ref. *59*].

The biological effects of SC-61213 have been evaluated using two isogenic strains of *C. neoformans* (*59*). One strain contained a single copy of wild-type *C. neoformans NMT*. The other strain contained *nmt487D* plus a recombinant DNA consisting of the human *NMT* ORF under the control of transcriptional regulatory elements from the *C. neoformans NMT* gene. Human *NMT* can complement the myristic acid auxotrophy produced by *nmt487D* and both strains have identical growth kinetics at 35°C. Because a single copy of *nmt487D* will not support growth at 35°C, survival depends on human Nmt. The advantage of using these two isogenic strains to evaluate a compound such as SC-61213 is that if an inhibitor is selective for the fungal compared with human enzyme *in vitro*, and if that compound also produces a greater degree of inhibition of growth of the fungal Nmt-producing compared with the human Nmt-producing strain, then an Nmt-dependent mechanism for its biological effect can be invoked. SC-61213 satisfied both criteria and was fungicidal (*59*).

These results show that an octapeptide substrate (GLYASKLS) containing seven peptide bonds and eight chiral centers can be reduced to a nonpeptidic inhibitor with one chiral center that is not only selective for fungal versus human Nmt but has Nmt-dependent fungicidal activity. SC-61213 also has fungicidal activity against *C. albicans* (*62*).

Isogenic *C. neoformans* strains, containing wild-type *C. neoformans* or human *NMT*, should be useful for future high-throughput screens of large chemical or natural product libraries that seek to identify Nmt-based fungistatic or fungicidal compounds with relatively little activity against the human acyltransferase.

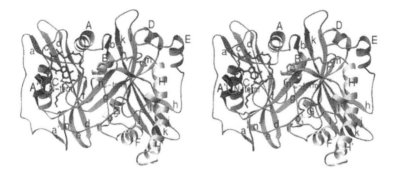

Fig. 4. Stereodiagram of Nmt1p with bound substrate analogs. Stereodiagram of *S. cerevisiae* Nmt1p with bound *S*-(2-oxo)pentadecyl-CoA (red) and SC-58272 (lavender). α-Helices, 3_{10} helices, strands, and loops are colored light blue, deep blue, green, and amber, respectively. Secondary structural elements are named according to Weston *et al.* (103). Strands are labeled with lower-case letters and helices with capital letters. Elements that are designed as primed (e.g., A') do not appear in the *C. albicans* apoenzyme structure.

III. Enzymology

A. STRUCTURE OF NMT WITH BOUND SUBSTRATE ANALOGS

Understanding the structural basis for Nmt peptide recognition and its chemical mechanism of catalysis should facilitate the design of potent and selective enzyme inhibitors. To this end, the structures of two orthologous Nmts have been solved by X-ray crystallography: *Candida albicans* Nmt without bound substrates (*103*) and *S. cerevisiae* Nmt1p with two bound substrate analogs (*104*).

Nmt1p was crystallized with a nonhydrolyzable myristoyl-CoA derivative, *S*-(2-oxo)pentadecyl-CoA (*105*), and the dipeptide inhibitor SC-58272. The 2.9-Å model of this ternary complex contains a total of 422 residues (Ala-34 to Leu-455) and is presented as a stereodiagram in Fig. 4. The enzyme is composed of a single compact domain with distinct binding sites for its two ligands.

The Nmt fold consists of a large saddle-shaped β sheet that spans the center of the protein. This β sheet is flanked on both of its faces by several helices. In the orientation shown in Fig. 5A, the sheet has two lateral walls and crossing strands in front. Each lateral wall is flanked by a large α helix (helices C and H). Figure 5B is a back view of the enzyme obtained by

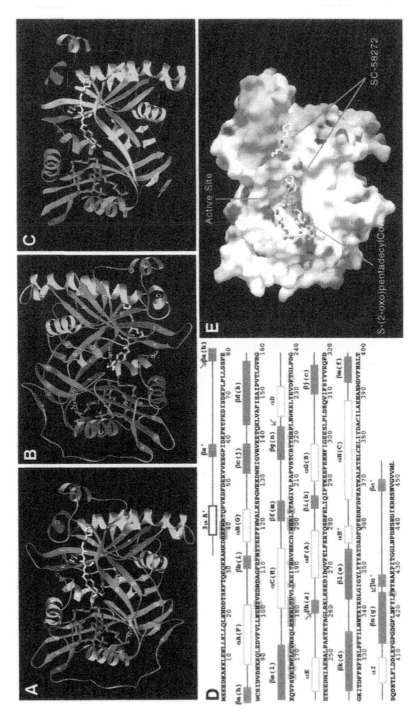

rotating the model shown in Fig. 5A 180° about its x axis. Four smaller α helices (A, B, F, and G) form a semicircular back face of the protein.

A remarkable feature of the Nmt fold is its internal pseudo twofold symmetry. The topology of the N-terminal half of the protein is similar to that of the C-terminal half (Fig. 5C). Each of the major secondary structural elements is represented in both symmetry-related halves. Figure 5C and D illustrates how the left side and crossing strand βg of the sheet are topologically equivalent to the right side and crossing strand βn. Similarly, the large helix C that flanks the left lateral wall of the sheet has a symmetry mate, αH, that flanks the right lateral wall of the sheet. This symmetry is also evident in the four small helices that form the back face: αA and αB are derived from the N-terminal half of the protein and are matched by αF and αG from the C-terminal half.

Phe-79 through Pro-223 encompass the secondary structural elements of the enzyme's N-terminal motif that have symmetry mates in the enzyme's C-terminal motif (Leu-261 through Phe-425). The C-terminal motif contains two additional secondary structural elements that are not represented in the N-terminal motif: αH' and αI (Fig. 5C and D). The N- and C-terminal symmetry-related motifs do not contain detectable primary sequence homology.

Elements of the primary sequence not encompassed by Phe-79 $\rightarrow$ Pro-223 or Leu-261 $\rightarrow$ Phe-425 do not have symmetry-related secondary structures. The sequence N terminal to Phe-79 contains a 3_{10} helix A' (Fig. 5A and D). The sequence between Pro-223 and Leu-261 contains αD and αE and forms a linker between the two symmetry-related motifs (Fig. 5A and

---

Fig. 5. The Nmt fold. (A) Ribbon diagram of Nmt1p with bound substrate analogs. This vantage point, showing the ligand-binding sites, is designated as the "front" view. (B) Rear view obtained by rotating the molecule 180° about the x axis relative to the orientation shown in (A). Note that the four helices, F, G, B, and A, enclose the space between the large walls of the β sheet. (C) The Nmt fold has pseudo twofold symmetry. The N-terminal symmetry-related half of the enzyme (Ala-34 to Asn-225) is colored green. The C-terminal symmetry-related half (Trp-226 to Leu-455) is colored yellow. (D) Primary sequence of Nmt1p showing secondary structural elements and defining the pseudo-twofold symmetry. Helices are designated by open boxes and strands as filled boxes. Elements are named as in Fig. 4. Many of these elements have symmetry mates. In these cases, the symmetry mate is indicated in parentheses: e.g., the symmetry mate of βa is βh. Residues involved in binding the myristoyl-CoA are highlighted in green while those that are involved in binding the peptide analog are highlighted in pink. Residues involved in binding both analogs are highlighted in blue. The red arrows indicate the boundaries of the sequences that encompass all the symmetry-related elements. (E) Surface (GRASP) diagram (106) of Nmt1p colored by electrostatic potential. The protein is in approximately the same orientation as (A). Deep blue represents the most positive regions. Deep red represents the most negative regions. Note that the peptide-binding groove is open ended to accommodate nascent peptide substrates.

D). The sequence C terminal to Phe-425 folds into the center of the enzyme so that the terminal carboxylate of Leu-455 can form part of the enzyme active site (Fig. 5A, D, and E).

The two interdigitated halves of the protein have distinct functions. The N-terminal half forms the myristoyl-CoA-binding site. The only contribution of the C terminus to this site is the side chain of Phe-425, which is positioned at the C-terminal end of βn (Fig. 5C and D). The C-terminal half forms the bulk of the peptide-binding site, although important contacts with the substrate (analog) are formed by αA, βB, and the intervening Ab loop of the N-terminal half (Fig. 5C and D).

The structure of *C. albicans* Nmt without bound ligands was originally reported at 2.8-Å resolution by Weston *et al.* (*103*). The model has subsequently been refined by the same group to 2.45-Å resolution [Protein Data Bank (PDB) accession 1NMT]. The orthologous *C. albicans* and *S. cerevisiae* N-myristoyltransferases have 55% sequence identity. The structure of *S. cerevisiae* Nmt1p with bound ligands (PDB accession 2NMT) is similar to the structure reported for the *C. albicans* apoenzyme: the overall root mean square (rms) deviation for the C_α atoms in these two models is 1.6 Å. Nonetheless, there are some notable differences between the structures. The first 59 residues of *C. albicans* apo-Nmt are not detectable, indicating that they are disordered. In contrast, electron density is visible beginning at residue Ala-34 of Nmt1p (equivalent to Glu-38 of the fungal enzyme). The additional ordered residues at the N terminus of Nmt1p include the 3_{10} helix A' involved in myristoyl-CoA binding (see below). The region immediately after αA in the *C. albicans* apoenzyme (Val-108 → Asp-112) displays considerable conformational heterogeneity. In contrast, the Ab loop (Val-104 → Asp-108) is well ordered in the *S. cerevisiae* ternary complex structure and is an important component of its myristoyl-CoA- and peptide-binding sites (see below).

1. Myristoyl-CoA-Binding Site

In the ternary complex structure, myristoyl-CoA is represented by *S*-(2-oxo)pentadecyl-CoA. *S*-(2-Oxo)pentadecyl-CoA differs from myristoyl-CoA only by the presence of an additional methylene interposed between the sulfur atom of CoA and the carbonyl carbon of the fatty acid. Thus, *S*-(2-oxo)pentadecyl-CoA has the same total number of carbons as pentadecanoyl-CoA, but the same number of carbons from its carbonyl to the ω terminus of its acyl chain as tetradecanoyl-CoA (myristoyl-CoA). Insertion of the extra methylene changes the reactive thioester bond of myristoyl-CoA to a nonreactive thioether. Isothermal titration calorimetric analyses (*39, 40*) have shown that *S*-(2-oxo)pentadecyl-CoA binds to apo-Nmt1p with an enthalpy (ΔH) similar to that of myristoyl-CoA (Table II). This

TABLE II

THERMODYNAMIC STUDIES OF BINDING OF ACYL-CoAS AND S-(2-OXO)PENTADECYL-CoA
TO APO-Nmt1p[a]

Ligand	$\Delta G°$	ΔH
Myristoyl CoA (C14:0)	−10.8	−24.8
S-(2-Oxo)pentadecyl-CoA	−11.7	−25.8
Tridecanoyl-CoA (C13:0)	−10.8	−18.3
Pentadecanoyl-CoA (C15:0)	−10.8	−18.8
Palmitoyl-CoA (C16:0)	−10.4	−15.1

[a] Values are expressed as kilocalories per mole and were determined by isothermal titration calorimetry as described in Bhatnagar et al. (39, 40).

suggests that the binding mode of the analog closely duplicates that of myristoyl-CoA.

The secondary structural elements contributing to myristoyl-CoA binding are the 3_{10} A′ and αC helices, strands βe, βf, and βn, and the fg, eC, and Ab loops (Fig. 5D). βe, αC, βf, and the intervening loops form a $\beta\alpha\beta$ motif that interacts with CoA in a manner similar to the way the $\beta\alpha\beta$ motifs in the Rossman folds of certain nucleotide-binding proteins interact with their ligands (107–109).

Binding of S-(2-oxo)pentadecyl-CoA results in burial of 850 Å^2 or 70% of its total solvent-accessible area. Four bends in the bound myristoyl-CoA analog produce an overall conformation resembling a question mark (Fig. 6A). The bends are located at the pyrophosphate group, at the C6p–C7p positions of pantetheine, at C1m of myristate, and at C5m–C6m (Fig. 6B; note the atom nomenclature—atoms in pantetheine are designated by the suffix "p" while atoms in the acyl chain are designated by the suffix "m"). As a result of this conformation, the adenine ring is nearly completely encircled by the pantetheine moiety and acyl chain, with extensive intramolecular contacts occurring between the adenine ring, pantetheine, and the fatty acid.

Both sides of the adenine ring interact with the protein (Fig. 6C). On the side facing the protein core, residues in helix αC (Thr-183 and Ile-187) and in strand βe (Leu-171) form the floor of the adenine-binding site. On the opposite side, hydrophobic residues contributed by the fg loop (Ile-208 and Leu-210) clamp the adenine base to the floor of the binding site. Interactions with the ribose of the nucleoside are limited to hydrogen bonds between His-38 and Trp-41 in the 3_{10} A′ helix and the 2′-hydroxyl of the pentose sugar (Fig. 6C).

The highly charged 3′-phosphate and pyrophosphate moieties of CoA are expected to be responsible for a large portion of the potential binding

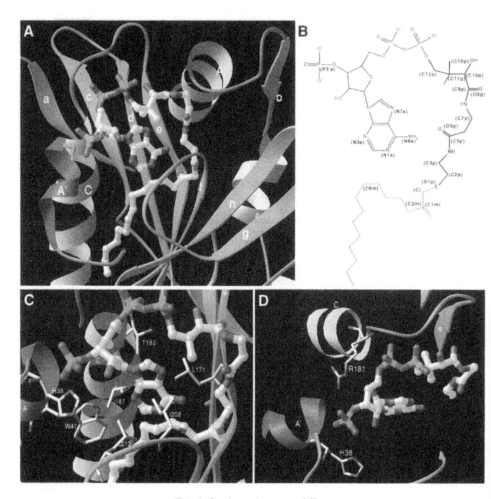

FIG. 6. See legend on page 268.

energy of CoA (*110*). In the myristoyl-CoA-binding site of Nmt1p, the CoA phosphates are positioned at the N termini of helices. The 3′-phosphate is positioned directly above the N terminus of the 3_{10} helix A′. The pyrophosphate is located above the N terminus of the large α-helix C (Fig. 6C and D). The macroscopic dipoles of these helices (positive at their N termini because of the regular orientation of backbone CO and NH groups) are utilized by the enzyme to bind the negatively charged phosphate groups. This is a common means employed by proteins for phosphate binding (see Ref. *111*).

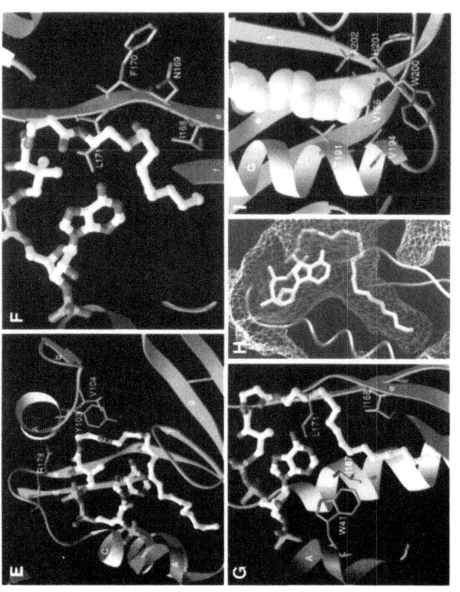

FIG. 6. (*continued*)

The 3'-phosphate is also neutralized through interactions with His-38 and by an ion pair with the guanido group of Arg-181 (Fig. 6D). Additional Nmt1p:pyrophosphate interactions consist mainly of hydrogen bonds between the phosphate oxygens and the amide nitrogens of residues in the eC loop (Ser-179 through Thr-183) and helix C. These hydrogen bonds and the main-chain atoms of the eC loop constrain the conformation of the pyrophosphate group so that a bend is formed (Fig. 6D and E).

The proximal part of the pantetheine chain runs in the direction imposed by this bend in the pyrophosphate group. The hydroxyl, O10p, is hydrogen bonded to the main-chain amide nitrogen of Cys-172 while the carbonyl oxygen, O9p, is surrounded by a dense network of polar or charged residues and appears to be involved in a π-stacking interaction with the guanido group of Arg-178 (Fig. 6E). The next atoms in the pantetheine chain (N8p and C7p) are stacked against the main-chain atoms of Tyr-103, and Val-104 of the Ab loop. C7p is also within van der Waals contact of the side chains of Val-104 and Tyr-103 (Fig. 6E). This contact imposes a bend in the pantetheine chain (the second bend in the acyl-CoA analog). As a result, the C7p-to-S1p half of the pantetheine chain is directed toward the

---

FIG. 6. Myristoyl-CoA-binding site of Nmt1p containing bound S-(2-oxo)pentadecyl-CoA. (A) Overview of bound S-(2-oxo)pentadecyl-CoA. The substrate analog contains four bends creating as overall conformation that resembles a question mark. Atoms are color coded as follows: oxygen, red; phosphorus, magenta; carbon, silver; nitrogen, dark purple; and sulfur, yellow. (B) Diagram of S-(2-oxo)pentadecyl-CoA showing component domains and summarizing atom nomenclature. The adenosine, 3'-phosphate, and pyrophosphate are shown in amber, the pantetheine in purple, and the acyl chain in orange. Atoms within the adenosine and phosphates are given an a suffix, those in pantetheine an p suffix, and those in the myristoyl chain an m suffix. The methylene (C') is inserted between the sulfur and carbonyl of myristoyl-CoA, producing a nonhydrolyzable derivative with a thioether rather than a thioester bond. (C) Interactions of the adenine and ribose rings with Nmt1p. Side chains are colored as described in (A). Note that the adenine ring makes extensive contacts with pantetheine and acyl chain. (D) Interactions of CoA phosphates with Nmt1p. The phosphate groups interact with the N termini of helices and hydrogen bond primarily with main-chain NH groups. (E) Interactions of the pantetheine chain with Nmt1p. Note that pantetheine contacts Tyr-103 and Val-104 of the Ab loop. Arg-178 appears to have a π-stacking interaction with the C9p–O9p carbonyl of pantetheine. (F) Region of S-(2-oxo)pentadecyl-CoA representing the reactive thioester of myristoyl-CoA. The carbonyl oxygen forms hydrogen bonds with the main-chain amide NH groups of Phe-170 and Leu-171 in βe. These hydrogen bonds are able to form because of a bulge in βe. The bulge is created by residues Asn-169 and Phe-170, which project from the same face of the strand. The NH groups of Phe-170 and Leu-171 form the oxyanion hole of the enzyme. (G) Hydrophobic side chains and the adenine ring force a bend at C-5–C-6 of myristate. (H) Outline of the internal surface of the myristoyl binding pocket of Nmt1p. (I) Floor of the acyl chain-binding pocket. The acyl chain is shown as a space-filling model. The floor is formed by the side chains of Val-166, Val-194, and Ala-202, and by the main-chain atoms of Trp-200 and His-201.

active site center (see Fig. 6F). The conformation of the pantetheine chain in this region is stabilized further by a water-mediated hydrogen bond between the O5p carbonyl oxygen and the N6a amine group in the adenine base (Fig. 6F).

The region of S-(2-oxo)pentadecyl-CoA representing the reactive thioester bond of myristoyl-CoA is positioned in front of a bulge in strand βe (Fig. 6F). The β bulge is formed where two consecutive side chains (Asn-169 and Phe-170) project from the same face of the strand, rather than alternating faces as in typical β strands. As a result, the carbonyl oxygen of S-(2-oxo)pentadecyl-CoA forms hydrogen bonds with the backbone NH groups of Phe-170 and Leu-171 (Fig. 6F). These backbone amide groups are not involved in other hydrogen bond interactions, as would typically be the case for backbone atoms in the core of a protein. This structural feature—a β bulge where the reactive thioester carbonyl forms hydrogen bonds with NH groups—constitutes an important enzymatic feature, the oxyanion hole (see below).

Steric constraints produce the third bend at C1m that takes the acyl chain around the adenine ring toward helix αC (Fig. 6G). At C6m, the acyl chain bends, as predicted from previous kinetic analysis of conformationally restricted acyl-CoA analogs with single, double, and triple bonds (41, 42). This bend is forced by the side chains of Trp-41 (from the 3_{10} A' helix), Ile-168 and Leu-171 (from βe), Ile-187 (from αC), Tyr-204 (from βe), and by the adenine ring itself (Fig. 6G). This fourth bend allows the acyl chain to insert itself into a deep narrow pocket located within the core of Nmt1p (Fig. 6H). The pocket is of limited depth and width, and is bounded on its sides primarily by hydrophobic residues. The side chains of Val-166, Val-194, and Ala-202, along with the backbone atoms of Trp-200 and His-201 in the Cf loop, form the floor of the acyl chain-binding pocket of the enzyme (Fig. 6I). The ω terminus (C14m) of the myristoyl chain is within van der Waals contact of this floor and closely surrounded by protein atoms. The fixed depth of this pocket appears to be an important component of the chain-length recognition machinery of the enzyme.

In summary, the ternary complex structure indicates that three features in Nmt1p act together as structural reference points against which acyl chain length is measured. These features are as follows: (1) the oxyanion hole, formed by the main-chain amide groups of Phe-170 and Leu-171, into which the thioester carbonyl must be inserted; (2) residues that force the bend at C6m; and (3) the floor of the pocket that accommodates C7m–C14m.

2. Peptide-Binding Site

The peptide-binding site in the Nmt1p ternary complex is occupied by SC-58272. A total of 71% (667 $\mathring{A}^2$) of the surface area of the analog is

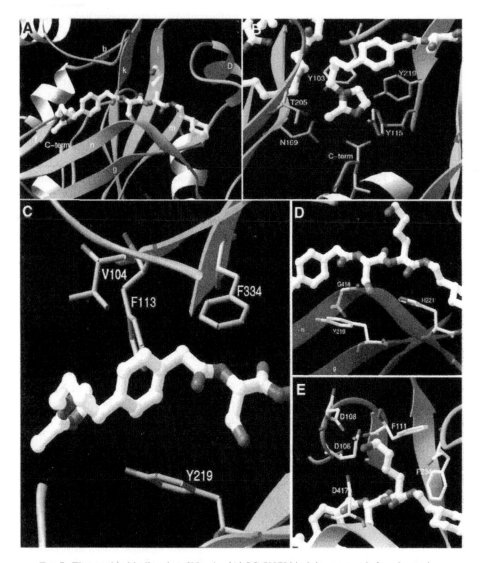

Fig. 7. The peptide-binding site of Nmt1p. (A) SC-58272 binds in an extended conformation. Atoms are colored as in Fig. 6A. The peptide-binding site spans much of the C-terminal half of Nmt1p. However, components of helices A and B, the Ab loop, and strand g from the N-terminal half of the enzyme make important contacts with the substrate analog. (B) Interactions of the 2-methylimidazole of SC-58272 with active site residues. The 2-methylimidazole, representing the N terminus of a substrate peptide, is located close to the portion of *S*-(2-oxo)penta-decyl-CoA that represents the reactive thioester bond of myristoyl-CoA. The N terminus of the peptidic inhibitor is in a pocket formed by aromatic side chains and is directly apposed to the C-terminal carboxylate of Leu-455. (C) Interactions of the phenylene ring of SC-58272

buried. SC-58272 is bound in a largely extended conformation (Fig. 7A). A prominent 90° bend (into the plane of Fig. 7A) is located between the 2-methylimidazole and the phenylene and directs the N terminus to the active site. Although the C-terminal half of Nmt1p forms the bulk of the peptide-binding site, the site is also composed of elements that are shared by the myristoyl-CoA-binding site—principally the Ab loop from the N-terminal half of the enzyme (Fig. 7A). In addition, the bound myristoyl-CoA analog itself forms part of the binding site for SC-58272 (Fig. 7B).

The imidazole ring is buried deeply within Nmt1p. Its NH nitrogen (N_γ), representing the Gly-1 ammonium of a substrate, is presented to the C-terminal carboxylate of Leu-455 (Fig. 7B). As discussed below, this finding has important implications for the catalytic mechanism. Other interactions contributing to the coordination of imidazole include a hydrogen bond between its N_γ and the hydroxyl of Tyr-115, plus hydrophobic contacts with the Ab loop (Tyr-103) and βb (Phe-113) (Fig. 7B and C). The aromatic imidazole is inserted into a pocket containing of a number of aromatic side chains (Tyr-103, Tyr-115, and Tyr-219; see Fig. 7B). These aromatic–aromatic interactions may explain, in part, why SC-58272 is bound with such high affinity even though it lacks the critical Gly-1 recognition element present in peptide substrates.

The region of SC-58272 between its 2-methylimidazole and serine is devoid of peptidic elements, and is composed of a flexible four-carbon chain followed by benzene (Fig. 7B). Thus, this region presumably demonstrates only some of the interactions that occur between a bound peptide substrate and Nmt1p. The flexible chain initially runs parallel to the C′-C3p portion of S-(2-oxo)pentadecyl-CoA (Fig. 7B) in a direction leading away from the catalytic pocket. The carbon chain then bends 90°, allowing the benzene ring to lie flat along a peptide-binding groove, where it is supported by a hydrophobic/aromatic surface formed by residues in the Ab loop (Val-104), the βb strand (Phe-113), and the βg strand (Tyr-219) (Fig. 7C).

As discussed earlier, serine and lysine are critical determinants for recognition of the peptide substrates of Nmt1p. The serine side chain in SC-

---

with Nmt1p. A favorable face-edge aromatic-to-aromatic interaction is evident with the side chain of Phe-113. (D) Interactions of the serine in SC-58272 with Nmt1p. Strands n and g are rendered in a semitransparent green to show the position of some of their component residues. The serine hydroxyl is hydrogen bonded to the side chain of His-221 and to the amide nitrogen of Gly-418. (E) Interactions of the lysine of SC-58272 with Nmt1p. The ε-ammonium group of lysine interacts electrostatically with Asp-106 and Asp-108 from the Ab loop, as well as Asp-417 from strand n.

58272 is completely buried: its hydroxyl is hydrogen bonded to the N_ε atom of His-221 and to the amide nitrogen of Gly-418, while its C_β carbon makes hydrophobic contacts with Gly-418 (Fig. 7D). The hydrogen bond formed between the serine hydroxyl and His-221 is critical for binding. Site-directed mutagenesis studies have established that substitution of this conserved histidine with alanine has a dramatic effect on K_m: e.g., for GNSGSKQH, representing the N terminus of Ppz2p, an N-myristoylated *S. cerevisiae* phosphatase, the K_m is 1480 μM and a V_{max} is 900 pmol/min/mg for Nmt221Ap compared with a K_m of 0.5 μM and a V_{max} of 1110 pmol/min/ mg for wild-type Nmt1p. Similar increases in K_m were also noted with other octapeptides (T. Farazi, J. Manchester, R. Bhatnagar, G. Waksman, and J. I. Gordon, unpublished observations, 1999).

The charged ε-amino group of the lysine of SC-58272 is surrounded by a large negative electrostatic field formed by the side chains of Asp-106 and Asp-108 from the Ab loop, and Asp-417 (Fig. 7E). Asp-417 also forms a hydrogen bond with the backbone NH of this lysine. The aliphatic portion of the lysine side chain is enclosed by the benzene side chains of Phe-111 and Phe-234 (Fig. 7E).

The part of the peptide-binding groove that contains the C-terminal 2-cyclohexylethylamide of SC-58272 is formed by the DE loop, the βg and βn strands, and has a mixed chemical composition. The groove is open ended (Fig. 5E), allowing the enzyme to accommodate a nascent polypeptide substrate as it catalyzes this cotranslational modification.

3. Ligand-Induced Conformational Changes and Nmt Kinetic Mechanism

As noted above, the Nmt1p ternary complex structure encompasses residues 34 through the C-terminal Leu-455: i.e., amino acids 1–33 are disordered. The region of *C. albicans* apo-Nmt corresponding to residues 34–55 of the *S. cerevisiae* holoenzyme is disordered, so that the 3_{10} A′ helix is missing. In addition, portions of the Ab loop are disordered in the fungal apoenzyme.

The 3_{10} A′ helix (Lys-39 through Pro-45) of Nmt1p interacts with both the 3′-phosphate and the myristoyl chain. While the positive dipole of this helix is important for binding the phosphate, the negative charge of the phosphate can also be viewed as stabilizing the helix. The absence of the A′ helix in the apoenzyme structure, its presence in the holoenzyme, and the interactions between the helix and 3′-phosphate provide strong evidence that CoA binding induces formation of A′.

The 3_{10} helix also has an impact on acyl chain binding through its Trp-41 residue. The indole side chain of this residue forms a substantial portion of the wall of the acyl chain pocket (Fig. 6G). Thermodynamic and kinetic studies have established that there is a functional coupling between changes in acyl chain length and interactions between the enzyme and the 3'-phosphate (40). The structural basis for this coupling appears to be the 3_{10} helix. Residues corresponding to the 3_{10} helix in Nmt1p are conserved in 11 orthologous Nmts [Fig. 8A and B].

In the original published 2.8-Å resolution structure of *C. albicans* apo-Nmt (103), Val-108 through Asp-112 (corresponding to Val-104 through Asp-108 in *S. cerevisiae* Nmt1p) were disordered and were not included in the model. In a subsequent 2.45-Å resolution refinement of this structure, these residues display conformational heterogeneity, indicating that they are flexible (Fig. 9A). In the Nmt1p ternary complex structure, this portion of the Ab loop makes contacts with both the pantetheine of the bound myristoyl-CoA analog (Fig. 6E) and with the dipeptide inhibitor. The latter contacts include the important electrostatic interaction between residues Asp-106 and Asp-108 and the lysine of SC-58272 (Fig. 7E). Residues in the Ab loop contribute substantially more buried surface area to peptide binding than to acyl-CoA binding: 94 Å^2 or 15% of the total buried surface area of SC-58272 versus 51 Å^2 or 6% of the total buried surface area of S-(2-oxo)pentadecyl-CoA. Therefore, one would expect binding of peptide to be more dependent on the surfaces of the Ab loop and that these surfaces would need to be presented prior to peptide recognition/binding.

The conformational differences between apo- and holoenzymes provide insights about the structural determinants of the ordered Bi–Bi reaction mechanism of the enzyme (Fig. 9B). Binding of myristoyl-CoA to the apoenzyme is the first step in the kinetic mechanism. We propose that this step induces formation of the 3_{10} A' helix and allows ordering of the Ab loop. Stabilization of the Ab loop, in turn, allows peptide to bind. Thus, the Ab loop functions as a switch that defines the order of substrate binding. Once myristate is catalytically transferred from CoA to the acceptor glycine amine of a nascent peptide substrate, the order of product release is dictated in part by the conformation of the Ab loop. CoA without an attached acyl chain is able to leave the complex. Once CoA departs, the Ab loop is able to revert to its flexible state. Moreover, the 3_{10} A' helix loses critical contacts with the 3'-phosphate that are required to maintain its secondary structure. Disassembly of the A' helix likely results in removal of Trp-41 from the acyl chain-binding pocket. Freed of its contacts with the Ab loop and a substantial portion of the acyl chain-binding pocket, the myristoyl peptide product is now free to be released.

This proposal concerning the role of the Ab loop in the ordered Bi–Bi reaction mechanism is based on two assumptions. First, the Ab loop is flexible in *S. cerevisiae* apo-Nmt1p just as it is in the orthologous *C. albicans*

A

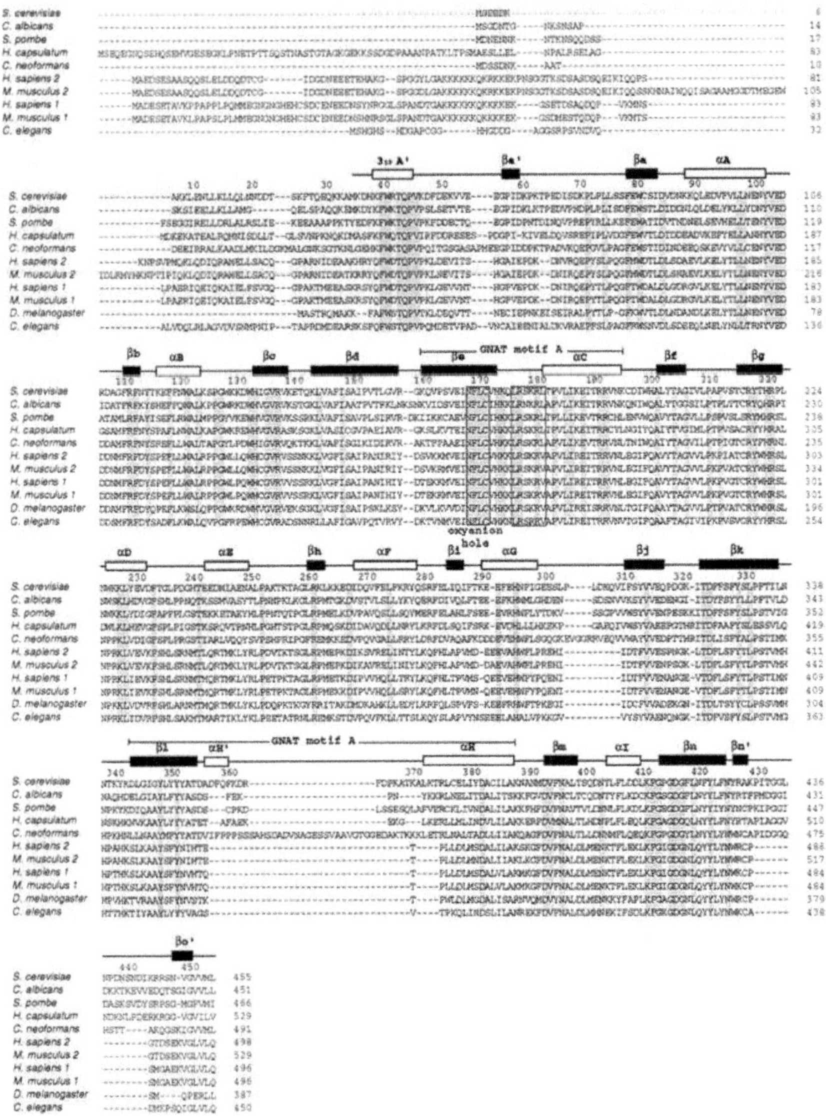
S. cerevisiae
C. albicans
S. pombe
H. capsulatum
C. neoformans
H. sapiens 2
M. musculus 2
H. sapiens 1
M. musculus 1
D. melanogaster
C. elegans
GNAT motif A
oxyanion hole

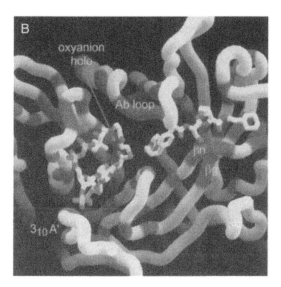

FIG. 8. Comparisons of orthologous Nmts. (A) Primary sequence alignments of 11 known orthologous Nmts. The alignments were generated using program CLUSTAL (*112*). Residues conserved in all sequences are highlighted with yellow. Secondary structure elements are indicated. Residues forming the β bulge containing the oxyanion hole are boxed in red. The sequences corresponding to GNAT superfamily motif A are indicated in both the N-terminal and C-terminal symmetry-related halves of the enzyme. The sequence corresponding to the highly conserved pyrophosphate binding submotif of GNAT motif A is boxed in green. See text for further discussion. (B) C_α trace of *S. cerevisiae* Nmt1p with bound substrate analogs. Regions of absolute sequence conservation among the orthologous Nmts are colored blue. Structural features important for substrate recognition and the reaction mechanism are also indicated.

apoenzyme. Second, formation of a binary myristoyl-CoA : Nmt1p complex is sufficient to stabilize the conformation of the Ab loop seen in the ternary Nmt1p structure.

4. Active Site and Catalytic Mechanism

Enzyme-catalyzed acyl transfer reactions typically proceed through the nucleophilic addition–elimination mechanism (*113*). Catalysis of acyl transfer from CoA to glycine through this mechanism has a number of requirements: (1) polarization of the C1m–O1m thioester carbonyl to make the carbon atom more amenable to nucleophilic attack, (2) stabilization of transition states and reaction intermediates, and (3) deprotonation of the glycine ammonium to generate a nucleophilic amine. On the basis of the structure of the ternary complex, it appears that polarization and transition

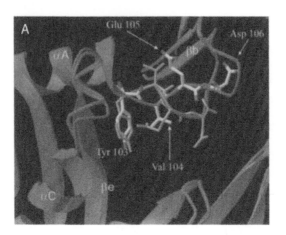

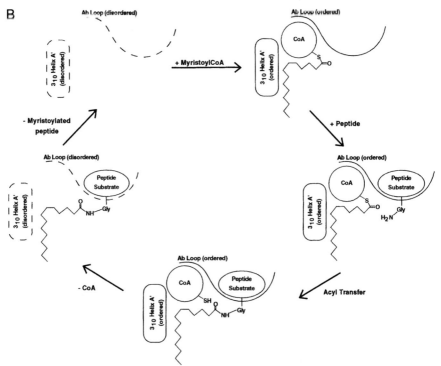

Fig. 9. Ligand-induced conformational changes define the Nmt ordered Bi–Bi reaction mechanism. (A) The structures of *C. albicans* Nmt and *S. cerevisiae* Nmt1p are superimposed as red and purple ribbons, respectively. The side chains of residues located in the Ab loop are drawn in yellow for *C. albicans* Nmt and colored by atom type for *S. cerevisiae* Nmt1p (oxygen, red; carbon, white; nitrogen, purple). Residue numbers are named with respect to the *S. cerevisiae* sequence. (B) Diagram describing the Nmt ordered reaction mechanism. See text for discussion.

state stabilization are achieved by proximity of the thioester carbonyl to the oxyanion hole and that deprotonation of the ammonium to an amine is mediated by the carboxylate of Leu-455 located in the active site.

The role of an oxyanion hole in any transferase, hydrolase, or other enzyme that breaks and/or forms peptide or ester bonds is to stabilize one or more tetrahedral intermediates. The general reaction involves a nucleophile (water, an enzyme side chain, or a transferase cosubstrate) displacing a leaving group through the nucleophilic addition–elimination mechanism (Fig. 10A). During this reaction, addition of a nucleophile to the trigonal (sp^2) carbonyl carbon results in formation of a tetrahedral (sp^3) carbon with covalent single bonds to the attacking nucleophile, the leaving group, a negatively charged oxygen atom (hence "oxyanion"), as well as a fourth substituent (Fig. 10A). Elimination occurs subsequently with reformation of a double bond between the oxygen and carbonyl carbon, resulting in restoration of trigonal geometry about the carbonyl carbon and expulsion of the leaving group. The transition state is thought to be close to the structure of the tetrahedral intermediate (114).

An oxyanion hole contributes to catalysis of the nucleophilic addition–elimination mechanism by stabilizing the tetrahedral intermediate, as well as the transition states for its formation and decomposition. This is achieved by using positive dipoles or charges to stabilize the negatively charged alkoxide of the intermediate, and by steric constraints favoring tetrahedral geometry about the carbon (Fig. 10B). In addition to stabilization of the intermediate, an oxyanion hole destabilizes the target carbonyl carbon to favor the nucleophilic addition by polarizing the bond to enhance the partial positive charge on the carbon and the partial negative charge on the oxygen. Such polarization is equivalent to increasing the single-bond character of the carbon–oxygen bond and distorting the carbon toward sp^3 tetrahedral geometry.

The first step in the nucleophilic addition–elimination mechanism is the nucleophilic attack. The pK_a for a free glycine amine is 9.8 (115). Thus, at physiologic pH, the glycine nitrogen at the N terminus of a nascent peptide would be expected to exist predominantly as the ammonium ion. NH_4^+ is a poor nucleophile but deprotonation to the amine by a base generates a good nucleophile (Fig. 10C).

The overall reaction catalyzed by Nmt can be summarized as follows (see Fig. 10D).

1. The myristoyl-CoA-binding site contains all the elements responsible for acyl chain length measurement and positioning such that the C1m–O1m carbonyl faces the main-chain amides of Phe-170 and Leu-171, which form the oxyanion hole. As a consequence, the carbonyl group becomes polarized with a partial positive charge on the carbon.

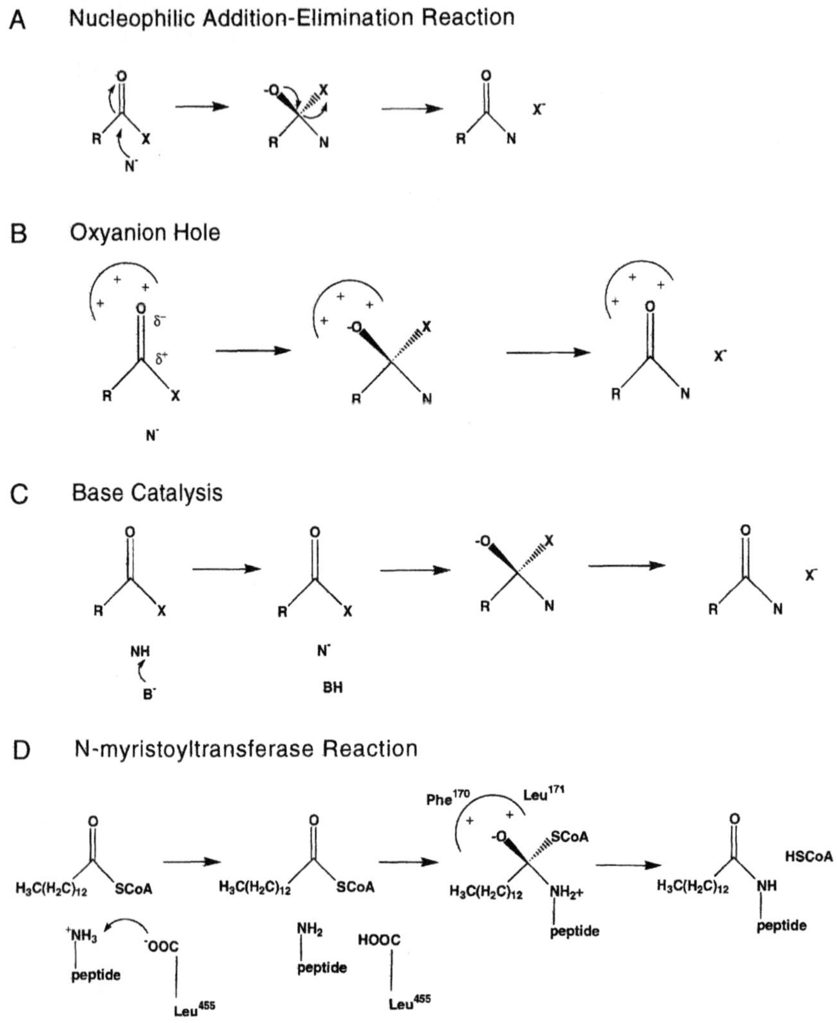

A   Nucleophilic Addition-Elimination Reaction

B   Oxyanion Hole

C   Base Catalysis

D   N-myristoyltransferase Reaction

FIG. 10. Nmt chemical mechanism of catalysis. (A) Generic nucleophilic addition–elimination mechanism. A nucleophile, N⁻, attacks the target thioester carbonyl, forming a tetrahedral intermediate, which collapses with elimination of a leaving group, X⁻. (B) Role of the oxyanion hole: stabilization of the tetrahedral intermediate formed during the addition–elimination reaction. (C) General base catalysis. B, General base. A base is necessary to deprotonate the generic nucleophile NH to its reactive N⁻ form. (D) Summary of the myristoyl transfer reaction catalyzed by Nmt.

2. Binding of myristoyl-CoA is also accompanied by ordering of the flexible Ab loop, resulting in formation of a fully competent peptide-binding site.

3. Peptide binding results in juxtaposition of the Gly-1 ammonium ion and the carboxy terminus of Nmt1p. This results in neutralization of both charges with proton transfer from the ammonium to the carboxylate. The resulting amine is now competent for nucleophilic attack on the thioester carbonyl. In the ternary complex structure of Nmt1p, the C1m carbon is 7.5 Å away from the N_γ of imidazole. This raises the possibility that the amine nucleophile of a bound peptide substrate must move in the active site toward the thioester carbonyl, which is already positioned in the oxyanion hole. This notion is supported by the finding that introducing proline at position 2 of an octapeptide substrate abolishes activity (45).

4. Acyl transfer occurs via a direct nucleophilic substitution reaction. Transfer proceeds through formation of a tetrahedral reaction intermediate that is stabilized by interactions between the alkoxide anion (O1m) and the positive potential of the oxyanion hole.

5. Acyl transfer is followed by release of free CoA, disordering of the Ab loop and 3_{10} A' helix, and release of myristoyl peptide. The C-terminal carboxylate must then be deprotonated so that it can act again as a general base in the next round of catalysis. Presumably, this occurs by solvent contact in apo-Nmt.

5. Evidence against Acyl-Enzyme Intermediate

Instead of proceeding through a single nucleophilic addition–elimination step with direct transfer of the acyl chain from CoA to the peptide, the Nmt mechanism could, in principle, proceed through two such steps. In the first step, nucleophilic addition–elimination would result from attack of a nucleophilic side chain of Nmt on myristoyl-CoA, with elimination of the CoA leaving group and formation of a stable acyl-enzyme intermediate. In the second nucleophilic addition–elimination step, the peptide glycine amine nitrogen would be the nucleophile, and the enzyme side chain to which the acyl chain is now attached would be the leaving group. Such a two-step mechanism is employed by serine proteases (114). However, it appears unlikely that the Nmt mechanism proceeds through an acyl-enzyme intermediate because mutation of all the nucleophilic side chains in the active site (Tyr-103 → Phe, Tyr-115 → Phe, and Thr-205 → Ala) fails to abolish N-myristoyltransferase activity (T. Farazi, R. Bhatnagar, G. Waksman, and J. I. Gordon, unpublished, 1999). However, even if such a two-step mechanism were to be used by Nmt, the roles and requirements for the oxyanion hole and general base catalyst would be as described for the one-step reaction.

6. *C-Terminal Carboxylate Acting as General Base Catalyst*

The role of general base catalysis in acyl transfer is best described for proteases. While serine proteases use the aspartate/histidine charge relay system and do not directly deprotonate nucleophiles with a carboxylate, both the aspartyl and zinc proteases represent examples of direct involvement of carboxylate side chains in general base catalysis of water addition to a peptide bond: e.g., Asp-32 in the aspartyl protease pepsin and Glu-143 in the zinc protease thermolysin (*114*).

There is no precedent for a backbone carboxylate acting as a catalytic residue. The carboxyl terminus is known to affect substrate binding and specificity in several enzymes including the Gal6 protease (*116*) and the aminoglycoside kinase APH(3′)-IIIa (*117*). In addition, the N termini of several enzymes are involved in catalysis. The N-terminal nucleophile hydrolase family of amide hydrolase and transferase enzymes (*118, 119*) features involvement of the side chain of the N-terminal residue (serine, threonine, or cysteine) as a nucleophile in a double-displacement nucleophilic addition–elimination mechanism. The primary amine group is thought to serve as a general base catalyst in these enzymes, abstracting a proton from the nucleophilic side chain during the acylation step, and from water during the deacylation step. Another example of general base catalysis by the N-terminal amine of an enzyme is provided by 4-oxalocrotonate reductase, in which the secondary amine of Pro-1 is thought to abstract a proton from a methylene adjacent to a carbonyl carbon as part of a concerted general acid–base-catalyzed mechanism (*120*). Nmt represents the first reported enzyme in which the C-terminal carboxylate group appears to have a catalytic role.

B. TWO *NMT* GENES OF MICE AND HUMANS

Alignment of the primary structures of 11 orthologous Nmts (Fig. 8A) indicates that sequence variation is greatest at their N termini. Unlike the *S. cerevisiae, C. albicans, C. neoformans,* and *C. elegans* genomes, which contain one *NMT,* the mouse and human genomes contain at least two *NMT* genes (*53*). The N terminus of each Nmt (Nmt1 and Nmt2) varies considerably within each mammalian species. However, the N terminus of each Nmt is conserved between species: i.e., mouse Nmt1 resembles human Nmt1. Work indicates that the N-terminal sequences of the human Nmts may regulate their interactions with ribosomes (*52*).

Defining the subcellular locations of each of the two mouse or human Nmts in different cell lineages at different stages of development or differentiation, and characterizing their substrate specificities, will undoubtedly help us understand their functions. Such an analysis will also likely help guide the development of safer therapeutic strategies directed at Nmts produced by eukaryotic pathogens, or at the human Nmts themselves, when the target is a virus or a transformed cell. All that can be said at present is that the two human Nmts are remarkably conserved beyond their divergent N-terminal domains. All substitutions appear to be conservative and do not involve residues corresponding to those in the *S. cerevisiae* Nmt1p ternary complex structure that are involved in key interactions with the myristoyl-CoA or peptide substrate analog.

C. NMT AS MEMBER OF GCN5-RELATED N-ACETYLTRANSFERASE SUPERFAMILY

Members of the GCN5-related *N*-acetyltransferase (GNAT) superfamily of proteins are represented in all kingdoms and include families of drug-resistance enzymes and protein and amino acid *N*-acetyltransferases, as well as proteins with no identified functions (*121*). Many of the enzymes in this superfamily catalyze the formation of an amide bond between an acceptor amine group and an acetyl group from the donor substrate acetyl-CoA. The superfamily is predominated by acetyl-CoA-dependent transferases. These include numerous histone acetyltransferases such as GCN5, which are known to be important in regulating chromatin structure and gene transcription.

The first structures of GNAT family members were reported in 1998 (*109, 122*). These structures have striking similarity to the Nmt fold (A. Murzin, personal communication, 1999; Ref. *108*). Each of the symmetry-related halves of Nmt is topologically equivalent to the core structure of the monomeric enzyme histone acetyltransferase from *S. cerevisiae* (Hat1p; Ref. *122*) and the monomer structure of the dimeric enzyme *Serratia marcescens* aminoglycoside *N*-acetyltransferase (SmAAT; Ref. *109*) (Fig. 11A and B). These similarities were unexpected, given the finding that Nmt has no appreciable pairwise homology to the GCN5 superfamily or to any members of current protein sequence databases, other than orthologous Nmts.

Four conserved sequence motifs were used initially to define the GNAT superfamily (*121*) and were modified slightly after the SmAAT structure was determined (*109*). The most highly conserved of these elements is motif A, which encompasses the first β strand and the α helix of the $\beta\alpha\beta$ acetyl-CoA-binding motifs in SmAAT and Hat1p. The structural correlate of motif A is clearly represented in the Nmt fold: it includes strand e (containing the β bulge, which forms the oxyanion hole of the enyzme), the eC loop, and the N terminus of helix C, which interact with the CoA pyrophosphate and

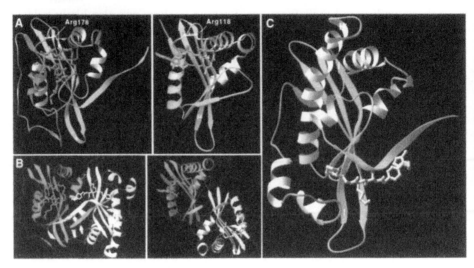

FIG. 11. Comparison of *S. cerevisiae* Nmt1p and *Serratia marcescens* aminoglycoside *N*-acetyltransferase (SmAAT). (A) Comparison of the N-terminal half of Nmt1p (*left*) and monomeric SmAAT (*right*). Myristoyl-CoA is shown in red in the Nmt1p structure. CoASH is shown in red in the SmAAT structure. The secondary structure elements forming motif A in SmAAT (S4, H4, and the intervening loop) and its correlate in Nmt (βe, αC, and the eC loop) are indicated. Note the similar orientations and interactions of Arg178 (Nmt1p) and Arg118 (SmAAT). (B) Comparison of the intact Nmt1p molecule (*left*) with dimeric SmAAT (*right*). (C) C-terminal half of Nmt1p oriented similar to the structures shown in (A). Secondary structure elements comprising the C-terminal half of Nmt correlate to the GNAT motif A (i.e., strand I and helix H, with the inserted helix H'). The motif is distorted relative to those shown in (A) and can no longer bind to CoA, although the β bulge is preserved. The peptide analog binds to this half of the enzyme at a site that is quite distinct from the GNAT CoA-binding site.

pantetheine moieties, plus the remainder of αC, which forms part of the acyl chain-binding pocket (Fig. 11A and B).

1. Conserved R/Q-X-X-G-X-G/A Pyrophosphate Binding Submotif of GNAT Superfamily

Within motif A of the GNAT superfamily, the most highly conserved sequence is R/Q-X-X-G-X-G/A. This sequence is represented as R(117)RQGIA in SmAAT and as Q(227)NKGHG in Hat1p. However, this submotif sequence does not appear in any known Nmt. Nonetheless, a structural alignment reveals that the corresponding region of Nmt1p is composed of the sequence L(187)RSKRL (forming the eC loop and N terminus of αC in the Nmt fold). The only identical residue in these regions of Nmt1p and SmAAT is Arg-188 of Nmt1p (corresponding to Arg-118 in

SmAAT; Ref. *109*). This shared arginine residue has a similar conformation in the SmAAT and Nmt1p structures. Its side chain appears to have π-stacking interactions with the carbonyl C9p–O9p of the pantetheine chain in both Nmt1p and SmAAT (Figs. 6E and 11A). This type of interaction may be recapitulated by glutamine in the R/Q-X-X-G-X-G/A submotifs of other GNAT members: e.g., in the reported Hat1p structure, the π cloud of the amide group of its Q227 side chain is positioned near the pantetheine carbonyl (*122*).

It is interesting that the two glycines in the highly conserved R/Q-X-X-G-X-G/A sequence submotif of GNAT superfamily members are replaced by a charged and a hydrophobic residue, respectively, in the Nmts. Although a distinct sequence, L-R-S/N-K/R-R-L/V, describes the Nmt structural homolog of R/Q-X-X-G-X-G/A (Fig. 8A,D), the two structures share a similar function: namely, binding of pyrophosphate and the proximal portion of pantetheine.

Arg-181 of Nmt1p, corresponding to the last X of R/Q-X-X-G-X-G/A, interacts with the 3'-phosphate of CoA and stabilizes the CoA conformation with 3'-phosphate pointing away from the center of the protein and the adenine ring pointing into the center. This CoA conformation is not present in the known GNAT structures, where the 3'-phosphate makes more contact with the protein than does the adenine ring (Fig. 11A). An important consequence is that in Nmt1p the adenine forms a portion of the binding site for myristate (Fig. 6A and G), whereas in SmAAT the adenine is solvent exposed and does not interact with the acetyl moiety of acetyl-CoA (Fig. 11A). These differences likely arose from the distinct requirements of these enzymes to accommodate a long versus short acyl chain.

The four conserved sequence motifs in GNAT superfamily members encompass 119–161 residues (*109*) and are ordered (N-terminal to C-terminal) as C-D-A-B. Motifs C, D, and B are less highly conserved between GNAT members. Nonetheless, each has an obvious structural correlate in each half of Nmt. In the N-terminal half of Nmt, motif C is represented as strand a and helix A, motif D as strands c, d, and the cd loop, and motif B as strand f. In the C-terminal half of Nmt, motif C appears as strand h and helix F, motif D as strands j, k and the jk loop, motif A as strand l and helix H (with the inserted helix H'), and motif B as strand m (see Fig. 5D).

2. Does β Bulge Function as Oxyanion Hole in Other GNAT-Related CoA-Dependent Acyltransferases?

The structural correlate of motif A in the N-terminal half of Nmt1p contains the oxyanion hole of the enzyme. Motif A in SmAAT and Hat1p, and its representative in the C-terminal half of Nmt, each contain a similarly oriented β bulge (Fig. 11C). Although Nmt is the only GNAT superfamily member where the bulge has been identified as the site of an oxyanion hole, we propose that the β bulge functions as an oxyanion hole in all other GNAT-related proteins that are CoA-dependent acyltransferases. The Hat1p structure with bound acetyl-CoA (122) supports this notion: its acetyl-CoA thioester carbonyl oxygen forms a hydrogen bond with the backbone NH group of Phe-220. This would be equivalent to the hydrogen bond between the thioester carbonyl of myristoyl-CoA and the NH group of Leu-171 in Nmt1p. [The conformation of bound CoASH in the SmAAT structure places the sulfur atom away from the β bulge of its motif A. However, CoASH is the product of the acetyltransfer reaction and lacks the reactive thioester carbonyl: its conformation would not necessarily be expected to resemble that of the acetyl-CoA substrate (109).]

While the N- and C-terminal halves of Nmt each contain all of the conserved GNAT superfamily structural motifs, the distinct functions of the two halves of Nmt illustrate how the fold may subserve functions other than acyl-CoA recognition. The β bulge in the C-terminal GNAT fold of Nmt is clearly not an oxyanion hole because this half of the protein does not interact with acyl-CoA. In fact, acyl-CoA binding to this region of the protein appears to be prohibited for two reasons. First, the structure between strand l and helix H is substantially different from both the conserved pyrophosphate-binding submotif of the GNAT family (R/Q-X-X-G-X-G/A) and the corresponding, highly conserved L-R-S/N-K/R-R-L/V submotif in the N-terminal half of Nmt. Second, helix I occupies the space where the adenine and pantetheine regions of CoA bind to the N-terminal half of Nmt (and to Hat1p/SmAAT).

SmAAT and Nmt both have ordered reaction mechanisms in which acyl-CoA binding precedes binding of the acceptor substrate (38, 109). If one of these acyltransferases forms a binary complex in which the thioester carbonyl is properly positioned within the oxyanion hole, it would seem easy for a nearby water molecule to hydrolyze the destabilized thioester bond. In the case of Nmt, general base catalysis by its C-terminal carboxylate would further accelerate this "wasteful" reaction. However, Nmt1p does not have detectable myristoyl-CoA thioesterase activity. Why? The distortion of βe that forms the oxyanion hole also appears in the C. albicans apo-Nmt structure (103). Moreover, the β bulge representing the proposed oxyanion hole in SmAAT does not require a thioester bond to maintain its structure (i.e., it is present even though the CoA sulfur is not nearby). These observations suggest that an oxyanion hole is represented in the binary complexes of these enzymes. One possible explanation for the lack of acceptor-independent acyl-CoA hydrolysis in Nmt is that its C-terminal backbone carboxylate does not act as a general base in a complex lacking a bound peptide substrate. If the pK_a of the backbone carboxylate group is near a normal solution state value (~3) in the binary complex, it may function to prevent hydrolysis of the thioester bond by the OH^- anion, rather than catalyzing formation of this anion.

IV. Prospectus

Because Nmt1p is the only member of the GNAT superfamily whose structure has been determined with both donor and acceptor substrate analogs, future studies of molecular recognition and catalysis in this enzyme should have importance beyond the field of protein N-myristoylation. Such studies should include analysis of the structures of orthologous Nmts complexed with authentic peptide substrates so that the basis for differences in their peptide recognition can be deduced. The answers will undoubtedly facilitate development of species-selective inhibitors. A related question is how Nmt "finds" its acyl-CoA and nascent polypeptide substrates. As noted above, work suggests that the divergent N-terminal sequences of Nmts may help define their subcellular compartmentalization; e.g., to the ribosome (52, 53). The cellular locations of the myristoyl-CoA pools accessed by Nmt are unknown, as are the devices employed to present this rare cellular acyl-CoA to Nmt. In addition, the mechanisms by which different N-myristoyl proteins are targeted to different cellular membranes need to be determined.

Although this review has emphasized the essential role of Nmt in yeast and fungal cell growth and survival, it will be important to explore the biological contributions of this acyltransferase and its substrates to the viability of other pathogens (including those whose genomes are being sequenced or are already known), to normal human development, to aging, and to the initiation and progression of neoplasia (e.g., Ref. 123). The publication of the mouse *NMT* gene sequences (53) should help investigators obtain answers to some of these issues because conditional, tissue-specific *NMT* knockouts can now be performed in mice with or without mutations known to lead to various pathologic states.

ACKNOWLEDGMENTS

We are grateful to our colleagues Charles McWherter, Balekudru Devadas, and James Sikorski for developing the species-selective inhibitors of Nmt described above. Work from the authors' laboratories cited in this review was supported in part by a grant from the National Institutes of Health (AI38200).

References

1. Aitken, A., Cohen, P., Santikarn, S., Williams, D. H., Calder, A. G., Smith, A., and Klee, C. B. (1982). *FEBS Lett.* **150,** 314.
2. Carr, S. A., Biemann, K., Shoji, S., Parmelee, D. C., and Titani, K. (1982). *Proc. Natl. Acad. Sci. U.S.A.* **79,** 6128.
3. Boutin, J. (1997). *Cell. Signal.* **9,** 15.
4. Bhatnagar, R. S., and Gordon, J. I. (1997). *Trends Cell Biol.* **7,** 14.
5. Resh, M. D. (1997). *Cell. Signal.* **8,** 403.
6. Hori, T., Nakamura, N., and Okuyama, H. (1987). *J. Biochem.* (*Tokyo*) **101,** 949.
7. Schjerling, C. K., Hummel, R., Hansen, J. K., Borsting, C., Mikkelsen, J. M., Kristiansen, K., and Knudsen, J. (1996). *J. Biol. Chem.* **271,** 22514.
8. Pietzsch, R. M., and McLaughlin, S. (1993). *Biochemistry* **32,** 10436.
9. McLaughlin, S., and Aderem, A. (1995). *Trends Biochem. Sci.* **20,** 272.
10. Murray, D., Hermida-Matsumoto, L., Buser, C. A., Tsang, J., Sigal, C. T., Ben-Tal, N., Honig, B., Resh, M. D., and McLaughlin, S. (1998). *Biochemistry* **37,** 2145.
11. Aderem, A. (1992). *Trends Biochem. Sci.* **17,** 438.
12. Shenoy-Scaria, A. M., Diefzen, D. J., Kwong, J., Link, D. C., and Lublin, D. M. (1994). *J. Cell Biol.* **126,** 353.
12a. Linder, M. E. (2000). "The Enzymes," Vol. XXI, Chap. 8. Academic Press, San Diego, California. [this volume]
13. Buser, C., Sigal, C. T., Resh, M. D., and McLaughlin, S. (1994). *Biochemistry* **33,** 13093.
14. Sigal, C. T., Zhou, W., Buser, C., McLaughlin, S., and Resh, M. D. (1994). *Proc. Natl. Acad. Sci. U.S.A.* **91,** 12253.
15. Zozulya, S., and Stryer, L. (1992). *Proc. Natl. Acad. Sci. U.S.A.* **89,** 11569.
16. Klenchin, V. A., Calvert, P. D., and Bownds, M. D. (1995). *J. Biol. Chem.* **270,** 16147.
17. Chen, C. K., Inglese, J., Lefkowitz, R. J., and Hurley, J. B. (1995). *J. Biol. Chem.* **270,** 18060.
18. Ames, J. B., Tanaka, T., Ikura, M., and Stryer, L. (1995). *J. Biol. Chem.* **270,** 30909.
19. Ames, J. B., Ishima, R., Tanaka, T., Gordon, J. I., Stryer, L., and Ikura, M. (1997). *Nature* (*London*) **389,** 198.
20. Braunewell, K.-H., and Gundelfinger, E. D. (1999). *Cell Tissue Res.* **295,** 1.
21. Dizhoor, A. M., Olshevskaya, E. V., Henzel, W. J., Wong, S. C., Stults, J. T., Ankoudinova, I., and Hurley, J. B. (1995). *J. Biol. Chem.* **270,** 25200.
22. Hughes, R. E., Brzovic, P. S., Dizhoor, A. M., Klevit, R. E., and Hurley, J. B. (1998). *Protein Sci.* **7,** 2675.
23. Olshevskaya, E. V., Hughes, R. E., Hurley, J. B., and Dizhoor, A. M. (1997). *J. Biol. Chem.* **272,** 14327.
24. Rothman, J. E., and Wieland, F. T. (1996). *Science* **272,** 227.
25. Goldberg, J. (1998). *Cell* **95,** 237.
26. Chang, W., Ying, Y.-S., Rothberg, K. G., Hooper, N. M., Turner, A. J., Gambliel, H. A., DeGunzburg, J., Mumby, S. M., Gilman, A. G., and Anderson, R. G. W. (1994). *J. Cell Biol.* **126,** 127.
27. Robbins, S. M., Quintrell, N. A., and Bishop, J. M. (1995). *Mol. Cell. Biol.* **15,** 3507.
28. Li, X., and Chang, Y. H. (1995). *Proc. Natl. Acad. Sci. U.S.A.* **92,** 12357.
29. Robinson, L. J., Buscon, L., and Michel, T. (1995). *J. Biol. Chem.* **270,** 995.
30. Feron, O., Michel, J. B., Sase, K., and Michel, T. (1998). *Biochemistry* **37,** 193.
31. Feron, O., Saldana, F., Michel, J. B., and Michel, T. (1998). *J. Biol. Chem.* **273,** 3125.
32. Van't Hof, W., and Resh, M. D. (1997). *J. Cell Biol.* **136,** 1023.

33. Arbuzova, A., Wang, J. Y., Murray, D., Jacob, J., Cafiso, D. S., and McLaughlin, S. (1997). *J. Biol. Chem.* **272**, 27167.
34. Towler, D. A., Adams, S. P., Eubanks, S. R., Towery, D. S., Jackson-Machelski, E., Glaser, L., and Gordon, J. I. (1987). *Proc. Natl. Acad. Sci. U.S.A.* **84**, 2708.
35. Rudnick, D. A., McWherter, C. A., Gokel, G. W., and Gordon, J. I. (1993). *Adv. Enzymol.* **67**, 375.
36. Chang, Y. H., Teichert, U., and Smith, J. A. (1992). *J. Biol. Chem.* **267**, 8007.
37. Klinkerberg, M., Ling, C., and Chang, Y. H. (1997). *Arch. Biochem. Biophys.* **347**, 193.
38. Rudnick, D. A., McWherter, C. A., Rocque, W. J., Lennon, P. J., Getman, D. P., and Gordon, J. I. (1991). *J. Biol. Chem.* **266**, 9732.
39. Bhatnagar, R. S., Jackson-Machelski, E., McWherter, C. A., and Gordon, J. I. (1994). *J. Biol. Chem.* **269**, 11045.
40. Bhatnagar, R. S., Schall, O. F., Jackson-Machelski, E., Sikorski, J. A., Devadas, B., Gokel, G. W., and Gordon, J. I. (1997). *Biochemistry* **36**, 6700.
41. Rudnick, D. A., Lu, T., Jackson-Machelski, E., Hernandez, J. C., Li, Q., Gokel, G. W., and Gordon, J. I. (1992). *Proc. Natl. Acad. Sci. U.S.A.* **89**, 10507.
42. Kishore, N. S., Lu, T., Knoll, L. J., Katoh, A., Rudnick, D. A., Mehta, P. P., Devadas, B., Huhn, M., Atwood, J. L., Adams, S. P., Gokel, G. W., and Gordon, J. I. (1991). *J. Biol. Chem.* **266**, 8835.
43. Lu, T., Li, Q., Katoh, A., Hernandez, J., Dffin, K., Jackson-Machelski, E., Knoll, L. J., Gokel, G. W., and Gordon, J. I. (1994). *J. Biol. Chem.* **269**, 5346.
44. Towler, D. A., Adams, S. P., Eubanks, S. R., Towery, D. S., Jackson-Machelski, E., Glaser, L., and Gordon, J. I. (1988). *J. Biol. Chem.* **263**, 1784.
45. Towler, D. A., Gordon, J. I., Adams, S, P., and Glaser, L. (1988). *Annu. Rev. Biochem.* **57**, 69.
46. Duronio, R. J., Towler, D. A., Heuckeroth, R. O., and Gordon, J. I. (1989). *Science* **243**, 796.
47. Wiegand, R. C., Carr, C., Minnerly, J. C., Pauley, A. M., Carron, C. P., Langner, C. A., Duronio, R. J., and Gordon, J. I. (1992). *J. Biol. Chem.* **267**, 8591.
48. Lodge, J. K., Johnson, R. L., Weinerg, R. A., and Gordon, J. I. (1994). *J. Biol. Chem.* **269**, 2996.
49. Ntwasa, M., Egerton, M., and Gay, N. J. (1997). *J. Cell Sci.* **110**, 149.
50. Zhang, L., Jackson-Machelski, E., and Gordon, J. I. (1996). *J. Biol. Chem.* **271**, 33131.
51. Duronio, R. J., Reed, S. I., and Gordon, J. I. (1992). *Proc. Natl. Acad. Sci. U.S.A.* **89**, 4129.
52. Glover, C. J., Hartman, K. D., and Felsted, R. L. (1997). *J. Biol. Chem.* **272**, 28680.
53. Giang, D. K., and Cravatt, B. F. (1998). *J. Biol. Chem.* **273**, 6595.
54. Duronio, R. J., Jackson-Machelski, E., Heuckeroth, R. O., Olins, P. O., Devine, C. S., Yonemoto, W., Slice, L. W., Taylor, S. S., and Gordon, J. I. (1990). *Proc. Natl. Acad. Sci. U.S.A.* **87**, 1506.
55. Kishore, N. S., Wood, D. C., Mehta, P. P., Wade, A. C., Lu, T., Gokel, G. W., and Gordon, J. I. (1993). *J. Biol. Chem.* **268**, 4889.
56. Rocque, W. J., McWherter, C. A., Wood, D. C., and Gordon, J. I. (1993). *J. Biol. Chem.* **268**, 9964.
57. Lodge, J. K., Jackson-Machelski, E., Toffaletti, D. L., Perfect, J. R., and Gordon, J. I. (1994). *Proc. Natl. Acad. Sci. U.S.A.* **91**, 12008.
58. Lodge, J. K., Jackson-Machelski, E., Devadas, B., Zupec, M. E., Getman, D. P., Kishore, N., Freeman, S. K., McWherter, C. A., Sikorski, J. A., and Gordon, J. I. (1997). *Microbiology* **143**, 357.
59. Lodge, J. K., Jackson-Machelski, E., Higgins, M., McWherter, C. A., Sikorski, J. A., Devadas, B., and Gordon, J. I. (1998). *J. Biol. Chem.* **273**, 12482.

60. Devadas, B., Zupec, M. E., Freeman, S. K., Brown, D. L., Nagarajan, S., Sikorski, J. A., McWherter, C. A., Getman, D. P., and Gordon, J. I. (1995). *J. Med. Chem.* **38,** 1837.

61. Devadas, B., Freeman, S. K., Zupec, M. E., Lu, H.-F., Nagarajan, S., McWherter, C. A., Kuneman, D, W., Vinjamoori, D. V., Getman, D. P., Gordon, J. I., and Sikorski, J. A. (1997). *J. Med. Chem.* **40,** 2609.

62. Devadas, B., Freeman, S. K., McWherter, C. A., Kishore, N. S., Lodge, J. K., Jackson-Machelski, E., Gordon, J. I., and Sikorski, J. A. (1998). *J. Med. Chem.* **41,** 996.

63. Nagarajan, S., Devadas, B., Zupec, M. E., Freeman, S. K., Brown, D. L., Lu, H.-F., Mehta, P. P., Kishore, N. S., McWherter, C. A., Getman, D. P., Gordon, J. I., and Sikorski, J. A. (1997). *J. Med. Chem.* **40,** 1422.

64. Sikorski, J. A., Devadas, B., Zupec, M. E., Freeman, S. K., Brown, D. L., Lu, H.-F., Nagarajan, S., Mehta, P. P., Wade, A. C., Kishore, N. S., Bryant, M. L., Getman, D. P., McWherter, C. A., and Gordon, J. I. (1997). *Biopolymers* **43,** 43.

65. Duronio, R. J., Rudnick, D. A., Johnson, R. L., Johnson, D. R., and Gordon, J. I. (1991). *J. Cell. Biol.* **113,** 1313.

66. Johnson, D. R., Cok, S. J., Feldman, H., and Gordon, J. I. (1994). *Proc. Natl. Acad. Sci. U.S.A.* **91,** 10158.

67. Meyer, K. H., and Schweizer, E. (1974). *J. Bacteriol.* **117,** 345.

68. Stearns, T., Kahn, R. A., Botstein, D., and Hoyt, M. A. (1990). *Mol. Cell Biol.* **10,** 6690.

69. Kahn, R. A., Clark, J., Rulka, C., Stearns, T., Zhang, C. J., Randazzo, P. A., Terui, T., and Cavenagh, M. (1995). *J. Biol. Chem.* **270,** 143.

70. Boman, A. L., and Kahn, R. A. (1995). *Trends Biochem. Sci.* **20,** 147.

71. Dohlman, H. G., Goldsmith, P., Spiegel, A. M., and Thorner, J. (1993). *Proc. Natl. Acad. Sci. U.S.A.* **90,** 9688.

72. Ashrafi, K., Farazi, T. A., and Gordon, J. I. (1998). *J. Biol. Chem.* **273,** 25864.

73. Schweizer, E., and Bolling, H. (1970). *Proc. Natl. Acad. Sci. U.S.A.* **67,** 660.

74. Singh, N., Wakil, S. J., and Stoops, J. K. (1985). *Biochemistry* **24,** 6598.

75. Schuller, H. J., Fortsch, B., Rautenstrauss, B., Wolf, D. H., and Schweizer, E. (1992). *Eur. J. Biochem.* **203,** 607.

76. Cok, S. J., Martin, C. G., and Gordon, J. I. (1998). *Nucleic Acids Res.* **26,** 2865.

77. Johnson, D. R., Knoll, L. J., Levin, D. E., and Gordon, J. I. (1994). *J. Cell. Biol.* **127,** 751.

78. Knoll, L. J., Johnson, D. R., and Gordon, J. I. (1995). *J. Biol. Chem.* **270,** 10861.

79. Knoll, L. J., Johnson, D. R., and Gordon, J. I. (1994). *J. Biol. Chem.* **269,** 16348.

80. Knoll, L. J., Schall, O. F., Suzuki, I., Gokel, G. W., and Gordon, J. I. (1995). *J. Biol. Chem.* **270,** 20090.

81. Werner-Washburne, M., Braun, E., Johnston, G. C., and Singer, R. A. (1993). *Microbiol. Rev.* **57,** 383.

82a. Hettema, E. H., van Roermund, C. W., Distel, B., van den Berg, M., Vilela, C., Rodrigues-Pousada, C., Wanders, R. J., and Tabak, H. F. (1996). *EMBO J.* **15,** 3813.

82b. DeRisi, J. L., Iyer, V. R., and Brown, P. O. (1997). *Science* **278,** 680.

83. Shiozaki, K., and Russell, P. (1994). *Cell. Mol. Biol. Res.* **40,** 241.

84. Stark, M. J. (1996). *Yeast* **12,** 1647.

85. Alepuz, P. M., Cunningham, K. W., and Estruch, F. (1997). *Mol. Microbiol.* **26,** 91.

86. Johnston, M. (1999). *Trends Genet.* **15,** 29.

87. Kanik-Ennulat, C., and Neff, N. (1990). *Mol. Cell. Biol.* **10,** 898.

88. Vallier, L. G., and Carlson, M. (1991). *Genetics* **129,** 675.

89. Tu, J., Vallier, L. G., and Carlson, M. (1993). *Genetics* **135,** 17.

90. Kennedy, B. K., Austriaco, N. R., Zhang, J., and Guarente, L. (1995). *Cell* **80,** 485.

91. Bitter, G. A., and Egan, K. M. (1984). *Gene* **32,** 263.

92. Nakafuku, M., Obara, T., Kaibuchi, K., Miyajima, I., Miyajima, A., Itoh, H., Nakamura, S., Arai, K., Matsumoto, K., and Kaziro, Y. (1988). *Proc. Natl. Acad. Sci. U.S.A.* **85,** 1374.
93. Kübler, E., Mösch, H.-U., Rupp, S., and Lisanti, M. P. (1997). *J. Biol. Chem.* **272,** 20321.
94. Pardinas, J., Pang, Z., Houghton, J., Palejwala, V., Donnelly, R. J., Hubbard, K., Small, M. B., and Ozer, H. L. (1997). *J. Cell. Physiol.* **171,** 325.
95. Powderly, W. G., Finkelstein, D. M., Feinberg, J., Frame, P., He, W., van der Horst, C., Koletar, S. L., Eyster, E., Carey, J., Waskin, H., Hooton, T. M., Hyslop, N., Spector, S. A., and Bozzette, S. A. (1995). *N. Engl. J. Med.* **332,** 700.
96. van der Horst, C. M., Saag, M. S., Cloud, G. A., Hamill, R. J., Graybill, J. R., Sobel, J. D., Johnson, P. C., Tuazon, C. U., Kerkering, T., Moskovitz, B. L., Powderly, W. G., Dismukes, W. E., National Institute of Allergy and Infectious Diseases Mycoses Study Group, and AIDS Clinical Trials Group. (1997). *N. Engl. J. Med.* **337,** 15.
97. Fichtenbaum, C. J., and Powderly, W. G. (1998). *Clin. Infect. Dis.* **26,** 556.
98. Georgopapadakou, N. H. (1998). *Curr. Opin. Microbiol.* **1,** 547.
99. White, T. C., Marr, K. A., and Bowden, R. A. (1998). *Clin. Microbiol. Rev.* **11,** 382.
100. Weinberg, R. A., McWherter, C. A., Freeman, S. K., Wood, D. C., Gordon, J. I., and Lee, S. C. (1995). *Mol. Microbiol.* **16,** 241.
101. Mitchell, T. G., and Perfect, J. R. (1995). *Clin. Microbiol. Rev.* **8,** 515.
102. McWherter, C. A., Rocque, W. J., Zupec, M. E., Freeman, S. K., Brown, D. L., Devadas, B., Getman, D. P., Sikorski, J. A., and Gordon, J. I. (1997). *J. Biol. Chem.* **272,** 11874.
103. Weston, S. A., Camble, R., Colls, J., Rosenbrock, G., Taylor, I., Egerton, M., Tucker, M. D., Tunnicliffe, A., Mistry, A., Mancia, F., de la Fortelle, E., Irwin, J., Bricogne, G., and Pauptit, R. A. (1998). *Nature Struct. Biol.* **5,** 213.
104. Bhatnagar, R. S., Fütterer, K., Farazi, T. A., Korolev, S., Murray, C. L., Jackson-Machelski, E., Gokel, G. W., Gordon, J. I., and Waksman, G. (1998). *Nature Struct. Biol.* **5,** 1091.
105. Paige, L. A., Zheng, G.-Q., DeFrees, S. A., Cassady, J. M., and Geahlen, R. L. (1998). *J. Med. Chem.* **32,** 1667.
106. Nicholls, A., and Honig, B. A. (1991). *J. Comput. Chem.* **12,** 435.
107. Rossmann, M. G., Liljas, A., Bränden, C.-I., and Banaszak, L. J. (1975). "The Enzymes," Vol. 2, p. 61. Academic Press, New York.
108. Modis, Y., and Wierenga, R. (1998). *Structure* **6,** 1345.
109. Wolf, E., Vassilev, A., Makino, Y., Sali, A., Nakatani, Y., and Burley, S. K. (1998). *Cell* **94,** 439.
110. Fierke, C. A., and Jencks, W. P. (1986). *J. Biol. Chem.* **261,** 7603.
111. Pflugrath, J. W., and Quiocho, F. A. (1985). *Nature (London)* **314,** 257.
112. Thompson, J. D., Higgins, D. G., and Gibson, T. J. (1994). *Nucleic Acids Res.* **22,** 4673.
113. Streitweiser, A., and Heathcock, C. J. (1985). "Introduction to Organic Chemistry," 3rd Ed. Macmillan, New York.
114. Creighton, T. E. (1993). "Proteins: Structures and Molecular Properties," 2nd Ed. W.H. Freeman, New York.
115. Vollhardt, K. P. C. (1987). "Organic Chemistry." W.H. Freeman, New York.
116. Zheng, W., Johnston, S. A., and Joshua-Tor, L. (1998). *Cell* **93,** 103.
117. Hon. W. C., McKay, G. A., Thompson, P. R., Sweet, R. M., Yang, D. S., Wright, G. D., and Berghuis, A. M. (1997). *Cell* **89,** 887.
118. Brannigan, J. A., Dodson, G., Duggleby, H. J., Moody, P. C., Smith, J. L., Tomchick, D. R., and Murzin, A. G. (1995). *Nature (London)* **378,** 416.
119. Tikkanen, R., Riikonen, A., Oinonen, C., Rouvinen, J., and Peltonen, L. (1996). *EMBO J.* **15,** 2954.

120. Stivers, J. T., Abeygunawardana, C., Mildvan, A. S., Hajipour, G., Whitman, C. P., and Chen, L. (1996). *Biochemistry* **35,** 803.
121. Neuwald, A. F., and Landsman, D. (1997). *Trends Biochem. Sci.* **22,** 154.
122. Dutnall, R. N., Tafrov, S. T., Sternglanz, R., and Ramakrishnan, V. (1998). *Cell* **94,** 427.
123. Magnuson, B. A., Raju, R. V., Moyana, T. N., and Sharma, R. K. (1995). *J. Natl. Cancer Inst.* **87,** 1630.

Author Index

Numbers in regular font are reference numbers and indicate that an author's work is referred to although the name is not cited in the text. Numbers in italics refer to the page numbers on which the complete reference appears.

A

Abe, K., 137, *152*
Abeygunawardana, C., 280, *290*
Ackermann, K., 6, *17*, 40, *45*
Adames, N., 173, 175, *210*
Adams, D. J., 95, 96, *102*
Adams, G. A., 216, *237*
Adams, S. P., 247, 269, 279, *287*
Adamson, P., 6, *17*, 118, *129*
Aderem, A., 243, 244, *286*
Adjei, A. A., 95, 96, *102*
Afonso, A., 86, 96, *98, 99*
Ahmadian, M. R., 82, *98*
Ahmed, N., 81, *97*
Ailm, M., 149, *153*
Aitken, A., 242, 249, 256, 269, *286*
Akino, T., 202, 203, *212*
Akopyan, T. N., 159, 162, 163, *210*
Aksamit, R. R., 199, 200, *212*
Albanese, M. M., 40, *45*, 48, 75, *79*, 86, 96, *98, 102*
Alberts, S., 95, 96, *102*
Alder, J., 87, *100*
Alepuz, P. M., 254, *288*
Aleshin, A., 22, *44*
Alexandrov, K., 137, 138, 147, *152*
Alland, L., 216, 222, *237*
Allen, C. M., 5, *16*, 36, *45*, 56, 57, 61, 62, *80*, 95, *102*, 110, *129*
Aloise, P., 7, *17*
Alvarez, C. S., 40, *45*, 48, 75, *79*, 86, 96, *98, 99, 102*

Alvarez, E., 219, *238*
Alzari, P. M., 22, *43*
Ambroziak, P., 160, 175, 177, 181, 183, 197, *209*
Ames, J. B., 245, *286*
Amit, A. G., 22, *43*
Amo, S. E., 87, 88, *100*
Anant, J. S., 136, 138, 139, 140, 141, 145, 146, 147, *152*
Anderegg, R. J., 132, *150*, 188, 192, 201, *210*
Anderson, R. G. W., 5, *16*, 246, *286*
Ando, S., 135, *151*
Andres, D. A., 3, *15*, 21, 22, 23, 36, 39, *43*, 50, 51, 54, 56, 60, *79*, 136, 138, 144, 147, *151, 153*
Angibaud, P., 87, 96, *100, 103*
Ankoudinova, I., 245, *286*
Anna-A., S. S., 188, 189, 196, *210*
Anraku, Y., 36, *44*, 57, 67, *80*
Anthony, N. J., 7, *17*, 40, *45*, 87, 88, *99, 100, 101*, 108, *128*, 145, *153*, 157, *209*
Anzano, M. A., 141, *153*
Aoyama, T., 88, *100*
Appelt, K., 22, *43*
Applebury, M. L., 188, *210*
Arai, K., 256, *289*
Arai, S., 42, *46*, 88, *100*
Arakawa, H., 42, *46*, 88, *100*
Araki, S., 135, 136, *151*
Arbuzova, A., 246, *287*
Arellano, M., 49, 57, *79*
Armstrong, J., 133, *151*, 191, *211*

291

K

Kabcenell, A. K., 135, 136, *151*
Kabsch, W., 82, *98*, 134, *151*
Kagan, R. M., 205, *213*
Kahn, B. B., 91, *101*
Kahn, R. A., 132, 134, *150*, 248, 249, *288*
Kaibuchi, K., 137, *152*, 256, *289*
Kaina, B., 92, *101*
Kalk, K. H., 22, *43*
Kalman, V. K., 120, 121, *129*
Kaltenbronn, J. S., 88, *100*
Kalvin, D., 87, *100*
Kamada, S., 77, *80*
Kamoi, T., 42, 46, 88, *100*
Kaminski, J. J., 86, *99*
Kamiya, Y., 132, *150*
Kammlott, U., 4, *16*, 20, 22, *43*, 48, 57, *79*
Kanik-Ennulat, C., 254, *288*
Kao, G., 94, *102*
Karlson, J., 94, *101*
Kataoka, T., 171, *210*
Katayama, M., 135, *151*
Kato, K., 157, 165, 185, *209*
Katoh, A., 247, 269, *287*
Kaufmann, S. H., 95, 96, *102*
Kawagishi-Kobayashi, M., 204, *213*
Kawakami, K., 42, *46*, 88, *100*
Kawamura, Y., 185, 186, *209*
Kawata, M., 106, *128*, 133, 134, 135, 145, *151*, 191, 202, 203, *211*, *212*
Kawata, S., 88, *101*
Kaziro, Y., 256, *289*
Keck, W., 22, *43*
Kellogg, B. A., 26, *44*
Kelly, J., 86, 96, *98*
Kennedy, B. K., 255, *288*
Kennedy, M. E., 216, 218, *237*, *238*
Kerkering, T., 257, *289*
Kerwar, S. S., 109, *128*
Khosravi-Far, R., 133, 137, *151*, *152*
Kikuchi, A., 135, 136, *151*
Kilic, F., 187, *210*
Kim, B. M., 87, 88, *100*
Kim, E., 132, *150*, 160, 164, 172, 175, 177, 178, 181, 183, 197, *209*
Kim, I.-G., 141, *153*
Kim, R., 81, *97*
Kim, S. H., 81, *97*, 106, *128*, 132, 134, *150*, *151*

King, D. S., 158, 161, *209*
King, I., 86, *98*, *99*
Kinsella, B. T., 108, *128*, 133, *151*
Kirschmeier, P., 6, 8, *17*, *45*, 85, 86, 88, 89, 90, 93, 96, *98*, *99*, 120, *129*
Kishihara, K., 95, *102*
Kishore, N. S., 247, 248, 249, 257, 258, 259, 260, 269, *287*, *288*
Kitada, C., 187, *210*
Kitada, D., 106, *127*
Klausner, R., 221, *239*
Klee, C. B., 242, 249, 256, 269, *286*
Klenchin, V. A., 244, *286*
Kleuss, C., 216, 235, *237*
Klevit, R. E., 245, *286*
Klinkerberg, M., 247, *287*
Kloog, Y., 199, *212*
Knapp, D. R., 216, 218, *237*
Knoll, L. J., 216, *238*, 247, 269, *287*
Knowles, D., 87, 93, 94, *100*, 120, *129*
Knudsen, J. K., 227, 228, *239*, 243, 247, *286*
Kobe, B., 141, *153*
Koblan, K. S., 40, 41, *45*, 84, 87, 88, 92, *98*, *99*, *100*, *101*, 157, *209*
Kodera, T., 42, *46*, 88, *100*
Koester, S. K., 89, 90, *101*
Kohl, E. M., 5, *16*
Kohl, N. E., 3, 4, 5, 6, *15*, *16*, *17*, 23, 24, 25, 27, 36, 37, 38, 40, 41, 42, *44*, *45*, 50, 55, 56, 57, 59, 61, 62, 63, 64, *79*, *80*, 84, 85, 87, 88, 89, 90, 92, 93, *98*, *99*, *100*, *101*, 111, 113, 116, *129*, 157, *209*
Kohler, D., 95, *102*
Kokame, K., 202, 203, *212*
Koletar, S. L., 257, *289*
Komano, H., 185, 186, *209*
Korfmacher, W. A., 86, 88, 89, 93, 96, *99*
Kornfeld, S., 216, 219, 221, 230, *237*, *238*, *239*, *240*
Kornhauser, R., 201, *212*
Korolev, S., 261, *289*
Koshland, D. E., Jr., 188, *210*
Kost, T. A., 3, *15*, 20, *43*
Kovar, P., 40, *45*, 87, *100*
Kowalczyk, J. J., 6, *17*, 40, *44*, 86, 90, *98*, 120, *129*
Koya, R. C., 77, *80*
Koyama, T., 78, *80*
Kral, A. M., 4, 5, *16*, 23, 24, 25, 36, 37, 38,

Subject Index

A

A-170634, farnesyltransferase inhibition, 87
Acylprotein thioesterase 1 (APT1)
 $G_s\alpha$ palmitate turnover role, 235–236
 signal transduction role, 234
 substrate specificity, 234–235
ADP-ribosylating factor (ARF)
 myristoyl-CoA : protein N-myristoyltransferase reporter assay, 248–249
 myristoyl-conformational switching,
  245–246
a-factor processing
 AFC1p processing of N-terminal
  deletion mutants, 175
  identification as Ste24p, 175
  overview of processing, 173–174
  truncated a-factor mutant studies, 176
  ubiquitin–*MFA1* fusion protein studies,
   176–177
 Axl1p, 173–174
 identification of processing mutants,
  167–168
 Ste23p, 173, 175
AFC1, *see* Isoprenylprotein endoproteases
Apoptosis, induction by farnesyltransferase
 inhibitors, 91–92
APT1, *see* Acylprotein thioesterase 1
ARF, *see* ADP-ribosylating factor

B

B1088, farnesyltransferase inhibition, 744
N-Benzyloxycarbonylglycylglycyl-(*S*-
farnesyl-C)-chloromethyl ketone
 (ZGGFCCMK), CXXX endoprotease inhibition, 164
BFCCMK, *see* N-*tert*-butyloxycarbonyl-(*S*-
 farnesyl-C)-chloromethyl ketone
BMS-193269, farnesyltransferase inhibition,
 74, 87
BMS-214662, farnesyltransferase inhibition,
 95–96
N-*tert*-Butyloxycarbonyl-(*S*-farnesyl-C)-
 chloromethyl ketone (BFCCMK),
 CXXX endoprotease inhibition,
 164

C

Chemotherapy, farnesyltransferase inhibitor
 combination therapy for cancer,
 93–94
Clavaric acid, farnesyltransferase inhibition,
 87–88
CXXX endoproteases
 activity properties of various membrane
  preparations, 158–160
 AFC1 in humans
  a-factor as substrate, 186
  gene cloning, 185–186
  natural substrates, 186–187
  sequence homology with other species,
   186–187
  tissue distribution, 186
 cloning of genes, 157, 160, 163
 inhibitors

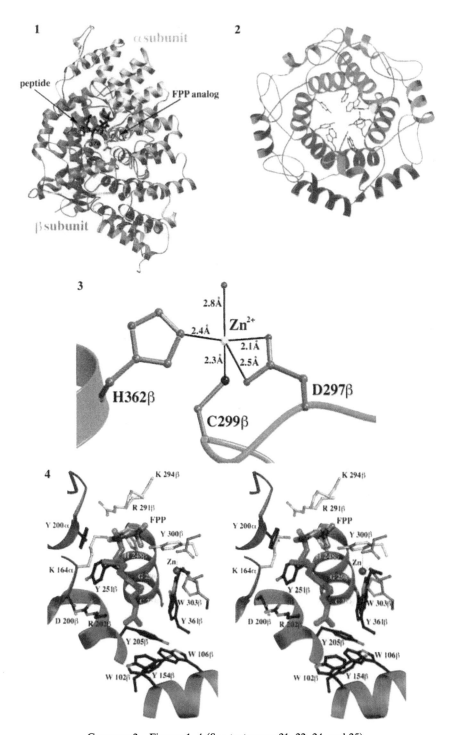

CHAPTER 2—Figures 1–4 (See text pages 21, 22, 24, and 25).

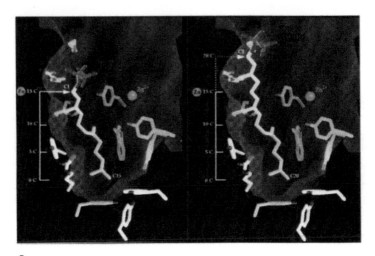

5

7

8

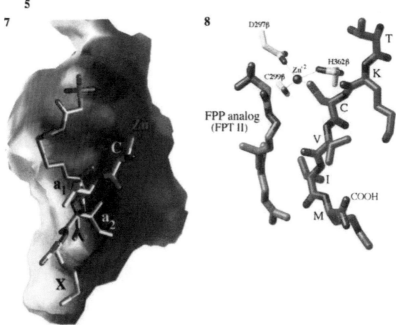

CHAPTER 2—Figures 5, 7, and 8 (See text pages 26, 29, and 30).

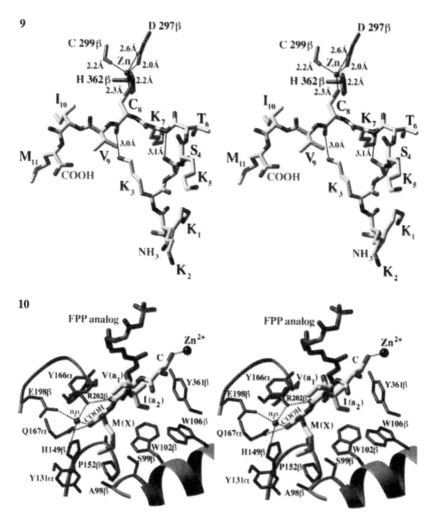

CHAPTER 2—Figures 9 and 10 (See text pages 30 and 31).

11

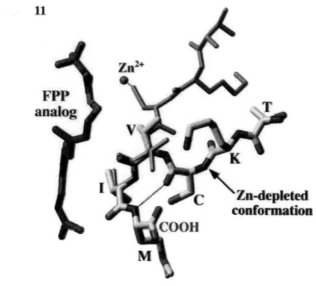

12

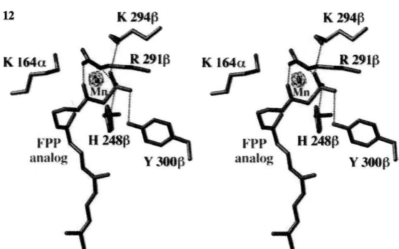

CHAPTER 2—Figures 11 and 12 (See text pages 32 and 35).

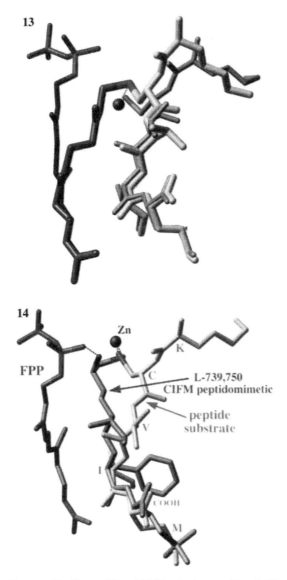

CHAPTER 2—Figures 13 and 14 (See text pages 36 and 40).

A

B

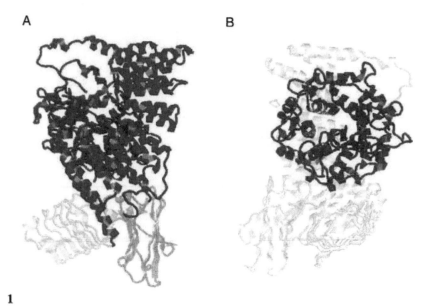

1

CHAPTER 6—Figure 1 (See text page 142).

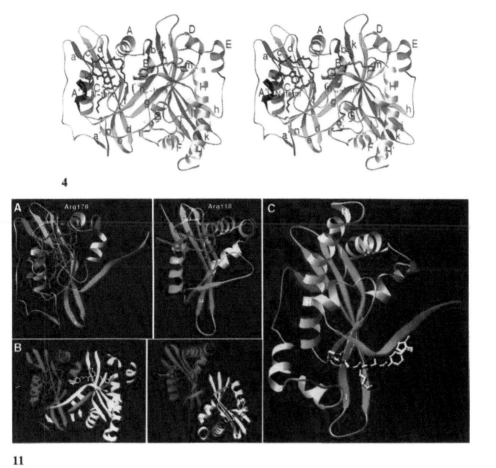

4

11

CHAPTER 9—Figures 4 and 11 (See text pages 261 and 282).

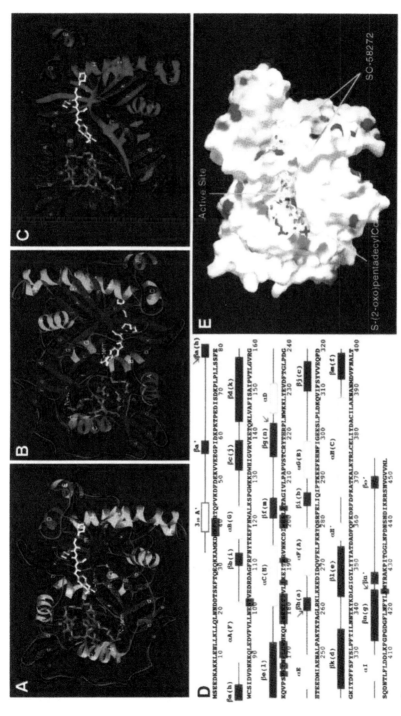

CHAPTER 9—Figure 5 (See text pages 262 and 263).

5

CHAPTER 9—Figure 6A–6D (See text pages 266 and 268).

6 (A–D)

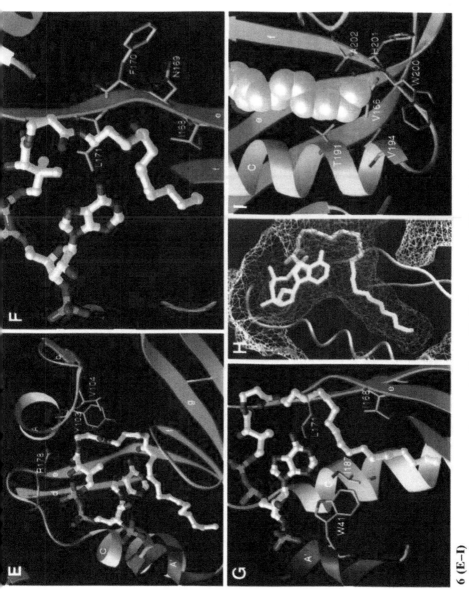

CHAPTER 9—Figure 6E–6I (See text pages 267 and 268).

6 (E–I)

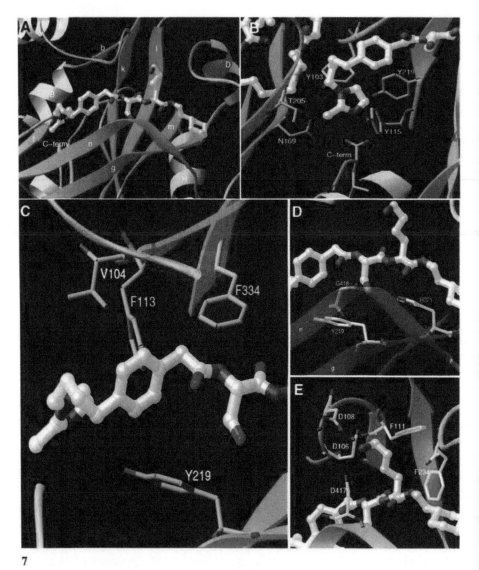

7

CHAPTER 9—Figure 7 (See text pages 270 and 271).

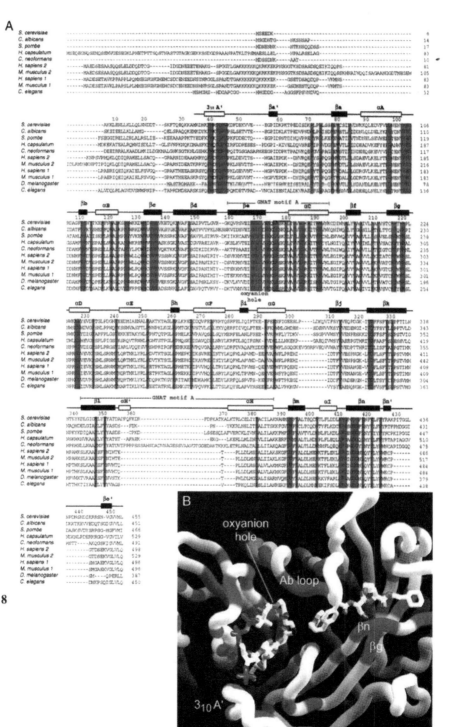

CHAPTER 9—Figure 8 (See text pages 274 and 275).

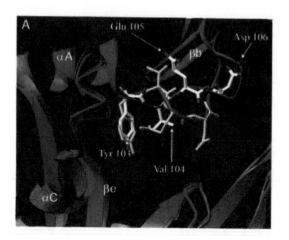

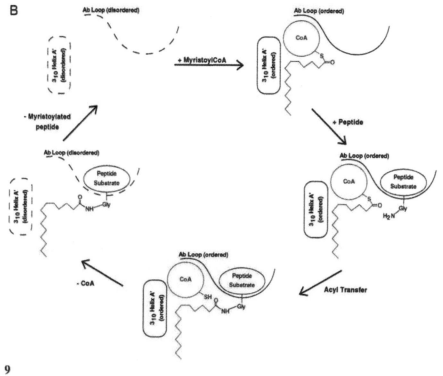

9

CHAPTER 9—Figure 9 (See text page 276).

ISBN 0-12-122722-7

Printed and bound by CPI Group (UK) Ltd, Croydon, CR0 4YY

08/05/2025

01864951-0001